Physicists Look Back

Studies in the History of Physics

Physicists Look Back

Studies in the History of Physics

Edited by

John Roche

Linacre College, Oxford

Adam Hilger, Bristol and New York

This compilation © 1990 by IOP Publishing Ltd

All rights reserved. No part of this publication may be reproduced, stored in a retrieval system or transmitted in any form or by any means, electronic, mechanical, photocopying, recording or otherwise, without the prior permission of the publisher. Multiple copying is not permitted under the terms of the agreement between the Committee of Vice-Chancellors and Principals and the Copyright Licensing Agency.

British Library Cataloguing in Publication Data

Physicists look back.
  1. Physics, history
  I. Roche, John
  530′.09

  ISBN 0-85274-001-8

Library of Congress Cataloging-in-Publication Data are available

The author has attempted to trace the copyright holder of all the material reproduced in this publication and apologises to copyright holders if permission to publish in this form has not been obtained.

Consultant Editor: **A J Meadows**

Published under the Adam Hilger imprint by IOP Publishing Ltd
Techno House, Redcliffe Way, Bristol BS1 6NX, England
335 East 45th Street, New York, NY 10017-3483, USA

Typeset by Mathematical Composition Setters Ltd, Salisbury, Wilts
Printed in Great Britain by J W Arrowsmith Ltd, Bristol

Contents

Preface		vii

Part 1. Method in the history of physics

1	A physicist looks at the resources of research libraries *Dennis F Shaw*	3
2	Preserving and making known the history of physics: the American Institute of Physics Center for History of Physics *Spencer R Weart*	29
3	Interviewing physicists and astronomers: methods of oral history *David H DeVorkin*	44
4	Some methodological problems in writing the history of CERN *John Krige*	66
5	Reflections of an amateur 'historian of science' *N Kurti*	78
6	Could we repeat it? *Ronald G Stansfield*	88

Part 2. Applying the history of physics

7	The role of history in physics teaching *A P French*	111
8	Historians and physics teachers: partnerships for packages *Brian Davies*	126
9	A critical study of the vector potential *John J Roche*	144
10	The photoelectric effect—a suitable case for surgery? *Stuart Leadstone*	169
11	Physics and society *Charles Boyle*	184

Part 3. Physicists look back

12	On attending to the instrument maker in physics history *Brian Gee*	205
13	Reflexions on crystals: instrumentation for crystal-structure determination *H Lipson*	226
14	From research to development: the early development of mechanical parts of electron probe instruments *D J Unwin*	237
15	Meteors and meteor showers: an historical perspective 1686–1950 *David W Hughes*	261
16	Solar variability and terrestrial weather *A J Meadows*	306
17	The early history of atmospheric ozone *C Desmond Walshaw*	313
18	Ionospheric research—some personal recollections *W J Granville Beynon*	327
19	The origins of ionisation and plasma physics *F Llewellyn-Jones*	348
20	The history of the discovery of weakly coupled superconductors *B D Josephson*	366

Notes on contributors   378

Index   383

Preface

There must be few physicists today who are unaware of the growth of the history of physics as a professional discipline. History of physics societies, conferences, books, journals, research centres, archives, libraries, educational centres and museums have proliferated in the past 20 years especially in the United States and in Britain and more recently on the continent of Europe and elsewhere.

Historical research in physics today may be broadly divided into three activities which are not mutually exclusive, namely the work of establishing sources, the social history of physics and the internal or technical history of physics.

The least glamorous but, perhaps, the most fundamental historical work being conducted at present is the careful collection and cataloguing of primary texts, manuscripts and apparatus and the recording of oral history interviews with senior physicists. Critical editions of this material provide a reliable foundation for scientific biographies, for detailed technical studies of the evolution of a particular topic and also for broad synoptic overviews of the history of physics.

The social history of physics has developed very rapidly in the past 20 years. It complements the more technical history of physics and is of considerable importance in that it reveals how the direction and emphasis of physics research is often influenced by factors which lie outside physics itself.

The main concern of the present text is with the history of physics concepts, instruments and research methods, that is, with the internal history of physics. Such research presents special difficulties which seem not to be so pronounced in the social history of physics or in the history of some other scientific subjects such as geology. The complex mathematical language employed and the subtlety of many of the concepts involved, especially since the nineteenth century, means that no historian can hope to carry out deep scholarly work in these subjects without a professional training in physics. Fortunately, many historians of physics today have just such a training. However, the more difficult the topic chosen the greater the maturity of understanding of the physics involved, and the greater the professional competence and general experience in physics required of the historian. This suggests that, beyond a certain level of difficulty, only an able and experienced professional physicist can hope to carry out research of the most penetrating kind in the internal history of physics.

Each generation has indeed produced one or two distinguished physicists, such as James Clerk Maxwell, Pierre Duhem, E T Whittaker and Abraham Pais, who have carried out serious historical research in physics. Nevertheless, scholarly research in the history of physics is hardly perceived today as an admired pursuit for a top-ranking physicist in mid-career, apart entirely from the question of how such a person could find the time to learn to do it and then to carry it out. Furthermore, how could the effort spent on such research be justified by the professional physicist, particularly within today's applications-oriented research environment? Is it not irrelevant to the serious work of the physicist and a waste of precious working time? Most physicists would probably accept that the history of science, like any other branch of history, is intrinsically important as part of the reflections of a civilised society on its own culture and on human culture in general. But detailed historical research is for historians, not for scientists.

Nevertheless, historical research, as part of the professional work of the physicist, can be justified.

In recent years the role of the history of physics in physics education has received increased attention in Europe, and its importance in this field has long been recognised in the United States. In 1983 an advisory council in science education was set up by the European Physical Society which has held a series of increasingly successful conferences on history in physics education. The British Society for the History of Science established an education section in 1987 with a similar purpose. These initiatives seem to be gathering momentum rapidly.

The applied history of physics also has a very important part to play in explanatory research in physics.

Prior to the seventeenth century physics was chiefly concerned with explaining the facts already known in nature. In the seventeenth century physics abruptly altered course and devoted its energies to discovering new laws, phenomena and devices, to the mathematicisation of nature and to the technological applications of physics.

Three centuries later physics has accumulated a vast body of

Preface

experimental and theoretical knowledge, not all of which is well understood. Explanatory research has not kept pace with exploratory research. Many physicists today will admit that they do not have a clear understanding of certain basic concepts, even in classical physics. At the same time everyone recognises that such understanding is the key to future progress. I would suggest, therefore, that a major challenge for the next generation of physicists is to clarify and consolidate foundations and to reconstruct each topic in physics, beginning with the various branches of classical physics, in such a way that physical explanation is given as much importance as mathematical and experimental explanation.

The history of physics has an enormously important role to play in the clarification of basic concepts. Most concepts in physics today, such as mass or potential, though posturing as single concepts, in fact represent a loose amalgamation of competing traditions, each with a different ancestry. Without resolving them into their component traditions such concepts are impossible to understand clearly. The historical study of foundations can help to isolate each tradition, make it understandable and assist in the reconstruction of a coherent explanation. This is a task which will require the efforts of some of the best minds in physics and is likely to occupy several generations. It should not be supposed that explanatory research is of marginal importance in physics. Given its comparative neglect for almost two centuries it could be as fertile and innovative as mathematical and experimental exploration. Indeed, without it I find it difficult to see how physics can make its next great leap forward.

The present volume attempts to encourage these developments at several different levels. The contributions have been selected from various meetings of the History of Physics Group of The Institute of Physics since its foundation in 1984. All of the contributors are professional physicists or physicists who have moved into other fields. It is hoped that this volume will guide the reader to interesting literature in the history of physics, indicate how to go about doing historical research, give some illustrations of the uses to which history can be put and provide a variety of examples of historical writing.

The first section is devoted to methodology. Dr Dennis Shaw, Keeper of Scientific Books at the Radcliffe Science Library, Oxford, gives a scholarly and comprehensive account of resources and services available in research libraries, to those interested in the history of physics. It covers three periods, before the nineteenth century, the nineteenth century and the present century. It concludes with an appendix by Mr R J Wyatt which provides a detailed listing of catalogues of major libraries with holdings in physics.

Dr Spencer Weart, who is head of the American Institute of Physics Center for History of Physics, describes the rich resources and many activities of the Center.

Dr David DeVorkin, who is Curator of the History of Astronomy at the Smithsonian Air and Space Museum gives a very thorough description of the techniques of oral history interviewing. This could almost serve as a *vade-mecum* for anyone interested in taking up this kind of research.

Dr John Krige, who leads an historical project at CERN, delves below the surface of historical research and illuminates the limitations on a comprehensive and objective institutional history posed by the nature of the archives available, the differing perspectives of large and small countries, the complexities of the subject, 'standard' views and personal preconceptions.

Professor Nicholas Kurti of Oxford throws further light on the dangers of assuming that received 'standard' views of historical events, or even of our own memories of research, are reliable. He does this through a series of examples ranging from Helmholtz in the nineteenth century to his own experiences at Oxford.

Mr R G Stansfield brings a wide range of experience both in physics and in the social sciences to bear on the interesting question of whether we could repeat the essentials of a past experiment. In doing so he takes us through a description of the laboratory culture of the 1930s, discusses the difficulties of replication of laboratory materials and local environmental conditions and comments on changes in meaning and nomenclature.

The second section of the volume deals mainly with the applications of the history of physics in the teaching of physics and in the clarification of concepts.

Professor A P French of the Massachussets Institute of Technology introduces this section with a broad discussion of whether and how history can be of service to physics teaching. Professor French's considerable experience as an educationalist in physics make this a valuable contribution to the current debate over the role of history in physics teaching.

Mr Brian Davies of The Institute of Physics also has a long experience as a physics educationalist. His well illustrated examples of what he terms 'history packages' will be particularly relevant to lower-school physics teaching.

My own contribution attempts to show how the study of the historical evolution of a topic, in this case the vector potential, can shed light on our present-day understanding of it.

Mr Stuart Leadstone of Atlantic College, Wales, continues a similar theme in his thorough-going study of difficulties in explaining the concept of the photoelectric effect. His research strategy is particularly interesting, involving as it does the systematic questioning of textbook authors.

Mr Charles Boyle, of Nottingham Polytechnic, surveys various aspects of the impact of physics on society and the influence of society on physics. In a period of increasing anxiety about the environment it is most important not to view physics or the history of physics in a narrow technical context and to be keenly aware of its larger implications. Mr Boyle's contribution is different from most of the others in that it is not primarily historical, but its importance and general relevance justifies and, perhaps, even demands its inclusion.

The final section of this book contains a selection of purely historical pieces. It begins with three articles on instrumentation, continues with five articles on experimental physics and astronomy and concludes with an article on theoretical physics.

Mr Brian Gee, formerly senior physics lecturer at Plymouth, provides a scholarly and reflective study of scientific instrument making in Britain and examines the relationships between the instrument maker and the physicist.

Professor Henry Lipson of Manchester gives an all too brief and self-effacing account of the development of instrumentation in the 1930s, and subsequently, for the determination of the structure of crystals.

Mr Don Unwin, formerly technical manager of research at Cambridge Instrument Company, gives a fascinating and technically detailed study of the development of electron microscopes at the company. This will delight experimentally minded physicists and historians with an interest in the 'internal' history of technology.

Dr David Hughes, of the University of Sheffield, surveys in detail the history of meteors from Aristotle to the present. His references effectively form a comprehensive bibliography of the subject.

Professor A J Meadows of Loughborough provides a brief but remarkably satisfying account of the still unsettled question of the influence of sunspot variations on earthly weather.

Dr C D Walshaw of Oxford gives a scholarly and comprehensive account of the history of ozone theory and research from the eighteenth century to 1921. Again there is a very full bibliography.

Sir Granville Beynon, of University College, Aberystwyth, has been a leading figure in ionospheric research for many years. In this survey of ionospheric research in the present century he includes glimpses of his pioneering work with Sir Edward Appleton.

Professor Frank Llewellyn-Jones of Swansea again provides a

professional scientific perspective, on a subject to which he has contributed so much, in his account of the history of ionisation and plasma physics in the early part of the present century.

The concluding article is by Professor Brian Josephson of Cambridge who, for the first time in print, describes his Nobel Laureate work on the discovery of the 'Josephson effect'.

To conclude this introduction I would first of all like to thank especially those contributors who submitted their contributions promptly and who have had to wait several years to see their work in print. I would also like to thank all contributors for their labours, sometimes considerable, in researching their articles and for cooperating with the editor during the process of editorial revision.

Several of the contributions, and that from Professor Josephson in particular, would not have been possible without the persuasive and organising skills of Mr Stuart Leadstone who ran a history of physics conference in Wales in 1986 and painstakingly transcribed tapes and edited the texts. Had he had the time he would have been joint editor of this volume.

Much of the editorial work on references was effectively a cooperative endeavour of the editor and the staff of the Radcliffe Science Library, Oxford, who enthusiastically tracked down obscure books, journals and dates. In particular I would like to express my gratitude to Mr P J Warren, Deputy Keeper, and to Miss Dianne Booth, Mr Theo Dunnet and Miss Jill Downing to mention but a few. Finally, I would like to thank Professor A J Meadows who first encouraged me to edit this volume and to Sean Pidgeon of Adam Hilger for his courteous patience over a long-delayed text.

John Roche
Linacre College, Oxford
15 October 1989

Part 1

Method in the History of Physics

1

A Physicist Looks at the Resources of Research Libraries†

Dennis F Shaw
(with an appendix by R J Wyatt)

Introduction

I have been a physicist for just over 40 years and for the last 12 of those years I have been the Keeper of Scientific Books in Oxford University, in charge of the university's rich collection of scientific literature. In this introductory chapter I want to pass on the benefit of some of the lessons I have learnt in those 12 years, in particular telling you about what some physicists do not seem to know about the resources that are available to them in general regarding the subject. Niels Bohr‡ (1885–1962) said that physics is the history of physics; you do not distinguish between them. The history of physics is not a different subject from physics; it is in reading about physics that one learns the history of the subject.

The plan of this chapter is to build upon those features of the literature and resources of research libraries that have been discovered since my coming to the Radcliffe Science Library.

† Based on a presentation given at the conference on the History of Physics for the Physicist, 2–4 July, 1986, Oxford.
‡ Personal communication.

Research libraries

Where are the research libraries? Many of them are in universities, many of them are owned by learned societies, others will be found in major museums, and there are the national libraries and the research establishments. Figure 1.1 shows the John Crerar Research Library of Science at the University of Chicago; it is one of the most modern buildings in the world of research libraries. The John Crerar Library is, in fact, quite an old library established by private foundation in Chicago in the middle of the nineteenth century. John Crerar (1827–89) was a Scotsman (not the only Scotsman to have invested in culture and learning in North America) and he left money to establish a library for the benefit of the citizens of Chicago. It has changed its location three times since its foundation in 1894 and the latest home shown in the figure is in the new Biomedical Center at the University of Chicago. This library houses one of the great science research collections in North America. Figure 1.2 shows the Radcliffe Camera, in the centre of Oxford, now used as a reading room by the Bodleian Library. This is where my own library, the Radcliffe Science Library, started its service in 1749.

Figure 1.1 The John Crerar Library of the University of Chicago. (By kind permission of the Trustees of the University of Chicago.)

The two buildings shown in figures 1.1 and 1.2, which are very different in architectural style and period, illustrate the fact that scholars have a very great concern for the homes where their research material is housed. It is a salutary consolation for scholars that they can still persuade their masters to pay for such facilities for learned societies and universities as well as for government and national institutions. Such magnificent buildings contribute an environment and create an atmosphere encouraging readers and visitors to appreciate that they represent an important institution housing a very significant part of our national and international culture.

Figure 1.2 The Radcliffe Camera, the Bodleian Library, Oxford. (By kind permission of the Curators of the Bodleian Library.)

Figure 1.3 shows one of the most modern reading rooms in Oxford, the Lankester Room in the Radcliffe Science Library where the Oxford University History of Science collection is now housed. The Lankester Room is beneath the lawn of the University Museum, and is lit entirely artificially. Readers often remark on the congenial atmosphere, which is said to be ideally suited to the needs of the student.

Figure 1.3 The Lankester Room, the Radcliffe Science Library, Oxford.

Research libraries in Europe

Thus we begin to answer the question, where are the research libraries? But those of us who do not know where they are may wonder where we can find them. If the library to which we have access has not got the essential information being sought and we want to find it in another library, what can be done about it? Unfortunately there is no reliable guide; this is one of the things lacking at present. If the library has a good reference librarian you can seek advice about where to find what you want. But if that channel of enquiry fails the next stage is to seek the information for yourself in another library. For this purpose it is helpful to know where the major collections of physics literature are housed. According to my assessment, the Universities of Oxford and Cambridge, University College, London, the Universities of Manchester, Birmingham, Leeds and Edinburgh, and Trinity College, Dublin have the most extensive collections in Britain and Ireland. This list is not exhaustive, and there are some collections, such as that at the Royal Society in London, which have important specialist source material. These university libraries in Britain and Ireland, which have the major physics collections, are of course supplemented by the national collections in the British Library, the Science Reference and Information Service (SRIS) and the British Library Document Supply Centre (BLDSC). One of the finest collections in the country of older literature on science and technology is housed at the Science Museum in London, but its coverage of modern physics is not nearly as good as it was 50–100 years ago. SRIS is the new name of the Science Reference Library in London which houses the collections of the former Patent Office Library. SRIS is complemented by BLDSC at Boston Spa in Yorkshire, which has one of the finest collections of modern scientific periodicals in the world. It is used by nations world-wide. Scientists in North America say they borrow more frequently from Boston Spa than they do from other American libraries, because the service is better and the coverage is greater. Finally, in this listing we must include the Royal Irish Academy in Dublin which, probably as a result of British influence, has a very fine collection of scientific periodicals, for which Irish scholars are eternally grateful.

Not all resource material for the history of physics can be found in the UK and Ireland. It may well be necessary to go further afield, particularly for foreign-language material. The guide to European Sources of Scientific and Technical Information (Harvey 1986) is a useful source of information on European holdings. Unfortunately, the author does not appear always to have read the small print about the libraries listed in the guide and, consequently, some of the observations about them are not strictly accurate.

Research libraries world-wide

For information world-wide, one of the finest reference sources is *The World of Learning* (1988). It has entries for universities, research establishments, scientific and technical institutions, libraries and museums. The descriptions of institutions give considerable detail which can be relied on for accuracy since the entries are provided by the individual institutions. Figure 1.4 reproduces part of the entry for West Germany to give an idea of the type of information obtainable from this source. The listing includes major libraries in main cities entered alphabetically. Each entry includes full postal address, date of foundation, details of holdings of books, periodicals, manuscripts, incunabula (books printed before 1501), drawings, engravings and maps and other audio-visual materials including microforms. The way to use this directory for locating source material in physics is to look for libraries which have large collections of serials and manuscripts and which specialise in scientific literature. The entries are not indexed by subject so that it is necessary to know what you are looking for. However, it does not take very long to discover, for instance, that in Frankfurt there is the Staats-und-Universität Bibliothek which has the best collection of biological periodical literature in West Germany, and this is the way to locate reference material that you cannot find by other means. The listing in figure 1.5 demonstrates the truly world-wide nature of this directory, showing part of the entry for Japan. The first entries under Tokyo are for the American Center Library, British Council Library and the Chuo University Library, listed in alphabetical order by name. The entry for Tokyo University Library shows it has the largest collection of scientific material in Japan, six million volumes, which is as large as the Bodleian collection.

Libraries

Aachen

Bibliothek der Technischen Hochschule: 5100 Aachen, Templergraben 61; tel. (0241) 80-4445; f. 1870; 960,000 vols; Dir Dr Ulrich Fellmann.

Oeffentliche Bibliothek: 5100 Aachen, Couvenstr. 15; tel. (0241) 20545; telex 832654; f. 1828; general information about the city and the region; regional history; 400,000 vols; Dir H. Frings.

Augsburg

Staats- und Stadtbibliothek: 8900 Augsburg, Schaezlerstr. 25; tel. (0821) 3242737; f. 1537; 400,000 vols, 3,540 MSS, 3,000 incunabula, 16,000 drawings and engravings; Dir Dr Helmut Gier.

Universitätsbibliothek: 8900 Augsburg, Alter Postweg 120; tel. (0821) 598802; telex 53830; f. 1970; 1,130,000 vols, 71,400 theses, 31,200 maps, 56,000 items of AV material and microforms, 1,000 incunabula, 1,500 MSS, 1,787 music MSS; Dir Dr Rudolf Frankenberger.

Bamberg

Staatsbibliothek Bamberg: 8600 Bamberg, Domplatz 8, Neue Residenz; f. 1803; contains 330,000 vols, including a special collection of old manuscripts (4,500), incunabula (3,400), and graphics (70,000); Chief Librarian Dr Bernhard Schemmel; publ. *Katalog der MSS.*

Universitätsbibliothek: 8600 Bamberg, Postfach 1549, Feldkirchenstr. 21; tel. (0551) 402-435; f. 1973; 695,000 vols on theology, humanities and social sciences; Dir Dr Dieter Karasek.

Figure 1.4 Part of the entry for West Germany from *The World of Learning.*

Libraries
Tokyo

American Center Library in Japan: 11th Floor, ABC Kaikan, 6-3 Shibakoen 2-chome, Minato-ku, Tokyo 105; Dir of Libraries JESSE REINBURG.

British Council Library: Iwanami Bldg, 1 Jimbo-cho, 2-chome, Kanda, Chiyoda-ku, Tokyo 101; f. 1953; 18,984 vols, 200 periodicals; Librarian (vacant).

Chuo University Library: 742 Higashi-nakano, Hachiojishi, Tokyo; f. 1885; 1,026,113 vols (422,595 in foreign languages), 10,860 periodicals; Librarian Prof. YASUNAO NAKADA.

Hitotsubashi University Library: Naka 2-1, Kunitachi-city, Tokyo 186; tel. (0425) 72-1101; f. 1887; 1,024,000 vols (including Kodaira branch); Librarian M. OKAWA; houses branch library for Institute of Economic Research; f. 1940; 234,000 vols; Dir S. FUJINO.

Imperial Household Agency Library: Imperial Palace, Chiyoda-ku, Tokyo; tel. (03) 213-1111; f. 1948; 80,102 vols; Librarian Mr YAMAMOTO.

International Christian University Library: 10 Osawa 3-chome, Mitaka-city, Tokyo 181; 279,300 vols (including 135,276 foreign), 2,000 periodicals; Dir Prof. MAKOTO SAITO.

Japan Meteorological Agency Library: 1-3-4 Ote-machi, Chiyoda-ku, Tokyo 100; f. 1875; 180,000 vols; Chief Librarian KIYOSHI KURASHIGE.

Figure 1.5 Part of the entry for Japan from *The World of Learning*.

When one has located the relevant library, one needs to find out exactly what it has in it. The most satisfactory way to accomplish this is to look up the library's catalogue, if it is available. The bibliography in Appendix 1.2 prepared by R J Wyatt lists the world's major research libraries with holdings in physics which have published catalogues.

If the catalogue has not been published in hard copy, it may be possible to access it on-line through a computer network. We await the introduction of the Open Systems Interconnection (ISO 1984, Open Systems Interconnection 1987) between major libraries which will enable the search for this sort of information to be performed on-line from the local library.

One of the finest history of science collections in the world is at the University of Oklahoma and the catalogue of this library was published by Roller and Goodman (1976). The title page of the first volume is reproduced in figure 1.6. Such a catalogue of a major collection is an excellent guide to the literature. Simply by opening it and browsing through it one can get a very good idea of the literature available on the history of any branch of science. Unfortunately, it has no subject index so the entries for physics cannot readily be identified. This is a shortcoming of many of these catalogues of celebrated collections, but it is compensated for partly by the existence of the subject index to the Royal Society catalogue of scientific papers 1800–1900 detailed below.

The Catalogue of the History of Science Collections of the University of Oklahoma Libraries

DUANE H. D. ROLLER
McCasland Professor of the History of Science *and* Curator, History of Science Collections

MARCIA M. GOODMAN
Instructor of Bibliography *and* Librarian, History of Science Collections

Volume 1

A - I

MANSELL 1976

Figure 1.6 Title page of Roller and Goodman catalogue for the University of Oklahoma collection.

History of physics before 1800

From the Renaissance until the end of the eighteenth century there were very few periodicals for the publication of discoveries in physics. The first specialist physics journal, the *Journal der Physik* started in 1790 to be followed by the *Philosophical Magazine* in 1798. The history of physics before 1800 has been well documented and will not be repeated here. A short summary of sources for the history of natural philosophy (the term usually used then to define physics) during the seventeenth and eighteenth centuries is given by Hackmann (1985).

Physics in the nineteenth century

During the nineteenth and twentieth centuries physics periodicals have assumed great importance and the historian must be able to find a way to study this wealth of literature in order to seek out original source material. The Royal Society (1867–1925) produced a definitive international catalogue of scientific papers from the nineteenth century. It was realised soon after it was completed and published that there was the need for a subject index.

The subject index of the *Royal Society Catalogue of Scientific Papers* (1908–14) was issued in three volumes. Volume 3, part 1, covers physics generalities, heat, light and sound, and part 2 covers electricity and magnetism. It is one of the most important guides to the literature of physics in the nineteenth century, if not the most important. The subject index, of course, is in terms of the language of physics in the nineteenth century. One will not, therefore, find entries for condensed matter nor its earlier description solid state physics. The user will have to look up solids, liquids and gases since that was the language used during the nineteenth century. This was a major project resulting from a collaboration between the staff of libraries in Cambridge, Oxford, University College, London, and the British Museum. (I believe the Science Museum contributed also. There were five or six librarians who spent their working lives going systematically through this index and picking out the subject entries.) It is a marvellous collective work and is typical of the sort of thing that scholarly librarians devote their lives to. Even today there are librarians in the Bodleian, and in other great libraries, who do this sort of thing. The Radcliffe Science Library does not have a large enough staff to be able to engage in this sort of activity. This is unfortunate since we have very rich resources and there are several important projects relating to the literature in various branches of science which would be worth undertaking. There is no doubt that guides to the literature, not only the published literature but also the private papers, are very greatly needed in the world of scholarship today and there are not enough scholars free to produce them.

The *Royal Society Catalogue of Scientific Papers* covers the major world journals published during the nineteenth century and, by way of illustration of its comprehensive coverage, the whole catalogue of scientific papers covers about 2700 journals, bulletins and society memoirs. The physics index includes materials from a selection of about 1300 physics journals world-wide. Nearly every major serial publication of a serious nature, containing any papers on physics, was included by the editors. The catalogue started in 1800 and it came out in four series finishing in 1902. When publication was completed the participating libraries got together to produce the index which took them 20 years to complete. It is unfortunate, but not surprising, that no manuscript material was included, only printed papers; but the importance of manuscript material was not recognised during the nineteenth century. Willem Hackmann (1985) has pointed out the change that has taken place in the sources for the history of physics from the Classical period through the Renaissance to the present day. Now we have come to realise how important it is that we should have access not only to the published work of physicists but also to their papers, particularly private correspondence (the papers that they do not publish). Many collections have now been catalogued and deposited in research libraries. Much of this work in England has been carried out in recent years by the Contemporary Archives Centre originally set up in Oxford by Professor Margaret Gowing. It has now become the National Cataloguing Unit for the Archives of Contemporary Scientists and is situated at the University of Bath.

There are details in these collections of private papers regarding the development of physics that we could never discover from reading published papers. This is one of the major lacunae in resources for the history of physics at the present time.

History of physics in the twentieth century

Although the history of nineteenth-century physics is becoming increasingly important it is during the twentieth century that most of physics, as we know it today, has been developed; and it is remarkable what a vast amount has been published. Figure 1.7 shows the growth of numbers of abstracts of physics published material in books, conference reports and journal papers considered to be of merit and containing original contributions to the subject, for the four major abstracting services: *Physics Abstracts*, *Physikalische Berichte*, *Referativnyi Zhurnal* and *Bulletin Signalétique*. The earliest of these was *Physics Abstracts* which started in 1898 as *Science Abstracts* but the amount of material produced, as is well known by the historians of physics, was not very significant until after World War I. There were of course very few physicists active then; in fact the significant growth in the number of research physicists did not take place until World War II. The output in 1920 of *Science Abstracts* was about 1800 items and in 1983 it was 90 000 items; this shows how great has been the growth of physics research during the twentieth century. *Physikalische Berichte* (now

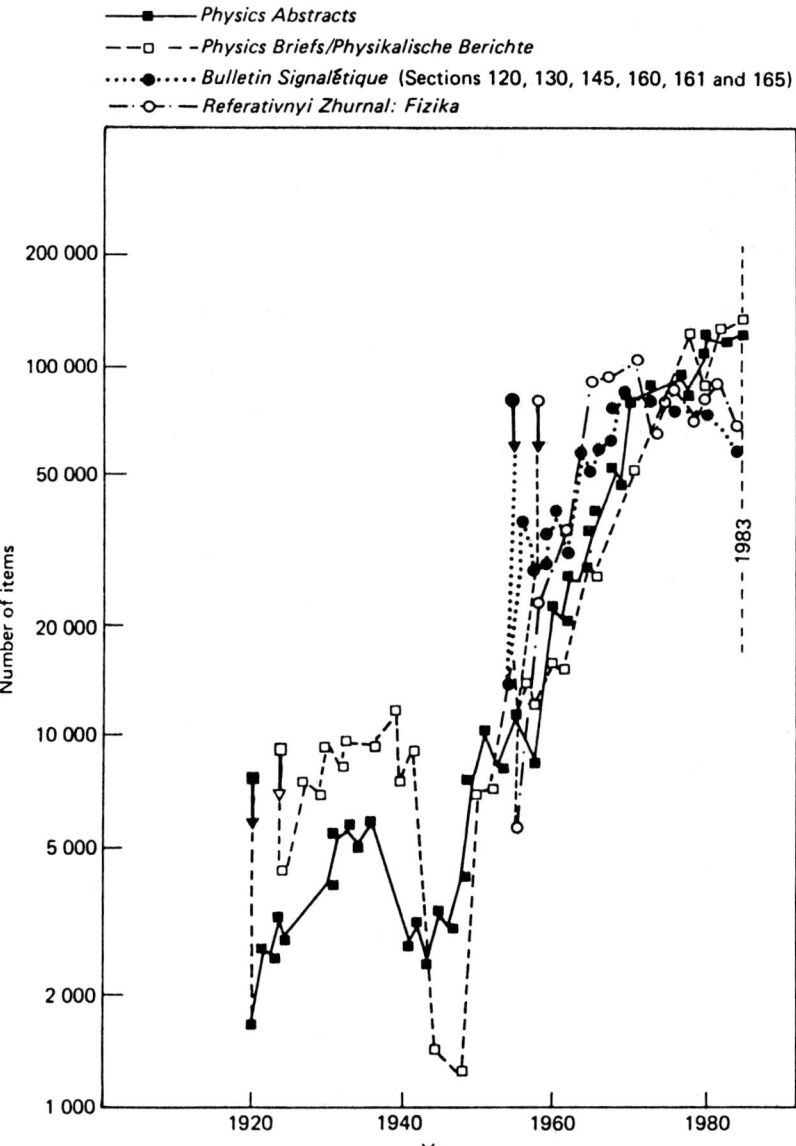

Figure 1.7 Growth of the four main abstracting services in physics, 1920–83.

entitled *Physics Briefs*) started in 1923 and was publishing rather more than *Physics Abstracts* at that time. Now it is just ahead and it duplicates a lot of the material that appears in *Physics Abstracts*. Why these two great organisations cannot get together, I do not know—except of course, that the output they would publish would not collect as much money from libraries as do the two separate publications, but there is a large amount of overlap between *Physics Briefs*, which is a joint American/German venture, and *Physics Abstracts*, which is produced by the UK Institute of Electrical Engineers in cooperation with the IEEE in America.

The French *Bulletin Signalétique* covers a lot of material not appearing in the other three abstracts services. *Referativnyi Zhurnal (Fizika)*, which is the printed version of the VINITI database held in Moscow, does not cover as much Russian literature as one might expect, in fact only about 20 per cent of the abstracts in this journal are from East European countries: the remaining 80 per cent, from the Western countries and China and Japan, are represented pretty well equally amongst the other three abstracting journals. The *Referativnyi Zhurnal* does not have very much other than English and Slavonic and Cyrillic language literature. Most of the literature, in Korean, Japanese and Chinese, is located either by the European or by the North American abstracters. This is probably a reflection of the skills of the staff employed by those organisations. There is no doubt that INSPEC (Information Services for the Physics and Engineering Communities) which publishes *Physics Abstracts* is the most important of the four physics abstracting services, because over 90 per cent of the world's physics literature is now published in the English language. One of the most important guides for twentieth-century scholars is the *INSPEC List of Journals and other Serial Sources* (1987). It now comes out about once every year. This publication lists the journals which are abstracted by the INSPEC services and it conveys a wealth of information of great interest to research physicists as well as to historians. But, it is important to appreciate that this is the list of journals which are now abstracted on to the INSPEC database and it does not include any information about the other journals which were abstracted into the hard-copy version before the INSPEC database was started since these data have not yet been added to the database. It is very much an instrument for locating contemporary information. It is of more value to technologists than scholars of the more established and ordered parts of physics. The entry reproduced in figure 1.8 illustrates the main features. This shows that the *Journal of the American Society for Information Science* (published in the USA) is fully abstracted and all information appears on the database from volume 21, no 1 January/February 1970. The Coden and ISSN given enabled direct entry into the database when searching for this particular title. The full title appears in italics and this is followed by any former title, in this case *American Documentation*. The next line gives the name of the sponsoring body and publisher together with full postal address. It is important to know the ancestors of current periodicals, which change their name more frequently than any other published materials: in the Radcliffe Science Library, with a collection of about 11 000 current periodicals, we get 500 title changes a year. This gives some idea of the rate at which scientific periodical titles change. The *Alphabetical Subject Guide to the INSPEC Classification*, published annually, is a very important additional resource to aid the searching of the literature.

INSPEC List of Journals and other Serial Sources   41

†**J. Am. Soc. Inf. Sci. (USA)**   (vol.21, no.1 Jan.-Feb. 1970—)
Coden: AISJB6   ISSN 0002-8231
Journal of the American Society for Information Sciences Formerly: Am. Doc. (USA)
American Society for Information Science, 1155 16th Street, N.W., Suite 210, Washington, DC 20036, USA. John Wiley & Sons Inc., 605 Third Avenue, New York, NY 10158

Figure 1.8 An entry from the *INSPEC List of Journals and Other Serial Sources*.

There are other subject classifications for physics and those listed in table 1.1 should be noted in particular. It would be much easier for us all if we could get agreement to use just one physics subject classification but attempts to date have been unsuccessful.

The classifications shown in table 1.1 are only slightly different, but one has to know which one is using. The *INSPEC Classification* (1988) is divided into four sections: physics; electrical engineering and electronics; computers and control; and information technology. There are about 3800 subject entries in the index and a small section on plasma is reproduced in figure 1.9. This example has been chosen to illustrate a very important feature of these classifications, namely the change in terminology with time. As an undergraduate I read about electrical plasma as part of my physics course, but it was called 'gas discharges' and the book that I used had the title 'Ionisation in Gases'. Now, in order to get 'gas discharges' one has to look up 'plasma'. There is no cross reference, there are two entries for 'gas discharges' in the index but they are both in section B (electronics and electrical engineering) and relate to applications. There are no entries in the physics section for 'gas discharges' and yet most of the phenomena concerning plasma are related to gas discharges so we have to be very aware of the fact that modern technology is not very well subject-linked to the older technology from which it has stemmed. Here is another feature lacking in

Table 1.1 Physics classification systems

DEWEY DECIMAL CLASSIFICATION (DDC)
(originated by Melvil Dewey and first published in 1876)

530	Physics (Theories, states of matter, physical units, etc.)
531	Mechanics
532	Mechanics of fluids
533	Mechanics of gases
534	Sound and related vibrations
535	Visible light (Optics) and paraphotic phenomena
536	Heat
537	Electricity and electronics
538	Magnetism
539	Modern physics (Molecular, atomic, nuclear physics)

UNIVERSAL DECIMAL CLASSIFICATION
(generally referred to as the UDC)
(Latest English edition published 1985 (BSI))

530.1	Basic principles of physics
531	General mechanics. Mechanics of solid and rigid bodies
532	Fluid mechanics in general. Mechanics of liquids (hydromechanics)
533	Gas mechanics. Aeromechanics. Plasma physics
534	Acoustics. Vibrations
535	Optics
536	Heat. Thermodynamics
537	Electricity. Magnetism. Electromagnetism
538.9	Physics of Condensed Matter (in liquid state and solid state)
539	Physical nature of matter

LIBRARY OF CONGRESS CLASSIFICATION
(was drawn up in the early 1900s and it is basically an enumerative scheme)
 (The sixth edition of the volume entitled *Science* was published in 1973)

The outline for physics is as follows:

QC 1–75	General
81–114	Weights and measures
120–168	Description and experimental mechanics
170–197	Atomic physics. Constitution and properties of matter, including relativity, quantum theory and solid-state physics
221–246	Acoustics. Sound
251–338.5	Heat
310.15–319	Thermodynamics
350–467	Optics. Light
450–467	Spectroscopy
501–766	Electricity and magnetism
501–718.8	Electricity
669–675.5	Electromagnetic theory

Table 1.1 (*Continued*)

676–678.6	Radio waves (theory)
716.6–718.8	Plasma physics
750–766	Magnetism
770–798	Nuclear and particle physics. Atomic energy. Radioactivity
793–793.5	Elementary particle physics
794.95–798	Radioactivity
801–809	Geophysics. Cosmic physics
811–849	Geomagnetism
851–999	Meteorology. Climatology
974.5–976	Meteorological optics
980–999	Climatology and weather

our information distribution services. Unfortunately, there is no-one around with the time or interest to devise, and publish, a concordance of subject index terms in physics, although it is urgently needed. We are relying very much at present on the knowledge of the people who are looking after the literature.

Memoirs and biographies

To finish this account of resources of research libraries for the historian of physics let me draw attention to the very important subject of the memoirs and the biographical descriptions of the life works of major physicists. Many of these publications are extremely important sources of information for historians of science, not only of physics. The most important of these in my opinion is Poggendorff's (1863) *Biographisch-literarisches Handworterbuch* (full title given in list of references) started in 1863. Comparing the contents of that biography with the contents of the *Dictionary of Scientific Biography* (1970–80) published in 16 volumes by the American Council of Learned Societies you will note that the latter has far fewer entries for scientists even though it spans the centuries from antiquity to 1980; but it gives very much longer biographical notes on key individuals. If, for example, one wishes to know about Michael Faraday's (1791–1867) work I recommend using the American biographical dictionary, rather than Poggendorff, because in the former there is an eight-page article summarising Faraday's life and his contributions to science whereas, in Poggendorff, there is a factual list of his publications and no commentary. If one wishes to check one of Faraday's papers without having access to the collected papers (which he collected together himself) then Poggendorff is the best reference—that is really Poggendorff's main function. In the

Plasma	A52
Plasma antennas	A5240F
Plasma applications	A5275
Plasma arc spraying	A5275, A8115, A8160
Plasma beam interactions	A5240M
Plasma collision processes	A5220F, A5220H
Plasma confinement	A5255
Plasma deposited coatings	A6855, A8115J, A8160
Plasma deposition	A8115J, B0520F
Plasma devices	A5275
Plasma diagnostics	A5270
Plasma diodes	B2380
Plasma electromagnetic wave propagation	A5240D, B5210H
Plasma elementary processes	A5220
Plasma equilibrium	A5255
Plasma-filled waveguides	A5240F
Plasma flow	A5230
Plasma heating	A5250
Plasma heating by laser beam	A5250J
Plasma heating by shock wave	A5250L
Plasma heating by wire explosion	A5250L
Plasma magnetohydrodynamics	A5230
Plasma oscillations	A5235
Plasma panel displays	B2380, B7260
Plasma production by laser beam	A5250J
Plasma production by shock wave	A5250L
Plasma production by wire explosion	A5250L
Plasma radiation	A5225P
Plasma sheaths	A5240K
Plasma shock waves	A5235T
Plasma simulation	A5265
Plasma sources	A5250
Plasma transport processes	A5225F
Plasma turbulence	A5235R
Plasma-wall interactions	A5240H
Plasma waves	A5235
Plasmons	A7145G, A7230

Figure 1.9 INSPEC classification for 'plasma'.

American Council of Learned Societies' publication one gets a full commentary on the contribution of the individual, related sources of information about it, the original works, secondary literature and even such facts as the location of portraits (in the case of Faraday in the Royal Institution of Great Britain and in the National Portrait Gallery). Such information is important and hard to acquire. By way of illustration a few years ago I had a letter from a German drug company asking the whereabouts of Robert Hooke's (1635–1703) portrait as they wanted a copy of it to hang in their new entrance hall.

Since the Hooke Science lending library is part of the Radcliffe Science Library I tried to find some information, but without success. I discovered that a portrait of Hooke was reportedly given to the Royal Society when Hooke was Secretary and the Royal Society still had possession of it when they were accommodated in Somerset House; but when the Society moved to Burlington House the portrait did not go with them. Perhaps it is now unrecognised somewhere in somebody's loft, which is unfortunate since there was only that one portrait ever painted of him. (I have an Eric Gill sculpture of him in the door of my room, but Gill did not know what Hooke looked like so he just produced an image of a Renaissance scholar. It looks just like any other Oxford don although Hooke was not an Oxford don—but Gill did not know that; I presume that he imagined he must be because Hooke worked with Robert Boyle in Oxford.)

Conference proceedings

The final source of information to which I wish to draw attention is the conference proceedings of the major physics associations. Physicist colleagues say that these proceedings and reports age quickly and are of little use to them after (say) two years. This observation is not strictly true since it refers to the use of such publications as a guide to present day physics research. For historians these publications have an intrinsic value. Probably the most important are the reports of the General Assemblies of the International Union of Pure and Applied Physics (IUPAP); but very few research libraries hold copies of these reports and the only reliable source is the ICSU Headquarters in Paris. The General Assemblies of IUPAP witness some of the most important deliberations of representative physicists that ever take place but it is the people who go to the meetings who get the reports, and the members of the committees reporting to these assemblies who have the literature. So I hope the archives of IUPAP are kept in good order and are well preserved for future historians. It occurs to me that this is information that really ought to be in research libraries. We must endeavour to collect this material in future. It would be interesting to know how many physicists refer to these when they want information about some important development in physics. Some of them, of course, hit the headlines—I suppose the most significant example of recent years is the phenomenon called nuclear winter which has been taken on board by the ICSU Scope Committee and their reports were published and distributed around the world, about 600 copies of every bulletin being sent out. This is an exception, and it is rare for an ICSU activity to get that much coverage. Probably, there is a wealth of information for historians in the collection of literature produced by ICSU and it may well be worthwhile to carry out a systematic search to locate it.

Conclusion

In conclusion, a few publications of continuing importance to physics historians are listed in Appendix 1.1.

I hope that this chapter has conveyed some idea of the scope of research libraries for studying the history of physics. For the future, it is important for librarians (and physicists) to look at the gaps in the coverage in order that the resources that are made available to readers are improved. So often scholars assume that if something is not in the library it does not exist, but I can assure you that that is not a real commentary on the world situation.

References

Gillispie C C (ed.) 1970–80 *Dictionary of Scientific Biography* 16 vols (republished in 8 vols 1981) (New York: Scribner)

Hackmann, W D 1985 History of physics in *Information Sources in Physics* ed. D F Shaw 2nd edn (London: Butterworth) pp 209–32

Harvey A P (ed.) 1986 *European Sources of Scientific and Technical Information* 7th edn (Harlow: Longman)

INSPEC 1987 *List of Journals and Other Serial Sources* (Hitchin: IEE)

INSPEC classification 1988 (Hitchin: IEE). The alphabetical subject index is now printed at the end of the classification

ISO 7498 1984 Description of basic reference model for open system interconnection

Open Systems Interconnection 1987 (Maidenhead: Pergamon Infotech) (State of the art report 15:5)

Poggendorff J C 1863– *Biographisch-literarisches Handworterbuch zur Geschichte der exacten Wissenschaften* (Leipzig: Barth) (currently published by Akademie-Verlag, Berlin)

Roller D H and Goodman M M 1976 *The Catalogue of the History of Science Collection of the University of Oklahoma Libraries* (London: Mansell)

Royal Society 1867–1925 *Catalogue of Scientific Papers* 19 vols in 4 series (London: Royal Society)

Royal Society of London Catalogue of Scientific Papers, 1800–1900. Subject Index (1908–14) 3 vols (Cambridge: Cambridge University Press)

The World of Learning 1988 1987 38th edn (London: Europa)

Appendix 1.1
Some useful publications for physics historians

Books

Brush, S G 1972 *Resources for the History of Physics* New England: University Press.

Frost A J 1880 *A Catalogue of Books and Papers Relating to Electricity, Magnetism and the Electric Telegraph, etc.* including the Ronalds Library (Soc. Tel. Eng.)

Heilbron J L and Wheaton B R (eds) 1981 *Literature on the History of Physics in the 20th Century* Berkeley Papers in the History of Science 5 (Berkeley: OHST, University of California)

Jayawardene A 1982 *Handlist of Reference Books for the Historian of Science* Science Museum Library Occasional Publications 2 (London: Science Museum)

Knight D M 1975 *Sources for the History of Science 1660–1914* (The Sources of History Ltd, Hodder and Stoughton; now distributed by Cambridge University Press)

Mehra J 1975 *The Solvay Conferences on Physics. Aspects of the Development of Physics since 1911* (Dordrecht: Reidel)

Weaver W D 1909 *Catalogue of the Wheeler Gift of Books, Pamphlets and Periodicals in the Library of the American Institute of Electrical Engineers. With Introduction, Description and Critical Notes by Brother Potamian* 2 vols (AIEE)

Weindling P 1983 Periodical literature and societies in *Information Sources in the History of Science and Medicine* ed. P Corsi and P Weindling (Guildford: Butterworth)

Wheaton B R and Heilbron J L 1982 *An Inventory of Published Letters to and from Physicists, 1900–1950* Berkeley Papers in History of Science, 6 (Berkeley: OHST, University of California)

Whitrow M (ed.) 1972–82 *Isis Cumulative Bibliography. A Bibliography of the History of Science formed from Isis Critical Bibliographies 1913–65*, 5 vols (London: Mansell); Neu J (ed.) 1980 continuing *Isis Cumulative Bibliography 1966–1975* (London: Mansell).

Journals

Adventures in Experimental Physics (Princeton) (1972–)
Annals of Science (London) (1936–)
Archive for History of Exact Sciences (Berlin etc) (1960–)
British Journal for the History of Science (London and Oxford) (1962–)
Centaurus (Copenhagen) (1950–)
'Histoire des sciences et des techniques' section of the *Bulletin Signalétique* (Centre National de la Recherche Scientifique Paris) (1947–)
Historia Scientarium (Tokyo) (1980–) (Formerly *Japanese Studies in the History of Science*)
Historical Studies in the Physical Sciences (Philadelphia) (1969–79)
History of Science (Cambridge) (1962–)
Inventory of Sources for History of Twentieth-Century Physics (ISHTCP) Office for History of Science and Technology of the University of California at Berkeley

Isis (Philadelphia) (1912–)
Isis Critical Bibliography (Philadelphia) (1913–)
Janus (Leiden) (1896, but present form started in 1957–)
Notes and Records of the Royal Society (London) (1938–)
Physis (Florence) (1959–)
Technology and Culture (Detroit) (1959–)
The Natural Philosopher (New York etc) (1963–4)

Appendix 1.2
Published catalogues of major research libraries with holdings in physics

Compiled by R J Wyatt
Radcliffe Science Library, Oxford

This list is by no means exhaustive. Catalogues of national libraries have been excluded, as have lists of current periodicals. For a list, by place, of printed catalogues published up to the late nineteenth century, see Stein H 1897 *Manuel de bibliographie générale* (Paris: Picard) pp 711–68

Universiteit van AMSTERDAM.
Tijdschriftenlijst. 1913.

Indian Institute of Science, BANGALORE.
Union catalogue of periodical holdings available in the libraries of Indian Institute of Technology at Madras, Kharagpur, Kanpur, Delhi, Bombay and at Indian Institute of Science, Bangalore. 1971.

Freie Universität BERLIN.
Alphabetischer Katalog/Monographien. Microfiche edition. Stand: Juli 1981. 1982.

Humboldt-Universität zu BERLIN.
Catalogus librorum qui in bibliotheca Universitatis Litterariae Fridericae Guilelmae Berolinensis adservantur. 1839.

Technische Universität BERLIN.
Katalog der Bibliothek der Königlichen Technischen Hochschule zu Berlin. 1900.
——I. Nachtrag. 1907.

University of BRISTOL.
Catalogue of the periodical publications in the libraries of the University of Bristol. [1940].

University of CALIFORNIA.
University of California union catalog of monographs catalogued by the nine campuses. 1963/67– . 1972– .

The University of California union list of serials. Prelim. edn, Dec. 1971. [1971?]
University of California union list of serials. [Microfiche]. 1976–9?
California academic libraries list of serials microform. [Microfiche]. 1980– .

(Berkeley).
List of serials in the University of California library. 1913.

Author–title catalog. 1963.

Serials key-word index. 1st edn 1973.
——2nd edn [1974].

(Los Angeles).
Dictionary catalog of the library of the University of California, 1919–62. 1963.

CALIFORNIA Institute of Technology.
Serials and journals in the Caltech libraries. 1979.

University of CAMBRIDGE.
Union catalogue of scientific libraries in the University of Cambridge: scientific conference proceedings, 1644–1972. 1975.

List of serials available in Cambridge University Library and in other libraries connected with the university. [Microfiche]. 1985– .

John Crerar Library, CHICAGO.
A list of serials in public libraries of Chicago and Evanston (comp. by Chicago Library Club). 1901.
——Supplement. 1903.
——Supplement. 2nd edn 1906.

Author–title catalog. 1967.

Classified subject catalog. 1967.

Technische Universiteit te DELFT.
Catalogus. 1867.
——Vervolg. 1873– .

Systematische catalogus. Deel II. Wiskundige Wetenschappen. 1907.

Lijst der periodieken. 2. uitg. 1909.

Trinity College, DUBLIN.
Catalogus librorum impressorum qui in bibliotheca collegii ... Trinitatis ... juxta Dublin, adservantur. 1864–7.

University of EDINBURGH.
Catalogue of the printed books in the library of the University of Edinburgh. 1918–23.

Bibliothèque Publique et Universitaire, GENEVA.
Catalogue de la Bibliothèque publique de Genève. 1834.

Catalogue de la Bibliothèque publique de Genève. 1875–87.

Niedersächsische Staats- und Universitätsbibliothek, GÖTTINGEN.
Verzeichniss. (Periodicals). 1883.
Katalog der Zeitschriften und Serien. ⟨Bestand 1946ff.⟩ Stand: 1.3.1968. [1968].

Göttinger Zeitschriften-Nachweis. 1933.

Göttinger Zeitschriften-Nachweis (GOZN). Verzeichnis der seit 1924 gehaltenen Zeitschriften und Serien-Werke. 1953.

Universität HANNOVER.
Katalog der Bibliothek der Königlichen polytechnischen Schule zu Hannover. 1868.

——Nachtrag von 1868 bis 1877. 1877.

Katalog. 1893.
——Nachtrag von 1893 bis 1904. 1904.

Kataloge der Technischen Informationsbibliothek (TIB) Hannover. Microfiche-Edition. Alphabetischer Katalog bis 1982. Konferenzkatalog bis 1983. 1983.

Zeitschriften-Verzeichnis. 1968.

HARVARD University.
Distributable union catalog. [Microfiche]. 1981– .

Periodical publications in Harvard science libraries [Microfiche]. [1981–].

Combined list of serials. 1983.

University of ILLINOIS.
List of serials in the University of Illinois library, together with those in other libraries in Urbana and Champaign, comp. by Francis K W Drury. 1911.

Leopold-Franzens Universität INNSBRUCK.
Catalogus. 1792.

University of IOWA.
A list of serial publications in the libraries of the University. 1911.

Serial publications in the University of Iowa libraries. 1972.

Linda Hall Library, KANSAS CITY, Missouri.
Serial holdings in the Linda Hall Library. 1973.

Universität Fridericiana KARLSRUHE.
Zeitschriftenverzeichnis der Universität Karlsruhe (T.H.), des Kernfor-

schungszentrums Karlsruhe und weiterer Forschungseinrichtungen im Raum Karlsruhe. 1968–69.

——Nachtrag. Alphabetischer Teil. 1970– .

KYOTO University.
Kyoto Daigaku Obun zasshi sogo mokuroku. Shizen kagaku hen. 1980.

Kyoto Teikoku Daigaku Fuzoku Toshokan yosho mokuroku. Catalogue of European books in the Kyoto Imperial University Library, 1897–1913. 1919.

Rijksuniversiteit te LEIDEN.
Catalogus bibliothecae publicae Lugduno-Batavae. 1640.

Catalogus Bibliothecae Publicae Lugduno-Batavae noviter recognitus. 1674.

Catalogus librorum tam impressorum quam manuscriptorum bibliothecae publicae Universitatis Lugdunobatavae. 1716.

——Supplementum ... usque ad 1741. 1741.

Lijst van periodieken te raadplegen in de bibliotheek der Rijks-universiteit te Leiden. 1903.

University of LIVERPOOL.
A hand-list of academies and periodical publications in the university libraries. 1913.

Finding list of scientific, medical and technical periodicals. 1966.

Finding list of periodicals excluding the humanities. 1971.

Imperial College of Science and Technology, LONDON.
Provisional list of periodical holdings. 1961.

Royal Society, LONDON.
Book catalogue of the library of the Royal Society. 1972.

University College, LONDON.
Catalogue of the periodical publications, including the serial publications of societies and governments. 1912.

University of LONDON.
Catalogue of the library of the University of London. 1876.

List of periodicals. 1956.

Union list of serials. [Microfiche]. 1979– .

MASSACHUSETTS Institute of Technology.
Serials and journals in the M.I.T. libraries. 1969–80?

Serials in the M.I.T. libraries. [Microfiche]. 1981– .

Ludwig-Maximilians-Universität MÜNCHEN.
Gesamtzeitschriftenverzeichnis der Universität München. 1970– .

Technische Universität MÜNCHEN.
Katalog der Bibliothek der Königlichen Technischen Hochschule zu München. 1881.
——Nachtrag. No. 1. Jan. 1881–June 1892. 1892.
——II. Nachtrag. 1908.

Bestand an deutschen und ausländischen Zeitschriften. 1939.

Gesamtzeitschriftenverzeichnis der Technischen Universität München (GZTUM). Stand: 1. Juni 1976. 1976.

OHIO State University.
Ohio State University periodical holdings. 1959?

University of OKLAHOMA.
The E De Golyer collection in the history of science and technology; a check list of the books and other materials. 2nd edn 1953.

A check list of the E De Golyer collection in the history of science and technology as of August 1, 1954. Compiled by Arthur McAnally and Duane H D Roller. [3rd edn]. 1954.

Short-title catalog of the history of science collections, University of Oklahoma Library. 1969.

The University of Oklahoma libraries serial holdings. 1968.

The catalogue of the history of science collections of the University of Oklahoma libraries. 1976.

Radcliffe Science Library, OXFORD.
Catalogue of the books in medicine and natural history contained in the Radcliffe Library. 1835.

Catalogue of books on natural science in the Radcliffe Library at the Oxford University Museum up to December, 1872. 1877.

Catalogue of books added to the Radcliffe Library, Oxford University Museum ... 1873–1926. 1874–1927.

Provisional catalogue of transactions of societies, periodicals and memoirs. 1866.

Catalogue of transactions of societies, periodicals and memoirs. 2nd edn 1871.
——3rd edn 1876.
——4th edn 1887.

Union list of serials in the Science Area, Oxford. Stage i. 1968.
——Stage ii. 1970.
——Stage ii. First Supplement. 1971.

Oxford University union list of scientific serials. [Microfiche]. 1983– .

École Polytechnique, PARIS.
Catalogue de la bibliothèque de l'École polytechnique. 1881.

———Premier supplément décennal. 1892.

Université de PARIS.
Catalogue de la Bibliothèque de l'Université de Paris. Section des sciences et des lettres (Sorbonne). 1905– .

Liste des périodiques français et étrangers qui se trouvent à la bibliothèque de l'Université (à la Sorbonne). 1883.

Academy of Natural Sciences of PHILADELPHIA.
Catalog. 1972.

PRINCETON University.
Alphabetical finding list. 1921.

Weizmann Institute of Science, REHOVOT.
Periodical holdings of the Weizmann Institute of Science. 4th edn 1969.

Università degli studi di ROMA 'La Sapienza'.
Catalogo delle pubblicazioni periodiche esistenti nella biblioteca della R. Università di Roma. 1932.

Kungliga Tekniska Högskolan, STOCKHOLM.
Katalog öfver Kongl. teknologiska institutets bibliothek, år 1848. 1849.

———Första supplementet ... 1849–53. 1854.

———Tillväxt-katalog, 1854–87. 1889.

———Tillväxt-katalog, 1888–1913. 1913–16.

TOKYO University.
Tokyo Daigaku gabujutsu zasshi sogo mokuroku. Wabun hen = The University of Tokyo catalogue of scientific periodicals in Japanese languages. 1985.

Eberhard-Karls-Universität TÜBINGEN.
Tübinger Zeitschriften-Verzeichnis. 1930.

———Neubearbeitung. 1965.

———Neubearbeitung. 1970–2.

———Neue Ausg. 1957.

———Nachtrag. 1957.

Norges Tekniske Høgskole, TRONDHEIM.
Samkatalog over tidskrifter i Norges tekniske høgskoles biblioteker. 1969.

Periodika i Norges tekniske høgskoles bibliotek. 1972.

UPPSALA Universitet.
Catalogus librorum impressorum bibliothecae Regiae Academiae Upsaliensis. 1814.

Technische Universität WIEN.
Katalog der Bibliothek des K.K. Polytechnischen Institutes in Wien. 1850.
——2. neu geordnete und verm. Aufl. 1868.

Systematischer Katalog der Bibliothek der K.K. Technischen Hochschule in Wien. 1901–9.
——Nachtragen. 1910–50.

Uniwersitet WROCLAWski.
Verzeichnis der von der Staats- und Universitäts-bibliothek und der Instituten der Universität gehaltenen Zeitschriften aus der Gebieten der Medizin und Naturwissenschaften. 1931.

Bayerische-Julius-Maximilians-Universität WÜRZBURG.
Gesamtzeitschriftenverzeichnis Würzburg. 1973.
——Supplement. 1975.

Eidgenössische Technische Hochschule ZÜRICH.
Katalog der Bibliothek des Eidgenössischen Polytechnikums in Zürich. 6. Aufl. 1896.

Verzeichniss der Bibliothek des Schweizerischen Polytechnikums. 2. Aufl. 1857.
——3. Aufl. 1859.
——4. Aufl. 1866.
——5. Aufl. 1876.
——Supplement zur 5. Aufl. 1887.

Zeitschriftenverzeichnis für Technik und Architektur, Mathematik, Physik und Chimie. [*c*. 1948].

Verzeichnis der in der Bibliothek der E.T.H. vorhandenen Zeitschriften aus den Gebieten der Technik und Architektur sowie der Mathematik, Physik und Chemie, nach dem Stand vom 1. April 1951. 1951.
——Nachtrag vom 1. April bis 1. August 1951. 1951.

Periodicaverzeichnis: Stand 1972.

2

Preserving and Making Known the History of Physics: The American Institute of Physics Center for History of Physics†

Spencer R Weart

The nature of the center

In 1961 the American Institute of Physics established a Center for History of Physics. By now the Center offers a great variety of resources and services. These resources and services will be described here, together with some of the things we ourselves have done with them. This is by way of inviting the reader to make use of what we have to offer.

But first of all, what exactly is this Center for History of Physics? In order to understand what we can do, it is necessary to know what we are. As an administrative unit, we are a division of the American Institute of Physics. The AIP is roughly but not entirely analogous to the British Institute of Physics. To be precise, the American Institute

† Based on a presentation given at the conference on The History of Physics for the Physicist, 2–4 July 1986, Oxford.

of Physics is a membership organisation whose members are *not* individual scientists, but rather societies: the members of AIP are the American Physical Society, the American Association of Physics Teachers, the American Geophysical Union and so forth—a number of leading American societies. Thus the Center ultimately reports to these societies.

AIP is a big business, with over 500 employees. It does housekeeping tasks such as handling dues collection for its member societies, and it markets the journals of the British Institute of Physics in the United States. More important, AIP handles many of its member societies' journals, ranging from the mighty *Physical Review* down to the little *Journal of the Society of Rheology*—on the library shelf, a year's worth of AIP publications occupy twenty feet. Included in this are many journals that do not belong to any of the member societies, for example, *Physics Today*, and journals translated from the Russian: these are owned by AIP itself. Since the AIP is a not-for-profit organisation, the revenue of the sales from its own journals is available to be spent for scientific and cultural purposes. It is this revenue that allows the AIP to support such activities as surveys of the demography of physicists; a news service; and even a history Center.

The AIP Center also draws financial support from grants for special projects, which we raise from government, private foundations and industry. Typically such grants account for a quarter to a third of our budget. And we get donations from hundreds of individuals, mostly physicists—we call them the Friends of the Center for History of Physics, and many of them are indeed not only donors but good friends. In fact our most important backing is not financial at all. We rely on many scientists, historians of science and science archivists, who cooperate in our work.

While the Center is located in the United States and chiefly deals with people in that country, our resources and services are available to anyone in the world. We certainly recognise that science is an international activity, and that the history of physics in particular has at least as much to do with Europe as with the USA. An important number of our friends and contacts are in Britain and other countries.

So much for what we are; next I shall describe what we do—and in particular what we do to help physicists and physics teachers.

The needs of the physics community were foremost in the minds of the founders of the Center, back around 1961—they were after all physicists themselves. Perhaps their greatest concern had to do with the famous problem of the public's image of physics, and of science in general. They noted that history textbooks and other repositories of popular culture gave little attention to twentieth-century science, except occasionally to view it as a mysterious fount of technology and especially of atomic bombs. Our founders hoped that the Center, by furthering accurate historical research, would exert an indirect effect on the general community of scholars, and so eventually on the public. The aim, then, was to help the public to understand the goals, methods and traditions of science. And this public was to include physicists themselves. Usually the Center's supporters have spoken in terms of physics students, who very much need to know the 'lore of physics' as one student put it. Older physicists felt that younger students had a very imperfect feeling for how physics has been done, and how it should be done, and that history could be a great help here. Personally, I don't think this problem is confined to young students; there is plenty of room for all of us to improve our understanding.

For, as we all know, much of our familiar lore is not history: it is myth. As the humorist said, the problem ain't what you don't know, it's what you do know that ain't so. The public, our students, and some of our colleagues, perhaps even you and I, know things about the history of physics that simply are not accurate either to the historical facts or to the real spirit of doing science. And if we believe anything as scholars, it must be that an accurate story is better for everyone than a myth.

In the long run, this accurate story must be created through scholarly research. The founders of the Center wanted to foster such research, and here they found a particularly troublesome problem. The raw data that must be in hand for scholarly research—such materials as the unpublished correspondence between modern scientists and the recollections within the memories of eminent figures—these were vanishing into the incinerator and the grave. In fact, as of 1961, virtually no physicist's recollections had been tape-recorded: we missed Einstein, for example. And in the entire United States at that time there was only one collection of papers of a prominent twentieth-century physicist that had been placed in a permanent repository. Physicists were unaware that their correspondence had any value for future historians; indeed, we learned not to tell physicists that their 'papers' should be saved (the usual archivists' lingo) because they would reply, 'but my papers are available in any good library'—they thought only their *published* papers were of permanent interest. Most archivists, on their side, shied away from the idea of preserving physicists' unpublished notes and correspondence: surely there would be nothing but a mass of equations that only a genius could understand?

This set the terms for the Center's primary task: to ensure the preservation of historical source materials. Closely following would be the second task: to make such materials available for use by anyone who might need them.

The result was a rapid accumulation of information resources. These were of such richness and interest that they began attracting not only professional historians of science but also journalists, film-makers and textbook writers. These groups included some physicists who doubled as writers or whatever. Meanwhile the Center sought ways to cut out the middlemen and make the materials directly available to ordinary working physicists and physics teachers.

Today the Center, located at the AIP headquarters building in midtown New York City, has a staff of five professional plus six secretarial and clerical assistants, under the direction of Joan Warnow—an archivist—and myself. The staff handle thousands of mail inquiries and hundreds of personal visits each year. I'll describe shortly what resources we make available. But I want to describe first how we ourselves have used the Center's resources. I hope it will promote awareness of some of the things that can be done with history.

First I'm going to describe how we have used our materials to support educational purposes. Later on I'll get to the matter of scholarly historical research. I think the chief long-term purpose of such research must be to instruct: that is, after one learns something from research, one ought to make the results known. So I'll start at that end first, the final result—public education and science teaching.

Educational activities

The AIP Center's most widely seen product is the Einstein exhibit. This drew on the Center's photograph collection—and further items we found in a search—to produce a set of 18 large printed posters, which could be hung along the walls of a corridor or mounted on an arrangement of folding stands. Our point was to portray Einstein not only as a theorist, but as a human being embedded in the scientific, political and cultural life of his times. For the text we relied, of course, on prior historical research, as interpreted for us by several leading scholars.

Some 80 copies of the exhibit, mounted on the folding stands, were circulated throughout the United States and abroad, going on display in colleges, libraries, museums, even state fairs and airport lobbies. In addition, over 400 sets of posters were sold to institutions that did their own mounting—I'm afraid they are now sold out, but many physics departments, in particular, were able to get their own copy. We sent out this exhibit in 1979, for Einstein's centennial, and after eleven years (in 1990) copies are still on display in some locations. Some time ago we got an inquiry from the State of Oklahoma Humanities Council, asking us for a replacement for a damaged part of the exhibit;

we were interested to learn that it was still in circulation there, and had visited a dozen locations in the state, mostly schools, in the previous year. The way the Oklahoma people circulate it is typical: they ask each school to pay them a small sum for administration, plus the cost of shipping the exhibit to the next location that requests it.

The Einstein exhibit has been seen by millions, probably tens of millions, of people, and we are told that many of them have spent the 20 or 30 minutes that are needed to read the entire text. In its impact we believe it has been at least as effective as a science television programme, and it was done for a fraction of the cost: roughly £100 000 at current prices, supported of course by grants. The real point here is that history can be an excellent vehicle for reaching people, and if the effort is supported by good photographs, it can give science a human face—no metaphor, the real thing.

Another example of an educational project drew on another of the Center's collections, our archives of the tape-recorded voices of scientists. I can think of nothing that can bring people closer to the life of science than letting them hear the scientists themselves, telling how they did their work. The Center experimented with this idea by putting together four tape cassettes (Hoddeson 1974). One was an 'anthology' of voices of famous figures, another took such voices and wove them together into a narrative, a third drew on an interview with a single Nobel prize-winner and the fourth used interviews with two young scientists. All four cassettes were field-tested in college and high-school classes, and an advisory panel of educators and media specialists recommended two of the cassettes for further development. The development they recommended was to build a package including illustrated transcripts, slides and teacher's guides. These guides include extensive background material as well as suggested teaching plans and student exercises. We got the help of an excellent secondary school teacher to help us prepare the guides: that's a substantial task in itself.

The whole package, published with the title, *Moments of Discovery*, could be made available at a low cost since its development was supported by foundation grants and the Friends of the Center for History of Physics (Warnow *et al* 1984). The package is divided into two units. One of these is called 'The Discovery of Nuclear Fission', and includes the voices of such figures as Rutherford, Einstein, Frisch and Fermi, drawn from recorded speeches and oral history interviews. These voices are connected by a narrator into a story that illustrates the complex path toward scientific discovery, in this case of fission and the chain reaction—not omitting some notice of the practical consequences. We were able to tell a full and accurate story because there has been much historical research into this subject; not only have scholars ploughed through the archives, but most of the leading participants have put their recollections on record. In fact many of these recollections were among the tape-recordings we drew upon, so that in this case historical research and educational purposes were mixed together from the start.

The second unit is called 'A Pulsar Discovery'. It is built on the experience of two young astronomers who made an important observation, the discovery of the first pulsar visible at optical wavelengths, during their first run on a telescope. They had used a tape-recorder to keep track of their observations—and the microphone was sensitive, so besides hearing them say 'This is run number 22' and the like, you can hear their exclamations of doubt and delight as the night proceeded. The unit includes this inadvertent recording—a striking real-life demonstration of the struggles and triumphs of extracting scientific fact from the jumble of observations. But extracts from this unique recording make up only a few minutes of the cassette; the rest is woven together from interviews that we conducted for the purpose with the two astronomers, plus narration that Philip Morrison kindly supplied.

I said the interviews were conducted expressly for this educational project, but they are also oral history interviews in their own right, and have been used as such by scholars. This is just the reverse of what happened with the fission interviews, which were done originally for scholarly purposes and were then turned to educational use. But actually this may be a false distinction: when people talk into a tape-recorder, it is to make the facts known—or anyway, their version of the facts—and their ultimate audience is not the historian, but everyone.

In any event we have put together these packages so that the scientists can, in a way, talk directly to students, through the cassettes. Each unit can be the basis of one or more days of classroom work, or can be used by individual students for extra credit. *Moments of Discovery* has been enthusiastically accepted by reviewers and teachers. We knew we were on the right track when we got some of the students' comments; one of them said, 'It helped me see that scientists can be human beings'. (Not *are*, perhaps, but at least *can* be!)

Again, we hope our experiment will inspire others to make similar use of historical source materials. As with the exhibit, we think the cost of making such units is low in comparison with the impact they can have. Still, I must confess that here too the initial cost—including marketing and so forth, before any income is received—could reach £100 000. In addition, such a package, and exhibits too, take a tremendous personal investment of interest and care; only those who have done such things can fully appreciate how many deadlines and difficult decisions are involved. In sum: not easy, not cheap, but well worth the effort.

Of course, not all use of historical source materials for educational purposes must be so elaborate. Much the same sort of thing can be done in still smaller pieces: for example, a few good photographs can do a lot in a textbook or when shown in class.

The Niels Bohr Library

For physicists who are interested in using historical materials for teaching or public education, the American Institute of Physics Center—or strictly speaking, the Niels Bohr Library, which is part of the Center—offers a wide range of resources, most of which are available even to those who cannot come to New York City.

Our most heavily used resource is the photograph collection. We have some 20 000 photographs of physicists, astronomers and others. Formal portraits are the most common type of picture, but we have also sought out informal snapshots showing scientists in their natural state, at the blackboard, in the laboratory, at a conference, or on a picnic or skiing vacation. This is what can really give a feeling for the human side of science. Our most interesting photographs are originals or copies from the personal photograph collections of eminent scientists—and I should mention here that these are not just Americans; a number of people in Britain and other countries have generously allowed us to make copies from their collections.

Our staff provides prints and slides at cost for non-commercial use in response to mail or telephone inquiries. It is not necessary to come to New York; given clear instruction, we can make a choice from among the many pictures available, or we can send Xerox copies for inspection, at a modest cost. As a further aid to those who cannot come to view the files, the Center has issued separate catalogues of our 600 photographs of Einstein and our 200 photographs of Niels Bohr (Warnow 1979).

Less frequently used but no less valuable is the library's collection of tape-recorded speeches and interviews. Here are the voices of some of the most famous modern physicists and astronomers. We have about 1000 tapes of speeches and the like given on public occasions, all catalogued by the name of the speaker, and another 800 or so oral history interview tapes. Again, these are not just from the United States but include eminent scientists of many countries, although naturally we have collected many more in the USA than abroad. Many of these tapes may be copied at cost for non-commercial use. For example, the Center has received a number of requests for a copy of the tape of the lively impromptu talk that Enrico Fermi gave in 1954, describing his pioneering work on the nuclear chain reaction (Fermi 1954).

So much for how we can provide direct aid in regard to instructional materials. This is not really our main business; it is more common for the AIP Center to reach physicists, physics students and the public in a still more indirect way. Our greatest concern is to provide aid to historians, who will then, we trust, make the results of their research known. Indeed we encourage historians not to write just for their fellow specialists, but to publish in magazines like *Physics Today*, magazines that reach physicists, including teachers and advanced students. Over the past two decades the number and quality of such historical articles have indeed markedly improved, providing a rich supply of information (Weart and Phillips 1985).

We also try to help people find their way to articles on the history of modern physics, in journals they don't commonly read, by providing a little bibliography—telling historians what is published in the physics journals, and telling physicists what is published in the history journals. This is the most popular feature of our Center for History of Physics *Newsletter*, which we issue twice a year. The *Newsletter* also reports on activity in the history of physics and related sciences. (The *Newsletter* is available to anyone free on request.)

Of course, before historians can publish they must actually do the research. We have now worked backwards to the heart of our work—and that is the task of making it easy for historians to get access to the materials they need. Of course, when I say historians, I include those physicists who themselves do historical work, at every level from amateur to professional. When their studies are serious enough to benefit from published or unpublished source materials, the AIP Center with its Niels Bohr Library is one of the first places they should look to. We welcome all scholars who can make use of our resources.

Further, since travel to New York and living expenses while there are costly, we have a program of grants-in-aid to help scholars who need to use the library. This is a very small programme, supported from our small endowment fund; we can provide up to $2000 reimbursement of travel expenses. Those who are either studying toward a degree in history of science, or who have some record of publication in the field can apply for funds: simply send a two-page letter describing the reasons for the visit, along with a curriculum vitae and a budget showing how the funds will be used.

However, many of our resources can be used at a distance. For example, the library not only contains much of the current literature on the history of physics, but also an extensive and unique collection of printed materials that are out of print. These must not leave the premises, but photocopies are made on request. We specialise in the period from around 1850 to 1950, and we try to collect materials not easily found elsewhere. For example, we have one of the largest existing collections of old textbooks and monographs, including their different editions; usually libraries do not retain such things, although it can be very instructive to see how new physics results were incorporated into successive editions of a text. We also collect printed materials that are not really books, such as instrument catalogues—these have interesting information on the prices and capabilities of commercially available apparatus, as well as wonderful illustrations. And we collect things like problem books and laboratory manuals. These have not yet found much use, for the history of science instruction is, so far, an unexplored area; but when someone does want to explore it, we have the resources.

Published materials are not the only thing that we can make available at a distance. We also do a brisk business by mail in loaning our microfilms of the unpublished notes and correspondence of scientists. With these a researcher can study the papers of people like Bohr, Rutherford and Hubble. Of course, many of these microfilms are also available at other locations, such as the Science Museum in London. But any *bona fide* scholar who has difficulty getting to a microfilm of the papers of a physicist (or astronomer, geophysicist and so forth) should ask us. We try to have a copy of every extant microfilm in our areas of interest. If we don't have the one required, we will try to get it, and then loan it out. This includes microfilms of doctoral dissertations—a fine resource that we would like to encourage historians to use more thoroughly.

Furthermore, if there exist historically important papers that are not microfilmed but should be, we would like to know of them. We will be glad to provide advice and perhaps funding so that the filming can be done. If the sum needed is not over $2000, a grant-in-aid may be applied for; for larger collections we may help to raise outside funds from foundations and the like. We feel strongly that documentation such as letters is the most reliable way to enter into the real history of science, and we want these materials to be used widely.

That does not mean, of course, that what was originally written as private letters can be opened to casual inspection or frivolous use. Like all archives, we will give access to unpublished materials only to people with *bona fide* scholarly or educational purposes; a potential user must fill out an application form, stating his or her purpose and listing two references who will vouch for them.

Oral history interviews

Much of what I've said of microfilms of unpublished papers is also true of the oral history interviews that the Center has conducted with leading scientists. Ranging up to 35 hours long and averaging over three hours each, the interviews give a penetrating and intensely personal view of scientific life.

Most of the interviews are fully transcribed and indexed. Some of the interviews are restricted, and may be used only with permission of the person interviewed, but most of them are unrestricted, and the transcripts can be loaned by mail, on the same terms as the microfilms of correspondence. To help scholars find out which of the over 800 interviews connect with their particular research questions, we have made abstracts and detailed tables of contents, and these too are available by mail. We also have a cross-index of names and institutions covering most of the interviews; a subject index is in preparation.

As mentioned by DeVorkin in Chapter 3 the quality of an interview greatly depends on the interviewer's preparation—several days of research should be done before each interview session—so the AIP Center has conducted its interviews in connected series within the framework of one or another extended project. This allows the preparation to be deeper than if we skipped from field to field. The purpose of these projects is not simply to gather interviews, but to preserve every kind of historical source material, and especially written materials such as correspondence. The chief projects in the past have covered atomic and quantum physics—this is the Archives for History of Quantum Physics project—and also the rise of nuclear physics, and modern astronomy and astrophysics. But many other areas from geophysics to health physics have also been touched upon.

Such work is currently being extended in several directions, such as the International Project in the History of Solid State Physics. This is nearing completion, resulting not only in over 100 tape-recorded interviews but also in published works by Lillian Hoddeson and others. We have done this work in close cooperation with a group in Britain including Ernest Braun and Paul Hoch and also with a group under Jürgen Teichmann at the Deutsches Museum in Munich. The AIP Center has been responsible for the bulk of the processing of the interviews. This project will result in a volume of historical monographs, to be published by Oxford University Press, arranged in chapters covering many aspects of the history of this field from the nineteenth century to about 1960 (Hoddeson *et al* 1990). This project has been the first sustained inquiry that has ever been made into the history of solid state physics—which is rather shocking actually, considering how tremendously important this field has been. We feel that in terms of social impact, and perhaps also intellectual achievement, solid state physics ranks among the most important fields of twentieth century science, and we would like to see still more historians working on it. To help stimulate this, we will later produce a catalogue of source materials. This will give access both to the interviews the project staff have conducted and the unpublished manuscript materials they have located.

By analogy with big science, we could call this a 'big history' project—involving over a dozen historians and archivists, hundreds of thousands of pounds, and much cataloguing and administration. It's the sort of thing that few institutions can, or indeed want to, do. Our advisors have always urged the Center for History of Physics to do such things, that is, things that we can do better than the usual university department structures.

Another project of this nature, now near its end, is the Laser History Project. This is sponsored by a number of scientific societies and groups including the AIP, and is under the direction of Joan Bromberg. The point, again, is to conduct interviews, to locate and catalogue manuscript materials and to write a monograph. The history of lasers, however, includes a new problem, since it is entirely in the post-World War II period. Like much of physics in this period, research and development was funded about half from the military budget, under conditions of secrecy. We therefore arranged for secrecy classified interviews and writing to be done by Robert Seidel, a historian who obtained the necessary Department of Defense Top Secret clearance and Department of Energy Q clearance. With the aid of military laser specialists who very much want their story to be told, he is finding it possible to declassify and make public the main results of his research.

We are taking a different approach in another current project, run by Finn Aaserud, a postdoctoral historian at our center. Aaserud is trying to preserve a historical record of the scientists who have been active in government policy advising, including military areas. As a special topic within this broad area, he is studying the history of the Jason group, a collection of very eminent younger physicists who have been advising the US Department of Defense since the 1960s. Aaserud has *not* sought security clearance, but we believe that publicly available manuscripts and open oral history interviews will allow him to reconstruct, and publish, the main outlines of what has been done in this area—an area, again, of great importance to all of us but scarcely explored by scholars. Here again it is the sort of thing that we can do better than others, simply because the reputation of the American Institute of Physics and its history programme, and our location *inside* the physics community, helps us gain the confidence of the physicists.

An even touchier project we are helping with is work by William Glen, who is recording the contemporary history of the controversy over the causes of mass extinctions in the geological record—the asteroid impact hypothesis. His interviews cover intellectual and personal conflicts that are entirely current: history in the making.

In all these projects, the interviewing is done chiefly in the United States, but interviews of key European scientists are also conducted, both when they visit the USA and when we come to Europe. We also try to locate relevant manuscript materials wherever they may be, without limiting ourselves to the USA.

More generally, much of the work on these projects is done outside the AIP Center in New York City. Of course, our staff (including myself) take part in the historical research and writing. But aside from that, the Center's role is to provide administrative support, to process the tape-recorded interviews and, not least, to help in raising grants from the federal government, industrial corporations and private foundations. We can be most effective, in fact, not by doing work ourselves but by helping and stimulating work by others.

Roughly half of our interviews were not done under our direction at all, but were donated by historians who conducted them during their own projects. We are glad to be known as a primary repository for such source materials. We can even support some interviewing by providing grants-in-aid to reimburse travel expenses, and by providing free transcription if a transcript is called for. If any tape-recorded interview is to be deposited with us, we do insist on two conditions. A brief abstract must be provided, describing what is on the tape. Without such a description to lead scholars to it, the tape would be buried on a shelf and forgotten. And of course we must have some evidence, either on the tape or better still in writing, that the person interviewed is willing to let other scholars hear the interview—with any restrictions clearly specified.

The archives

Interviews are some of our most fascinating materials, but still not the most important. Our greatest concern in all work is to save not just personal recollections but also historical 'hard data'—the unpublished correspondence and other papers of individuals, and the records of organisations. Center staff help to get these saved at the most appropriate repository, typically at the institution where most of the work represented in the papers was done. That is where people will look for the papers, and that is where they can be near the administrative files and the papers of colleagues that help tell the whole story. In short, we act as disinterested brokers, trying to get scientists or their families together with the appropriate archivists.

When this system was set up 28 years ago, it was unprecedented. Archivists usually had little way of knowing that they were neglecting to preserve papers of important scientists, or if they did find out about a collection of a particularly famous figure, they would compete with one another to get it. Our method of cooperative advice was devised by people quite ignorant of archival matters—the physicists who founded our Center—following their own traditions. I'm glad to say that the archival community has come to recognise the value of the cooperative approach, and that it is spreading rapidly.

The Center itself does not always place papers elsewhere; our Niels Bohr Library does sometimes serve as a repository for papers that do not have any more appropriate home.

On a number of occasions we have not only taken in and organised collections, but then used this as an argument to shame institutions into establishing an archival programme, so that they could recover the papers of their great scientists. We then ship them the collections, in one recent case 20 years after we acquired it. Our work is a bit different from that of Britain's National Cataloguing Unit for the Archives of Contemporary Scientists at the University of Bath but there are also many parallels in our approaches.

I should add that we at the AIP Center are interested not only in American papers but in any important set of historical source materials that need preservation. In Britain there is, of course, the fine Cataloguing Unit just mentioned and we rely on their work and their reports. But for other regions, anyone who knows of papers that ought to be preserved is urged to get in touch with the AIP Center. We'll do what we can to help get materials preserved.

In recent years the effort to save unpublished papers has run into a problem: archivists traditionally deal with the files of an individual, but since about 1940 science has become increasingly a matter of teams. Center staff under my colleague Joan Warnow have attacked this problem through extensive studies of how records are preserved—or lost or destroyed—at institutions such as the vast US Department of Energy laboratories. I should add that we were greatly encouraged by the example of Harwell in particular and the UK Atomic Energy Authority in general, which under Margaret Gowing had already developed a programme for saving records of such institutions. As a result of our own efforts, several major US laboratories have established similar programs. Now Joan Warnow and staff have begun a project to study a still more complex problem: how to save a historical record of teams that involve people from a number of different institutions. There are many physics projects nowadays that involve several universities, a government laboratory, industrial subcontractors and more. Whether a coherent historical record can be assembled for such a team remains to be seen, but we will at least study the issue.

Happily, so much is being successfully preserved that a new problem arises: how are historians to find what they want amid all that paper? The AIP Center from the beginning has had a programme to record the location of materials—whether in the United States, in Britain or anywhere else—in our International Catalog of Sources for History of Physics and Allied Sciences. Anyone seeking the unpublished papers of physicists, astronomers and so forth will find this catalogue a useful tool. Every year we receive, spontaneously or in response to our inquiries, information on over one hundred new deposits of papers. We report the chief items in our *Newsletter*. Since 1981, new information on papers in repositories is entered on a computerised database, and we are currently working to put our 1000 older records on the database as well. At present, if you need help locating papers you must write to us to ask about them. But we have raised funds so that in a few years the information will be available on-line for those who subscribe to the main US library database network. We also hope to publish in print a fully indexed summary of the International Catalog. In the meantime one part of the data, a guide to the archival collections in our own Niels Bohr Library, will be published shortly (Warnow 1990).

I have not covered all our activities or resources, but these are certainly the main ones. And how successful has the AIP Center been? It is not easy to document just how much such a programme has accomplished; so much of the work is invisible, since it aims to stimulate others rather than to perform on its own. What gives me the most confidence that we are on the right track is the fact that the AIP's example has been noted by groups in other scientific disciplines. During the past decade in the USA several history of science organisations have been founded in areas outside physics, largely inspired by our Center. The largest of these is the Center for the History of Chemistry, at the University of Pennsylvania; then there is the Center for the History of Electrical Engineering of the Institute of Electrical and Electronics Engineers, just around the corner from us in New York City; and the Charles Babbage Institute for the History of Information Processing, at the University of Minnesota. All these centres carry out most of the kinds of work I've described—exhibits, historical research and publications, oral history interviewing and other preservation of source materials, archival survey and research and cataloguing.

Naturally we're pleased that others concur in the importance of what we're doing—but we do not think that the American Institute of Physics Center is a model that can or should be imitated everywhere.

We can do some things that others cannot, and some other institutions can do things that we cannot. Still, we feel that many of the particular programmes we have could be taken up by one or another group in their own national context. On the other hand, among our programmes are a number that it would be pointless to duplicate, such as the International Catalog of Sources for History of Physics, and the photograph collection; here we have received close cooperation from groups and individuals around the world, and we look forward to more of the same. Although our sponsor is an American organisation, we very much hope that physicists the world over will consider it *their* Center for History of Physics as well. That means we are glad to serve you ... and we also feel free to ask for your help. This is the spirit that is found at the great accelerator laboratories and telescopes, the international cooperation of science. It can function just as well in history of science.

References

Fermi E 1954 *Address to the American Physical Society, 30 January 1954* (transcript published in Weart and Phillips)

Hoddeson L 1974 The living history of physics and the human dimension of sciences *Phys. Teacher* **12** 275

Hoddeson L, Braun E, Teichmann J and Weart S (eds) 1990 *Out of the Crystal Maze: Chapters in the History of Solid State Physics* (New York: Oxford University Press) in press

Warnow J N (comp.) 1979 *Images of Einstein : A Catalog* (New York: American Institute of Physics). (The *Catalog of Photographs of Niels Bohr* is available on loan only.)

—— 1990 *Guide to the Archival Collections in the Niels Bohr Library at the American Institute of Physics* (New York: American Institute of Physics) in press

Warnow J N *et al* 1984 *Moments of Discovery* (New York: American Institute of Physics)

Weart S R and Phillips M (eds) 1985 *History of Physics : Readings from Physics Today* (New York: American Institute of Physics)

3

Interviewing Physicists and Astronomers: Methods of Oral History†

David H DeVorkin

The data of science and its history

In science, and the history of science, the collecting of information and the doing of intelligent and creative things with it is at the heart of the enterprise.

Also as in science, there are many forms information might take in presenting its history. Traditional resources are the printed word intended both for peer and public consumption, and the privately written word meant not for public consumption but for communication between colleagues and peers, friends or family. Both the published and archival records are essential sources for historical research. Without the wealth of letters, reports, internal memoranda, minutes of meetings, laboratory notebooks, records of observations, photographs and diaries left behind by scientists and their institutions, modern historians would only be able to recapitulate the published record. While this might be appropriate in preparing a scientific review paper, it does not make for complete or satisfactory history. If, as

† Based on a presentation given at the conference on The History of Physics for the Physicist, 2–4 July 1986, Oxford.

Michael Hoskin and Owen Gingerich (1980) argue, 'The primary duty of the historian of astronomy [or of any scientific enterprise] is to illuminate his science as a creative human activity of the astronomical community of the time', then the historian must use all avenues available to reconstruct both the process of doing science and the social and cultural context within which the science was done.

In short, the historian, as with the scientist, is limited to the data at hand. The scientist takes every opportunity to gather data already available from earlier efforts, and then broadens the database by contributing observations of his or her own. The historian too takes every advantage available; and in these modern times when the paper trail has been replaced by ephemeral electronic conversations and paper shredders, the historian of modern science must seek additional ways to reconstruct or, in fact, create anew the historical database. At the very least, the modern historian has to find means of unravelling the complexities of modern science, and devising a plan of attack that uncovers original materials.

Oral history as an additional resource

Oral histories in science are formal interviews, usually taken with an audio tape-recorder and now sometimes with video technology when the situation requires documenting visual information. They are, in short, structured conversations between historians and those who have played a part in creating the history of the subject matter of interest to the historian. They may be useful for gaining specific information about a person's experiences, training, personal background, ideas, opinions, biases, relations with others; anything that helps to do, among other things, one or more of the following tasks:

(i) Answer specific questions that may clarify issues the historian is currently interested in.

(ii) Uncover issues and events the historian is unaware of in his or her general interest area.

(iii) Provide general documentary material that will be useful to future historians—a personal statement set down for posterity by a person of historical importance, or who played a part in events of historical importance, or who was in a position personally to view events of significance.

(iv) Obtain a collective body of knowledge about a discipline, institution or scientific body that goes beyond the published record to the processes of creating that discipline, or events that changed the character of the discipline, or institutions involved.

(v) Serve as a vehicle for historians entering new areas of study where no synthetic work is available as a guide, or which is so complex that the historian lacks the perspective even to know what proper questions to ask.

In addition to these five major tasks, characteristics of either 'focused' or 'archival' oral histories (project-oriented versus programme-oriented studies), there are several important parallel goals of oral histories that often provide materials exceeding the value of the interviews themselves:

(i) The identification of original documentation not already in competent archival hands. Collections of working notes, log-books, photographs, correspondence, internal minutes and original documents come under this category.

(ii) The identification of the material artefacts of the scientific enterprise that otherwise might be lost, cannibalised, or destroyed through neglect. These range from prototype detectors, optical, mechanical and electronic systems, to otherwise off-the-shelf devices that somehow played a role in normal research as well as in discovery, or played a role or represented an episode in instrument development that was later abandoned, but represents a valuable object lesson.

In both cases, engaging in oral history interviews with those who, through personality or circumstance, have preserved these types of records and objects provides some semblance of control over the preservation of the contemporary archival and artefact record by simply finding out what exists before it is destroyed or lost.

As point (v) above hints, properly designed oral histories are a learning process both for the historian and the subject. Oral history helps the historian gain knowledge of the existence of original documentation not yet in the hands of archivists or curators, and helps to increase the historian's appreciation for the archival record at hand. But oral histories may also act as a mode of gaining knowledge of a discipline or activity unfamiliar to the entering historian. They are, in effect, formal methods of gaining intimate and personal contact with scientists, engineers, project managers, administrators and other players in the scientific process. If well designed and executed, they are an effective *entré* into the scientific world for the historian, who must gain the confidence of his or her subject not only to obtain a candid history based upon personal recollections, but to seek out supporting documentation for those recollections, usually best available from the interview subject.

If the historian is sufficiently effective at communicating his or her interests and needs during the interview process (before, during and after the actual interviews), the scientist will become sensitised to the nature of contemporary historical enquiry. It is not unusual for practising scientists to be surprised at the questions posed by historians. Even though they may have taken the effort to read modern history of science, they might not be aware of the historiographical processes embedded in the doing of that history. Many sincerely feel that the equipment they used, the notes they made, or the letters they wrote during the course of scientific investigation are not worthy of preservation. To many, the published record of their efforts is all that matters; this to them is the scientific legacy. They are surprised and a bit bewildered to learn of the historian's argument that the scientific process is not transparent in the published record (see Rudwick (1985)). Motives behind specific problem choices, social and cultural influences, and the many and complex interactions the scientists had with their peers, institutions and patrons all mould the scientific landscape and are necessary ingredients of the historical record.

The pitfalls of interviewing

Despite the potential of oral history interviewing as a new source of historical information, as an *entré* to the scientific community and as a means of identifying original materials, many historians remain cautious in cultivating the art and practice of oral history. Some historians prefer to leave oral history alone, and some argue that the products can be dangerous and misleading, since they are apt to be misused by historians who are either uncritical or unaware of their limitations. When they are used as *entré* especially, pitfalls lurk everywhere, and the historian would do well to avoid at all cost using such documents without meticulous confirmation.

Oral histories are imperfect documents. At the very least, there is a 'Heisenberg uncertainty principle' working since the oral history process is a collaborative one between historian and scientist, and by being collaborative, inevitably changes both participants to some degree, however so slight. In the mere act of historian meeting scientist, and making the scientist aware that his or her opinions and recollections will be preserved and may be exploited by future historians, scientists may be prompted to adopt a public image, even a mask, if you will, that reflects what they want to have remembered about themselves, their life and their accomplishments.† Most scientists

---

† This tendency has been noted in a most fascinating study that contrasts the oral historian and the psychiatrist (Lomax and Morrissey 1989).

are willing to cooperate because they see it as an opportunity to set the record straight, or to provide a recollection of the past in a particularly personal light. Knowing the historian is watching, or is asking specific questions, scientists are likely to tailor responses to meet both the historian's expectations and their own, to the best of their ability.

Underneath the intentions of the scientist, memory is faulty to start with, and imperfectly designed questions posed by historians stimulate improper responses, and thereby foster distorted visions of history. In fact, there is good reason to suppose that the mere act of asking a question influences the reply. It is not unusual to find that an historian, already deep into his or her subject, may have a broader and quite different perspective on a scientist's life than the scientist being interviewed, especially if that scientist did not work in isolation but within a larger structure or organisation, as most do today. But the historian's picture is necessarily idiosyncratic; it is a reflection of the historian's interests and preconceptions. Thus, if the picture of that era, and where the scientist fitted in, is transferred to the scientist by the historian during the process of the interview, or during preparations for it, what biases or world views the historian might carry as intellectual baggage might as a result flavour the memory of the scientist who is being interviewed. Similarly, if the scientist has had the opportunity to read published memoirs by colleagues, either serendipitously or furnished to him or her by the historian, it is not unlikely to find that the scientist's oral history recollections, sincerely thought to be based upon personal experience over time, stem more from reading the recollections of others. Memory is tricky, and anyone depending upon it has to be terribly vigilant.

For example, recollections of time-related sequences of events, of who did what and when, or of how something happened, often are innocently garbled and in every instance should be verified. In one instance, an interview subject placed himself at a particular event when an original telegram showed he was several hundred miles away when it happened. His story was right, only he was not there. He no doubt heard about it later. Two contestants for priority for a discovery in the 1930s, still sensitive about it in the 1980s, each have recollections that favour themselves, but the archival record shows that what probably happened is a conflation of the two recollections that supports neither side. To cite only one other example: a third-party recollection of a controversy between two scientists reversed their roles in a manner favouring a colleague.

The interviewer and researcher hoping to generate and/or use oral history must be prepared to deal with such situations.

The key to good oral history

Oral history may well be an excellent vehicle as a means to an end, but in no way should it be considered an end unto itself. The successful use of oral history depends upon a sensitivity to what it is—what limitations are at play—but most of all, to what it is not. Oral history interviews are not replacements for original documentation such as correspondence, laboratory notebooks, manuscript sources, or other evidence contemporary to the historical period of interest. They should never be generated as primary source material for direct citation without exhausting all other avenues first. And when they have to be used directly in citation or inference, source notes must accompany them stating clearly that the conclusions derived were based upon recollections. And explicit qualifications ('according to Professor XXX, ...') must be built into the narrative to clearly identify quotations from oral history as opinion, not as fact.

The key to good oral history is sensitivity to its limitations. While questions about priority, who did what and why, and other typical issues cannot be avoided, different lines of inquiry stemming from more recent interests in the history of science often find oral history a most useful tool. Taking an anthropological or sociological view helps: the recollections of an interview subject ten, thirty or forty years after the fact (or even one week!) may well not be trusted as fact. Nevertheless, they *are* recollections, they are the intellectual baggage this person has been carrying around, they are the collective memory and mythology of a community of elite workers and they do represent the history one wishes to express more or less as fact, with full sincerity and honesty. It is not a question, however, of trust or sincerity: it is a question of having the proper regard for what memory is capable of providing, and what has conditioned that memory. The persistence of memory for some types of experiences or impressions, and its absence for others, should alert the historian to attitudes or issues embedded in the interview subject's psyche that may lead to more fruitful lines of inquiry if pursued in new and fresh ways. Considering the existence of an historical uncertainty principle, the act of taking an oral history injects the historian into the historical record. The ultimate user of that information must always be conscious of the fact that the interview subject is aware that an historian is asking him or her questions. Knowing this and appreciating what results from it can reveal a great deal of useful information about the interview subject, or about the norms of shared impressions the person expresses about his or her research group, workplace, chosen discipline and world. As such, oral history becomes an invaluable means for capturing a deeper perspective on the human condition in science, if used correctly and with cautious reserve.

Preparing for oral history interviewing

Whether an oral history is part of a general archival programme whose goal is to collect information on a particular discipline or major institution, or is part of an individual historical research project focused upon a particular issue or question, certain basic steps in preparation are essential. Assuming that specific problem areas or themes have been identified, either from exploratory interviews, examining review papers, studying citation patterns, or merely the specific interest and experience of the historian, the first problem is to establish priorities for whom to interview.

In subject- and institution-related projects there is usually little question about who should be interviewed. The sample is usually small and directly identifiable. If the subject sample or institution is vast, or if the goal of the project is to survey a large population of scientists, then priorities are essential.

First, age and status in the community are obvious criteria to consider. Citation patterns in major review journals can identify high-priority candidates, as can questionnaires sent to a select list of scientific patrons. These steps reveal what the community thinks about itself. Other criteria come from the particular interests of the historians involved, as well as a structured attempt to preserve the recollections of those who may have an unusual perspective on the science of their times. Finding the latter is difficult; the best way is to be as familiar with the community as possible.

Once the person or persons to interview are identified within a specific problem area there will be several types of data to collect about that person, the institutions and groups involved. The AIP commonly has taken the following steps.

Ingredients for biographical profile

(i) *Who's Who* citations, testimonials presented on the occasion of the interview candidate receiving an award, or upon retirement.

(ii) Curriculum vitae with significant published papers (books, patents, reports, etc) noted by interview subject which he or she feels best represent his or her career contributions.

These are to be compared with:

(i) Citation analysis of all publications. For example, the Institute for Scientific Information (ISI) based in Philadelphia compiles a *Science Citation Index* (SCI) which is a running account of the citations to articles appearing in major scientific journals. It identifies which authors cite the works of the interview subject, and where the citations appear. The yearly compilations should be distilled by tabulating the number of citations each paper receives year by year, identifying the total, any self-citations by the author, and the year of maximum citation. Highest cited papers and papers that are self-cited are likely candidates for discussion, especially if they are not previously mentioned by the interview subject as being perceived as important to understanding his or her career.

(ii) The most frequently cited papers should be read and digested, and, if of sufficient interest, those papers citing the subject's work should be examined to see why they are being cited. At least note who is citing the subject and prepare questions about those people.

(iii) Biographical information should be collected on major professors and colleagues, as well as on the tenure of major students, identified partially through examination of published papers, noting collaborators, etc.

Institutional profiles

(i) Obtain institutional histories of those places the subject worked at for any significant amount of time. Corporate histories, observatory and laboratory yearly reports, news items in professional and trade publications and alumni magazines are all good sources. In the case of an extended archival programme, it is useful to obtain likely institutional profiles before the interview process commences because many of the people to be interviewed may well have been trained at, or worked at, a variety of institutions.

(ii) Collect staff histories and tabulate this information in graphical form whenever possible: identify the growth of department or observatory or laboratory as a function of time and the tenure and positions of staff as a function of time. This material might refresh the memory of the interview subject about who he or she worked with, who was contemporary with him or her and who came at what time. Materials that quickly refresh time frames are quite valuable to have during the interview and often serve to get around sticking points. When an interview subject persistently forgets time sequences or dates, even though they are trivial and can be reconstructed later, this sometimes creates a feeling of unnecessary despair during the interview.

Committee and panel histories

Determine the basic purpose of major panels, committees and boards the subject served on and what they actually accomplished. Try to obtain membership histories, and chronologies of major issues each panel faced, especially during the tenure of the subject. As in the case of the institutional profiles, this information might best be gathered coherently for all significant boards and panels within the general subject matter area of interest to the oral historian prior to beginning any one oral history. Significant links and issues are often uncovered by reviewing such material, and it is best to have it at hand well before any interview so that the historian can prepare effective questions about the function of particular panels and boards.

All information obtained should be read and applied to the preparation of specific questions for each interview subject. Beyond standard biographical questions, separate questions should be designed to draw out the interview subject's status within his or her department or institution, relations with colleagues, teachers, students, administrators and family, based upon what work was accomplished during the tenure of the interview subject at each location. If a department chair retired, the historian should be interested in the process of selecting the new chairman; if the size or structure of the institution changed in any way, questions should be developed to find out why, and if these changes had any effect upon the career of the interview subject. Highly cited papers should be read in some detail, not only for their content, but for some insight into the types of questions to create to find out if the interview subject considers the paper to be successful, why it was cited, how he or she came to research and write the paper, and what the review process was like getting the paper into print. If the paper has co-authors, the interview subject's relation with the others must be determined. Reasons for self-citation and speculation on why others cite the paper (naming names from the *Science Citation Index*) are valuable pieces of information to gather from the interview subject.

Other sources

(i) Examine oral histories of people who had been colleagues or contemporaries of the interview subject. Use these to gain a perspective on possible reactions by the interview subject to questions posed during the interview. If oral histories are available of those who worked in allied areas of research, knowing their opinions beforehand of the value of the new interview subject's work prepares the historian, who, however, *should make every effort to design questions that do not reveal the sources from which questions had been obtained*. Previous oral histories are most useful when the historian wishes to obtain perspective on appropriate questions to ask. If a previous interview subject had something definite to say about a colleague or event, or a piece of research, the historian must design unbiased questions which are close to those asked of the previous subject. Otherwise the two responses cannot be evaluated collectively, which is sometimes the best use of oral histories.

(ii) Review papers, unpublished notes, letters and other archival records such as photographs, film footage and artefacts are often the best guides for specific questions. As these become available, and as time allows, the knowledge of such primary material is both helpful for generating intelligent questions, and is of critical importance in learning more about the significance of the material itself, especially the artefacts. Examining photographs during the interview helps to stimulate and focus memory, and in turn the elements of the scene can be identified: names, dates, instruments, the nature of the occasion and who took the photographs. All of this should be documented.

Based upon the foregoing research, questions generated can then be put on large index cards, and arranged in their order of use during the interview. Putting questions on cards allows for flexibility. If the interview subject wants to go off in another direction during the interview, cards can be shifted easily to accommodate the new direction. As questions are answered, the cards can be dropped.

While the historian ideally wishes to prepare carefully for each interview by some method such as that identified here, quite often interview subjects present themselves without warning. They are in town for a meeting, are passing through and decide to stop in, or for some reason or other are made available on short notice; so short in fact that the historian has time only to load his or her tape-recorder and photocopy a *Who's Who* entry. Of course, the historian will have a good reason for making the effort at short notice, and usually if experienced, has a set and ready arsenal of questions to ask that cover useful territory for the 'archival' or general record. If an historian has a well defined set of interests, has conducted numerous interviews and general research in the subject matter area, and has also identified clearly what problems have to be addressed and solved, then 'off-the-cuff' interviews are possible and profitable, but never preferable to deliberate planning.

But much depends upon the personality of the subject, his or her willingness to be expansive and helpful, and the degree to which the interviewer is adept at stimulating this expansiveness. Even so, adequate preparation before an interview, a structured plan for what is to be obtained during the interview, and sufficient familiarity with the field and general research area so as to be adept at taking advantage of opportunity when it arises, all are ingredients of a successful archival oral history programme.

The archival interview

The date and site of the interview should be identified as far in advance as possible. Depending upon the accessibility and age of the interview subject, initial interviews should not be too long or exhausting. They should be designed to gain the confidence of the subject, and to further establish useful lines of discussion. Usually two to three hours is optimum.

The site for the interview should be at the convenience of the subject. He or she should be at ease, but ideally also close to published material, working notes, correspondence, and even his or her laboratory if possible. While one might thus expect that interviewing at the worksite would be the best, because familiar surroundings sometimes stimulate memory, this is not always the case. Personality, preparedness and readiness of the interview subject are more significant factors. And there are problems with interviewing at worksites. The interview site should be as free from distractions as possible. Interviewing a scientist at the peak of his or her career, at the worksite, sometimes creates a very choppy record as the subject is unable to concentrate, being interrupted constantly by daily responsibilities. Those who are close to or at retirement age usually have schedules more amenable to reflection and contemplation. Priorities for choosing interview candidates stem naturally from circumstances such as these, but if a younger and highly active person is to be interviewed, a best time would be beyond normal working hours, possibly during weekends.

The interview is best conducted by one person, the historian, although two and even sometimes three people have been present in one session if the interview subject is willing. But clear ground rules must be established among the interviewers beforehand on who leads with the questioning, and how the others participate.

Ground rules have to be identified at the outset of the interview session. Standards developed at the American Institute of Physics in New York for interviewing physicists and astronomers and followed at the Air and Space Museum include:

(i) The interview subject retains literary and editorial control over the product.

(ii) The interview will be taped and transcribed, edited by the interviewer or his or her staff, and then sent back to the subject for comment and possible revision. The fully edited interview will then be retyped, with table of contents, and be made available subject to restrictions on access, reproduction and citation identified by the interview subject.

(iii) The interviewer tells the subject what will be happening: the interviewer will usually be taking notes during the session to aid the transcriber, and to develop further questions.

(iv) The direction of the interview is fully determined by the subject, but the interviewer is interested in specific areas and will strive to cover them, subject only to limitations set by the subject.

(v) The interviewer should state that the session can be interrupted at any time and for any reason, or that the tape-recorder can be turned off or put on pause at any time.

(vi) The interviewer should be clear and direct about his or her specific interests, as well as about the general archival nature of the interview (if this is the interviewer's mission as well). If there is any question in the interview subject's mind about the intent of the interviewer's questions, they should be brought up immediately.

Most interviews for archival preservation as well as for specific historical projects start with biographical profiles. Questions such as 'I know you were born in 1917 in Chicago, but I would like to know more about your family, your father and mother...' draw out the interview subject, put him or her at ease, and help to convince him or her that we are interested in his or her life, and that our questions are not narrow, closed yes or no enquiries, but are open and designed to get subjects to explore their own lives in any way they may choose.

As the initial session develops, try to be sensitive to the nature of the answers. 'Body language' and certain character traits often reveal dynamics that are invisible on audiotape, and require a certain facility in the interviewer to bring out what the person is really trying to say. Look for qualifiers in the subject's responses, and then ask about them, whenever possible. If the subject seems to be losing track of an issue or specific recollection, try to draw him or her back to the initial question. If the subject responds saying 'I'm getting to that...' let them do so, but be vigilant and quietly persistent to a degree determined only by how important the hoped-for information might be. Vigilance is proper, rigid adherence to a line of questioning is not. One never knows the course a reply will take, and often diversions reveal fruitful avenues of enquiry that could never have been anticipated or prepared for.

Good oral historians are always interested in tangents and diversions the interview subject takes, even if they are not the immediate target of the interview session. We know that someone in the future might well be interested in this material, and all effort must be made to stimulate expansiveness. Highly focused interviews and rigid criteria for discussion stifle opportunity. Would it not be sad to have had the opportunity to interview someone who was an active member of the Royal

Astronomical Society during the debates between Eddington, Jeans and Milne in the late 1920s and early 1930s, and to be either unaware of those debates when you interviewed that person, or not respond when they were mentioned! It takes both good preparation and a willingness to explore unexpected opportunities that make for good oral history. Even if a new subject is not fully familiar and you do not know all sides of an issue, if the opportunity presents itself, it is best to grab it, especially if the interview subject is elderly.

Always try to be sensitive to 'throw-away' statements by the interview subject. For example, if the Royal Astronomical Society debates were simply mentioned, and the interviewer did not fully appreciate their importance, or their substance, the fact that they were mentioned warrants a quick follow-up, or a note to follow up soon. At the right time, ask: 'You mentioned debates between Eddington and Jeans, did you actually witness them?' 'Why did you recall them just now?' 'What were they about?' 'Were they really debates?' 'Did you participate?' 'Where might I learn more about them?' 'Who is still around to talk to about them?'

While questions such as these may not help the historian write a particular research paper, they might aid the future historian immensely. This is one of the key differences between focused interviews and archival interviews.

The art of knowing how to ask questions is acquired through experience, although reading oral histories others have taken helps too. At its best, oral history is a carefully crafted art. There are some rules, to be sure, but on the whole, each interviewer must find a personal formula that works for him or her. We have mentioned a few examples. Here is a review and expansion:

(i) Ask open questions that cause the interview subject to reflect on events, issues and processes.

(ii) If there is a possibility that the interview subject has not fully understood the question, do not hesitate to rephrase it, but also do not belabour the point if the subject cannot recall or understand what you are after. Be flexible.

(iii) Try to ask only one question at a time. In the rush of the moment, interviewers often ask a string of related questions, and the subject responds only to the last one.

(iv) Be aware of qualifiers, and follow up by asking why they were used. If, for example, an interview subject notes that an experiment was 'almost perfect' ask why 'almost?' 'What stood in the way of its perfection?' 'Did others have this same opinion?' 'Where did it fail to meet your expectations and needs?'

(v) Ask 'why' questions, or 'why do you suppose...' questions. The latter asks for an opinion, and this makes it clear that the oral history interview documents opinions and recollections which are not necessarily facts. 'Why' questions are prompts to open up the interview subject. 'Why did you choose Dr X to be on your research team?' 'Why did you decide to (go to) (leave) (attend) that (school) (department) (meeting)?' 'Why did you choose to perform your research in that way? Why not another (specify an alternative, if you know of one)?'

(vi) Be sensitive to 'throw-away' remarks, small tangents and diversions. They often reveal important issues. Ask open questions: 'Please tell me more about your contact with that fellow?' 'Do you want to spend some time talking about this?' 'Why did you just mention that?'

(vii) Be on the lookout for any evidence of original documentation that might aid the discussion. Ask: 'Do you have a letter or any record of your contact with "xxx" that might help us to better understand this point?' 'Have you made a point of preserving your letters, correspondence, notes of meetings? Have you ever kept a diary? Where are these documents?' Unless these documents are critical to the discussion at hand, and are easily available, the interview should not be interrupted too frequently to get at them. But the tape-recorder can easily be put on pause for a few minutes, or even left running while the subject searches his or her files. Often perceptive and revealing comments emerge while the subject looks through his or her papers. Though these may be random comments, subsequent editing of the oral history transcript will make them accessible, and provide useful material for follow-up interviews.

If original documentation is readily available during the interview, ask the interview subject to read passages and then to comment on them, given the value of years of hindsight and insight. Asking the right questions during the commentary often results in fascinating revelations. Typical questions include: 'Was this an unusual letter or memorandum?'; 'Were its contents and directives already determined or understood verbally or through previous correspondence, before the execution of the document?'; 'What prompted the writing of the document in the first place?'; 'Are there other letters or documents that might help me better understand the significance of the material presented here?'

Questions such as these often reveal the process that brought the particular document into being. In specific cases, memoranda or letters brought to interview sessions as examples to raise during the questioning were found to be prepared after much invisible activity, and that the document was merely a reflection of the understanding that all parties had agreed to, rather than a part of the process of agreement.

Further, many documents were not prepared by the person signing it; this is especially true for documentation of recent vintage coming from large organisations. In several cases, letters and work orders were written by the recipient upon request by their superior. In a few cases, those creating the material freely admitted that they were expressing a collective opinion, subject to review, and that if they had been free to do so, a different message would have been conveyed.

The most common reaction of a subject upon reading letters or original documentation is that it stimulates their memory, and disciplines them somewhat in speculating upon what actually was transpiring at the time the document was created.

(viii) Be ready to reveal your tentative hypotheses, analyses or theories about the historical issues you are studying, but reserve them for later parts of the interview, or until the interview subject really needs to know. Ask questions that allow the interview subject to comment, but take care not to bias his or her responses. Try to ask a question based upon your hypothesis in such a manner that it does not unduly influence the response. Questions such as 'Did you do the experiment in this manner because you were out of money?' could intimidate or even anger a sensitive individual if asked in an abrupt or insensitive manner. If asked with a certain empathy, possibly using a famous example known by both historian and interview subject to set the stage for the reply, then the interview subject may be more at ease to respond with useful candour about their lack of support. Conversely, the subject may be defiantly proud of their perseverance, in spite of the lack of funding, and succeeded. After all, the subject is being interviewed presumably because he or she has been a successful contributor to their chosen profession.

The last type of question requires some additional discussion. Always be prepared to hear that your scenario about a particular process or event, and questions you have developed to test it, will be countered by the experience of the interview subject. If the interview subject responds that 'You have it all wrong...' it should be an invitation to ask him or her to set the record straight. Solicit commentary that supports or rejects your hypothesis, but realise that the technique is suspect at best, and should not be used if you think a subject can be convinced of your version of their history. It is better to hold back, ask open questions and not indulge in hypothesis unless you are desperately stuck.

It is disconcerting during an interview to be forced into asking leading questions, and to have the interview subject constantly agree with the wording of the question, as if to be done with it. From this one can only conclude that it is far too easy to 'lead the witness', to suggest that something happened in a particular way, and then to find that the respondent is only too happy to agree, for whatever reason.

Be careful to qualify any speculation. Preface your speculation by admitting 'I am not sure about this... but, could "xxx" have happened in the following way?' 'Does this sound right to you?' 'I have read in "xxx" that you said "yyy", did you in fact say this, and was this because "zzz"?' or 'Dr "X" noted to Dr "Y" that "zzz" happened in the following way...Does this sound right to you, because earlier you said...?'

This last example raises another issue: is it threatening to an interview subject to reveal that original documents you have read show that his or her memory is faulty? Tough problems such as these have to be treated delicately. Often it is wise not to reveal all that you know in too blunt a fashion, especially if the issue is not an important one to you and you are concerned that it might stifle the expansiveness of your interview subject. But it is equally important to remain honest and open. One way out of this delicate situation is to stop the tape-recording and reveal the question openly in confidence, and then ask how the interview subject wishes to treat the question. Another method suggested by Spencer Weart of the American Institute of Physics is to save controversial subjects and questions for the end of the interview session or sessions. Although they will be out of context, they will not interfere with all of the other responses you hope to obtain. In any event, avoid hiding intentions.

As long as the interview subject knows he or she is not compelled to answer your questions, and the historian demonstrates this by not being too obstinate or tenacious (although the latter is necessary in moderation) controversial issues should be entertained somehow, if they exist and the historian is interested in them. If willing to do so (for archival purposes primarily), the historian should clearly state that editorial and use restrictions can vary within an interview: parts of interviews can be 'closed' for 'xxx' number of years, or restricted, while others remain more open to use. If the interview subject knows this in advance, or is reminded of it during the interview, he or she will often relent. But it makes the information less accessible, and does not ensure that the interview subject will respond. The most common reaction by a subject is to declare ignorance of the issue at hand.

(ix) Finally, unless this session will be the only chance to talk with this person, the historian should try to be very sensitive to their limitations. As noted before, an initial session, and indeed follow-up sessions, are best limited to only two to three hours. Longer sessions are tiresome and should be avoided. In addition, follow-up interviews conducted after editing and processing the earlier ones are usually better because the historian has had a chance to review the material once again, read the subject's initial responses and prepare follow-up questions to clarify points and fill in gaps. The subject, reading the interview, will be able to do the same, and will know more about what to expect.

Without question the best way to end an initial session is to discuss with the subject what will be covered in the next session, and what types of documentary materials the historian would like to have at hand to prepare for it. Often the subject is able to find this material, and so both become better prepared for the next session in what becomes an iterative process.

Whenever possible, the historian should develop oral history interview sessions as iterative sets not only during the interview process, but during the writing or publishing phase. Once enough data have been gathered, and the historian is in the process of preparing a paper or book, the interview subject should be given a chance to review the rough drafts even though he or she might not be quoted. If needed, follow-up interviews can then be arranged to determine the subject's reaction. Here all sources have to be revealed, but now each source will have the chance to respond on an equal footing.

The product of the process

The initial product of a well-funded archival oral history is an edited, retyped transcription of the conversation, as close to the way it took place as possible, but with emendations provided by the interview subject after reflection. The text might include a table of contents, citations to major papers mentioned, clarifications of acronyms or esoteric terms, fully documented photographs and an index. The amount of processing is a function of the amount of time and money available. The tape itself is normally not used for direct citation, but to recover instances that may not be clear in the transcript, or to provide the historian with a sense of what the interview subject sounded like.

Since the finished product must be treated as if it were a private and confidential letter between colleagues, or as if it were a personal diary, it must be handled with a degree of care and control equal to that given to original data in an archive. For preservation and proper execution of editorial use restrictions, oral history interviews must be placed in established, responsible archives. This is an expensive process, and to justify it the product must be as reliable as possible.

The information gathered in any oral history interview must be regarded as provisional, subjective and in need of confirmation. Every effort must be made to find ways to check on what is said in an oral history before anything is used, either as a direct quotation or for general background. Therefore, since oral history is no replacement for the preservation and use of original archival documentation, it should be used as a technique to get at the original record, if that record is not already available.† The historian/interviewer should try to be as helpful as possible when depositing interview transcripts in an archive, providing information on where original documentation exists, or upon why the interview was taken. Conversely, established archives often accept materials only if they are accompanied by documentation generated by the historian/interviewer that help to illuminate why the interview was taken, what his or her motives and goals were and any cautions about the use of the material.

Non-archival interviews taken by historians interviewing only for specific personal interests might remain only on the tape. Most historians, unless they have decided to serve broader archival interests and have been successful in finding the support to do so, have worked in this manner in the past. The tapes can always be transcribed later, but the subject may not be available for comment, and thus the value of the further iterations is lost. In this latter case, if the interviews are deposited at an archive for future use by others, the importance of full documentation is even greater.

As with the data the scientist creates and uses, the data founded or created by historians, especially through oral histories, must be carefully weighed against all available forms of evidence. The best time and place to do this is during the interview process. Thus, locating original documentation becomes as much a product of the oral history process as the interview itself. During the interview, when documents are discussed that are not in the published literature, the interviewer should at least ask if the subject knows where they might be found. But the least obtrusive way to obtain information on primary documentation is during the editing of the transcript when the historian notes what is desired, and makes a point of requesting this material from the interview subject, or at least a statement that the documents are safe, identifiable and accessible in some permanent manner. If they are not, the interview process is an ideal time to identify and preserve such records. After the interview subject has read the interview, made comments and suggestions, and the transcript is put into final form, the product of the process must be augmented by asking for original documentation, especially to support or clarify important aspects of the recollections in the interview.

† To be sure, historians are keenly aware of the fact that original documents oftentimes contain the same pitfalls oral histories do, and for many of the same reasons (see Rudwick 1985).

This is an important follow-up process that promotes continual relation-building between the historian and the scientist. Most of all, it sensitises the scientist to the value of preserving the proper records. It is not surprising that a scientist might take the care to preserve memorabilia, such as notices of appointment or other milestones in his or her career. While these should be preserved, letters with colleagues are often far more useful in fleshing out the process of experimentation, speculation, discovery and elaboration. The historian will make this point usually by example, asking for information on documents that are naturally most useful to the historical record.

The art of oral history

The success of an interview is a combination of three ingredients: adequate preparation, the flexibility of the interviewer, and the personalities of the interviewer and subject. If the subject has a compelling wish to get a point on record, and sees the interview as his or her opportunity, then no amount of preparation will aid or detract from this, and the interviewer should be flexible enough to let the subject have their say. But a familiarity with that subject matter area might well help to draw out the interview subject beyond his or her 'prepared' or set statement. Even without a familiarity with the subject matter, the interviewer may simply be sensitive enough to the zeal of the interview subject to ask why he or she wished to have this particular statement on the record in the first place, or if this statement might give others pause, or be countered by a colleague. A general sensitivity to what the subject is anxious to communicate, and a curiosity to find out why, requires no preparation; only a receptive state of mind and a willingness to ask objective and moderately blunt but tactfully placed questions. Lack of interest is deadly.

On the other hand, deeply penetrating questions, based upon long hours of preparation, may not be too useful. Aking a subject to recall events and relationships 30 to 40 years cold is asking a lot. And even if the historian's expectations do not include critical and sharp replies to such questions, the mere act of questioning in this manner sensitises an interview subject to their limitations and diminishes their confidence. It takes a very special person to correctly recall events, impressions and experiences over a lifetime. Recalling experiences from the 1940s by someone still active today is quite an accomplishment. The interviewer must respect this and not try to intimidate. In many ways, the historian is closer to the subject matter than the interview subject, and has to carefully bring the latter back in time. Sometimes *not* knowing too much about a subject makes the interviewer less demanding and to some, less threatening. But above all, sensitivity to the limitations of memory is an essential requirement for an interviewer.

The interviewer can only try to aid a subject's ability to recall things, in the ways we have noted here, using what documents are available. But questions might also be asked in various different ways, not to trip up the subject, but to try to find the key that unlocks a particular memory. If a subject is not able to recall something, the interviewer should move on to other topics without troubling with it. The recollection may come later, but dwelling on it will only stifle the enterprise. The interviewer must above all try to promote the subject's willingness to concentrate, to participate in a fresh way, in reliving history. The right social chemistry or circumstance conducive to reflection is essential. How it is obtained is not a matter for textbooks or papers. It comes only from experience and striving for sensitivity.

Often one would wish to know about an interview subject's life and possess greater insight into his or her scientific legacy upon walking into the interview. But in my experience the interview subject is usually delighted to find that the 'interrogator' wished to be educated, and was educable. There have been times, on the other hand, when too much knowledge was a problem. Insight gained after several hundred hours of interviewing in a particular field sometimes can be intimidating, as can original letters or other primary documents that made them realise how much they had forgotten. Most interview subjects seemed to take this realisation philosophically and were not threatened, but a few required reassurance that it was, after all, the historian's chosen interest to ferret out these old things, and all he or she is after is an objective record; at least the historian should ask to hear their side of a story, if indeed there were sides to take. The fact is, historians have made it their life's work to think about the past, and therefore naturally have at hand more about someone's history than that person is ever expected to be able to draw directly from memory.

The art of the oral historian then, is to somehow use his or her specialised knowledge to stimulate accurate memories from the interview subject and to reconstruct with the interview subject the human dimension of the enterprise. The tools of the art, beyond a clearly defined and structured mission, specialised knowledge and questions that are based as much as possible upon original documentation and both subject and institutional profiles, include a sensitivity to the interview subject, and an ability to determine how to approach someone not in a threatening or fault-finding manner, but in a direct, objective and professional way.

The objective, simply put, is to secure recollections that are as accurate and useful as possible, though they may well be biased. Future users of these oral histories must be equally sensitive to their limitations. The art of oral history is as much in discriminating their use as in their production. They must be appreciated for what they are: personal visions of history.

Acknowledgments

The procedures reviewed here were developed by staff at the American Institute of Physics Center for History of Physics in New York City, where the author learned the art of oral history in the late 1970s. Although the present discussion is based upon a paper given at the Oxford Summer School, commentary and criticism provided by colleagues at the Summer School, and in the intervening years, have proven critical in the present revision.

References

Hoskin M A and Gingerich O 1980 On writing the history of modern astronomy *J. Hist. Astron.* **11** 145–6

Lomax J W and Morrissey C T 1989 The interview as inquiry for psychiatrists and oral historians—convergence and divergence in skills and goals *The Public Historian* **11** 17–24 (on p 21)

Rudwick M J 1985 *The Great Devonian Controversy* (Chicago: University of Chicago Press) pp 434–5 (and references therein)

Appendix 3.1
Selected Readings and Sources

General sources on oral history technique include:

Moss W W 1974 *Oral History Program Manual* (New York: Praeger)

Dexter L A 1970 *Elite and Specialized Interviewing* (Evanston, Il: Northwestern University Press)

Davis C, Back K and MacLean K 1977 *Oral History: From Tape to Type* (Chicago: American Library Association)

Baum W K and Dunnaway D (eds) 1984 *Oral History: An Interdisciplinary Anthology* (Nashville, TN: American Association for State and Local History)

Baum W K 1977 *Transcribing and Editing Oral History* (Nashville, TN: American Association for State and Local History)

For a discussion of the problems of oral testimony, see: Henige D 1982 *Oral Historiography* (New York: Longman)

For the pitfalls of gaining access to a specific cohort, see: Lerner D 1956 Interviewing Frenchmen *Am. J. Sociol.* **62** 187–94, which offers a fascinating account of techniques developed to gain access to members of the French elite, and how interview styles are developed. Applies as well to the study of any elite group.

A recent study that heightens the concerns of any oral historian to the pitfalls of gaining access through establishing a false trust is Janet Malcolm's two-part essay 'Reflections' (*The New Yorker* 13 March 1989 pp 38–73; 20 March 1989 pp 49–82).

Very helpful insight in placing oral history testimony in context with the established historical record can be found in: Weiner C 1988 Oral history of science: a mushrooming cloud? *J. Am. Hist.* **75** 548–59.

On methodology and process in studying the history of modern physics, see: Kuhn T S, Heilbron J L, Forman P and Allen L 1967 *Sources for History of Quantum Physics: An Inventory and Report* (Philadelphia: American Philosophical Society). This report introduces the Archives for History of Quantum Physics microfilm collection, deposited at a number of sites in the United States and in Copenhagen.

In contrast to the above, which was a finite project, programmatic interviewing closely linked to the central interests of thematic history centres is well described in: Aspray W and Bruemmer B (eds) 1986 *Guide to the Oral History Collection of the Charles Babbage Institute* (University of Minneapolis: Charles Babbage Institute Center for the History of Information Processing). This describes the collection, and identifies research themes.

Large-scale oral history projects in modern astronomy and astrophysics have been conducted at the American Institute of Physics in New York City, and at the National Air and Space Museum, Smithsonian Institution, Washington, DC. On the AIP project, see: Weart S R and DeVorkin D H 1981 Interviews as sources for the history of modern astrophysics *Isis* **72** 471–7. This includes hints for identifying selection criteria for an archival oral history program that surveyed astronomy at mid-century, as well as a name list of astronomers interviewed on the project, which totals over 400 hours with over 100 people. The fully transcribed interviews are on deposit at the Niels Bohr Library of the AIP in New York City. See also: Weart S R and DeVorkin D H 1982 The voice of astronomical history *Sky and Telescope* **63** 124–7. On the NASM project, see: DeVorkin D H, Collins M and Mayr S 1985 *Space Astronomy Oral History Project Catalog* (Washington: Smithsonian). This is the first report of an extensive study of space scientists, mainly those with astronomical backgrounds who were active in early space astronomy and space physics. See also: Collins M and DeVorkin D H 1986 Space astronomy oral history project *National Air and Space Museum Research Report 1985* (Washington: Smithsonian), 203–8, and DeVorkin, D H 1987 Thoughts on oral history *American Geophysical Union Newletter* **3** November 22–4.

In addition to its astronomical documentation project, conducted in the late 1970s, the American Institute of Physics Center for History of Physics has undertaken oral history studies in many areas of physics, including solid state physics, and continues today as the most active documentation centre for American physics. (See Chapter 2 of this book.)

4

Some Methodological Problems in Writing the History of CERN†

John Krige

CERN, the European Organization for Nuclear Research, situated just outside Geneva, was officially established in October 1954. From modest beginnings it has mushroomed into a huge research facility endowed with a variety of high-energy accelerators, a staff of some 3500 people supplemented by as many visitors, and a budget in 1986 of almost 750 million Swiss francs (about £270 million) distributed between its 14 member states. Some 16 per cent of this sum is provided by the United Kingdom. And although serious doubts have recently been raised in the United Kingdom over the wider effects of their substantial contribution to CERN, notably in the so-called Kendrew report, that same report concluded that 'CERN [was] at present [i.e. in June 1985] the leading particle physics laboratory in the world and [was] an inspiring example of what [could] be achieved by European collaboration'.‡

† Based on a presentation given at the conference on The History of Physics for the Physicist, 2–4 July 1986, Oxford.
‡ High Energy Particle Physics in the United Kingdom, report of a group set up by the ABRC and the SERC to review UK participation in the study of high energy particle physics, Advisory Board for the Research Councils, Science and Engineering Research Council, June 1985, p. 87.

Problems in Writing the History of CERN

The idea of writing a history of CERN was mooted soon after the laboratory was established. Indeed as early as 1961 CERN published a brief account of its 'origin and beginnings' written by one of the organisation's pioneers Lew Kowarski (1961). The scheme was carried further in the early 1970s, and Professor Margaret Gowing was asked to write an official history of the laboratory—a project which faltered primarily for lack of adequate resources. Another initiative taken soon thereafter led to the setting up of a Study Team for CERN History, whose main members are Armin Hermann (funded from the Federal Republic of Germany), John Krige (United Kingdom), Ulrike Mersits (Austria) and Dominique Pestre (France).†

The first members of the team started work in offices adjacent to the CERN archives in October 1982. The first volume of the study, which covers the period up to October 1954, appeared in 1987 (Hermann *et al* 1987). The manuscript for the second volume, which deals with the subsequent period up to 1965, was completed in summer 1989.

My aim in this chapter is to discuss some of the methodological problems encountered thus far in writing the history of CERN. I should say at once that it is not my purpose to present a quasi-philosophical exegesis on historical method. Nor do I want to suggest that the problems raised are necessarily unique or peculiar to the writing of CERN's history—far from it. Indeed my aims are more modest, my approach essentially pragmatic: I simply want to describe a few of the many difficulties, properly called methodological, which we have encountered in our day-to-day practice as historians, some of which, I should add, we have not been able to solve.

The chapter is divided into three roughly equal parts, each treating rather different issues. In the first I briefly describe the main archival sources we have used, and discuss some of the ways in which our solution has biased our history of CERN. The various phases in the growth of the organisation are explained in the next section, and the diversity of methodological approaches to each is stressed. Finally, I discuss some of the problems we have encountered in dealing with what we call the 'standard views' of CERN's history. Characteristic features of those views are identified, and the difficulties in dislodging them are explored.

---

† These members are funded respectively by the Volkswagen Foundation, the Joint Committee of the Science and Engineering Research Council and the Social Science Research Council, the Austrian Foundation for the Advancement of Scientific Research and the Centre National de la Recherche Scientifique. There were also two part-time team members, Lanfranco Belloni (Italy) and Laura Weiss (Switzerland).

The problem of archival bias

I want to start by discussing the scope of the archives and to explain some of the choices we have made in dealing with the huge amount of material potentially relevant to our project.

At the centre of our study lies the collection at CERN itself. This now comprises some 2500 boxes of papers, about half of them deposited by the offices of the Director of Administration and of the Director-General. Of the remainder, about a thousand boxes contain the 'personal' papers of some of the leading scientific staff and of the secretaries of some divisions. There are also some hundred boxes containing papers prepared for, and the minutes of, the CERN Council and its committees—the Committee of Council, the Finance Committee and the Scientific Policy Committee—as well as hundreds of tapes of many of the meetings themselves. In the very early years, say 1950–1, the archive is somewhat patchy; from about 1952 onwards it provides a comprehensive record of many aspects of the organisation's life. I should add that we have been granted access to the whole of it with the minimum of formalities.

This by no means exhausts the source material. One of the more important aspects of CERN is that it is an intergovernmental institution, in fact the first major intergovernmental research laboratory dedicated to basic science. The political and scientific relationships between the member states and the laboratory are therefore of interest. To study them one has to extend one's search to national archives and collections of papers—to study them 'in full' would require searching in over a dozen European countries, from Scandinavia to Spain and from Great Britain to Greece.

Granted the limitations of time and money, some way clearly had to be found to reduce the source material to manageable proportions. We have done this in three ways. Firstly, we have set the upper limit to our study at 1965, the year in which the construction of the first of the new generation of CERN accelerators got under way. Secondly, we have decided to limit specifically national dealings with CERN to the considerations which led the country to join the organisation in the first place, so imposing a cut-off in 1954. Finally—and this is reflected in the composition of the team—detailed studies of the domestic decision-making processes are only being made for the four major contributors to the CERN budget, the Federal Republic of Germany, France, Italy and the United Kingdom.

Though unavoidable these choices have inevitably introduced an element of archival bias into our work, particularly as far as the national studies are concerned. By concentrating on the four most powerful European states we have a rather good idea of how they negotiated with one another, how they perceived CERN, what their aims were in joining it and what they expected out of it. We have little or no idea of the attitudes of the less influential European states—of The Netherlands, of Switzerland, of Norway, or of Greece, for example. Their perception of the situation was certainly different: as Milan Kundera (1985) warns us,

> People suppose that the little countries necessarily imitate the big ones, but that is an illusion. In fact they're quite different. A little guy's outlook is different from a big man's. The Europe made up of little countries is *another Europe*; it offers another perspective and its culture is often completely at odds with the Europe of big countries.

By not studying these 'little countries' in detail we lose that perspective. We lose, for example, an insight into how the big four collectively wielded their power at this time, of how they imposed their conception as to what CERN should be like on the smaller states. In the absence of this critical point of view we subtly reinforce and legitimate the positions adopted by the major (west) European states, and to the extent that they succeeded in shaping CERN along lines that suited them (in terms of kind of equipment, level of budget, appointment of leading staff, etc), we subtly reinforce and legitimate the *status quo* as it was in the early 1950s, and indeed beyond.

There are other ways, too, in which the archival bias I have identified can dull our awareness of the factors shaping the behaviour of the major powers. To illustrate, let us consider the evolution of the United Kingdom's relationship with CERN.† Very briefly, early in 1951 when British scientists were first officially informed that there was a project afoot to equip Europe with the most powerful accelerator in the world, the bulk of them were relatively indifferent, even hostile, to the scheme, and offered only limited technical support. Later in the year, Sir James Chadwick and Sir George Thomson, in particular, went further, and allied themselves and their country with an alternative suggestion made by Niels Bohr and Hendrik Kramers, who had proposed that the whole venture be rethought. The terms of the debate changed again during the second half of 1952, notably after CERN decided to double or treble the design energy of the big accelerator by incorporating the newly discovered strong-focusing principle into it. A younger group of physicists centred at Harwell insisted that they have access to such a device, and the major issue became whether Britain should go it alone and build the machine herself or whether she should collaborate with the European venture. The latter course was adopted

† For a more detailed study see chapters 12 and 13 in Hermann *et al* (1987).

towards the end of the year, though not without misgivings and reluctance in certain quarters.

A striking feature of Britain's attitudes in this period is that they were *just what one would expect* in the light of her well known lack of enthusiasm for European ventures in the post-war period and of the periodic crises in her dealings with CERN. And just because there is nothing surprising here it is all too tempting to leave matters at that, to think that one has 'understood' the British case, to assume that no further explanation is called for because it fits into what one already knows, because Britain's attitude to CERN seems to be yet another manifestation of a general and persistent reluctance to become closely involved with Europe.

Matters become complicated, however, when we recognise that Britain was not alone in having doubts about CERN membership—Bohr in Denmark, Kramers in The Netherlands, Scherrer in Switzerland, among the scientists, as well as leading scientific administrators in at least the first of these countries, as well as in Sweden, all harboured doubts of one kind or another. This is another example of the different perspectives of the small countries, thought not them alone, for even in France and in Germany, who may generally be regarded as strong supporters of the scheme, the venture was not without its critics. At the very least this forces us to ask whether Britain's reluctance to join CERN was not *also* due to certain features of the project itself, and of its articulation with the United Kingdom's specific situation. And, indeed, this was so. As first presented the project seemed to some to be crazily ambitious and impractical; at least until the spring of 1952 the scientific establishment in Britain felt that the country's needs would be more than satisfied by the new domestic machines whose construction had been authorised after the war. And even when Britain's physicists decided that they needed access to big accelerators they, and probably they alone, as the leading Europeans in machine building at the time, had the experience, the manpower and the money to build the device themselves.

To conclude: doubtless there is something in the conventional wisdom, in the view that Britain was behaving true to type in being stand-offish about joining CERN. However, knowing that other countries also had their doubts helps break the grip of this somewhat glib explanation, and paves the way for a subtler analysis of the peculiarities of the British case. One does not need to study some of the similarly reluctant smaller countries in detail to appreciate this. But if we had had the opportunity to do so our treatment of Britain would have been that much more refined, and we would have been able to assess the relative weights of the several kinds of factors shaping her attitude more thoroughly. From what we know already I would say that a general lack of enthusiasm for European ventures in Britain was actually of relatively *little* importance in this regard.

The need for methodological pluralism

The next point that I want to draw attention to is that there is no *one* way of studying the launching and development of an organisation like CERN. Different periods in its history require different kinds of treatment, different kinds of histories. If one believes otherwise, if one tries to impose the same approach across the several phases of its evolution, one will fail to capture the richness and complexity of the object of study. In short, one has to be methodologically flexible.

Let me be more explicit. The history of CERN has passed through three distinct phases. During the first, which lasted from around the end of 1949 to early in 1952, the idea of building a collaborative European organisation in nuclear science was spelt out and refined by a number of experimental physicists with the support of some leading scientific administrators and government officials. Their efforts resulted in UNESCO calling an intergovernmental 'conference on the organization of studies relating to the possible establishment of a European nuclear physics laboratory', which met in December 1951 and again in February 1952. This conference decided to set up a Council of Representatives of European States whose stated aim was to plan an accelerator laboratory and to organise other forms of cooperation in nuclear science. This body, customarily called the provisional CERN, lasted for some 18 months from May 1952; its activities constitute phase two of CERN's history. Phase three commences with the official establishment of CERN (Conseil Européen pour la Recherche Nucléaire) in October 1954, when the Council of the European Organization for Nuclear Research held its first session.

The mode of development, the forces at work, the determinants of evolution during each of these phases had their own distinctive features. In the earliest period, the so-called prehistory, the project was in the hands of a rather small group of people, who kept in close contact with each other, and who for a variety of reasons, had the freedom to shape the project along the lines that they saw fit.† Though linked to various national and international bureaucracies, this group was able to steer its way through the pores in these structures. By virtue of its size, its compactness (notwithstanding internal conflicts) and its relative autonomy, we have been able to study this group in

† This phase is extensively described by Dominique Pestre in chapters 2–6 of Hermann *et al* (1987).

some detail. Our account of the first phase of CERN's history is thus essentially chronological, exhaustive and centred around the main actors and the decision-making process.

This situation was to change steadily during phase two and beyond, with the establishment of the provisional Council.† One of its first tasks was to set up a number of study groups, two of them dedicated to designing the accelerators, one to planning the laboratory to house them and a fourth devoted to theoretical studies. Essentially then the European laboratory project was disaggregated into its constituent components and institutionalised. Its several facets became steadily independent of one another, and each developed according to its own rhythm, a rhythm determined by the nature of the very different tasks at hand and the way in which the various groups in question dealt with those tasks. If before we were dealing with a compact group of people, now we are confronted with an increasingly complex organisational structure, comprising semi-autonomous units articulated with each other in intricate ways to form the whole. To deal with this situation we have decided to work on two distinct levels: the one narrative and chronological, giving a general overview of developments during the provisional period, the other primarily analytical, in which we present in-depth studies of particular and relatively short-lived *events* (like the choice of a site for the laboratory or the appointment of its first Director-General).

The gradual transformation of our object of investigation as we move from phase one to phase two is completed in phase three, when we are dealing with CERN itself. In October 1954, when the CERN Council first met, the organisation comprised four divisions, the office of the Secretary-General, and a handful of administrators. The total staff was 120, and some 3.7 million Swiss francs had been spent since May 1952. At the end of 1965—the terminus of our study—CERN comprised eleven divisions, eight of them scientific, overseen by a Director-General advised by a Directorate. Total staff numbered over 2000 and expenditure for the year (on the basic programme) was almost 133 million Swiss francs.‡ To deal with this new situation we have elaborated the main division initiated in our study of phase two of CERN's history, the division, that is, between narrative account and in-depth analysis. Very crudely, we present first a general chronology of the development of CERN to enable us to have a feel for the way the

---

† See John Krige, chapters 7 and 8 in Hermann *et al* (1987).
‡ For the data in this and the previous sentence see respectively the Final Report to Member States presented by the Secretary-General of the Interim Organization, document CERN/GEN/15, 30/6/55, and CERN's Annual Report for 1965, both in the CERN archives.

organism as a whole evolved. Thereafter, if before we studied *events* in depth, now we study *topics* in depth, topics like the development of CERN's financial policy, or the evolution of its organisational structure, or the physics done on the giant strong-focusing proton synchrotron. This then is what I mean when I say we have had to be methodologically flexible and supple. And to conclude this section, I want to discuss an issue narrowly related to the above: the question of how one chooses what to omit and what to include in one's history.

As is to be expected in the light of what we have just discussed, the kinds of choices we have had to make have been very different depending on which stage of CERN's history was at issue. For the 'prehistory', for example, we had a manageable amount of documentary material. We were therefore in a position to write an 'exhaustive' account of events. Having adopted this approach—a choice, I should add, which we sometimes regretted afterwards—the main decisions then concerned the level to which the analysis was pushed, how deeply one explored a particular decision, or behaviour pattern, or whatever. By contrast, when studying the organisation proper (phase three) the main problem presents itself differently: the volume of material is large, and the number of topics that one can study and the ways of approaching them are virtually limitless. The choices one then makes are determined by a number of factors: the recognition that some things simply cannot be left out of any book dealing with the history of CERN (like the laboratory's scientific policy), the sources available, the time and money at one's disposal, the professional background and interests of the historian and so on. By implication then, and for quite mundane reasons, reasons far removed from philosophical debates about the relativity of any historical work, there can be no such thing as the definitive history of CERN: the study of its genesis and development is an open-ended task.

The 'standard views' of CERN's history

To illustrate what I mean by this heading I shall begin with an example. Towards the end of 1952, as the British government was coming round to the view that it would be wise to join CERN, they invited the Secretary-General of the provisional organisation, Edoardo Amaldi, over to London to clarify a number of outstanding issues.† On the afternoon of the 11 December, Sir John Cockcroft, the Director of Harwell, and Sir Ben Lockspeiser, the Secretary of the DSIR, took Amaldi to see Lord Cherwell, Churchill's highly influential 'science

† See also § 13.5.1 in Hermann *et al* (1987).

minister'. Amaldi described this encounter as follows in a lecture given in 1972, i.e. twenty years later (Amaldi 1977):

> Lord Cherwell appeared to be very clearly against the participation of the U.K. in the new organization. As soon as I was introduced to his office he said that the European laboratory was to be one more of the many international bodies consuming money and producing a lot of papers of no practical use. I was annoyed and answered rather sharply that it was a great pity that the U.K. was not ready to join such a venture which, without any doubt, was destined to full success, and I went on by explaining the reasons for my convictions. Lord Cherwell concluded the meeting by saying that the problem had to be reconsidered by His Majesty's Government [...]
>
> A few weeks later Sir Ben Lockspeiser wrote an official letter asking that the status of observer be given to the U.K. in the provisional organization, and the D.S.I.R. started to regularly pay 'gifts', as they were called, corresponding exactly to the U.K. share calculated according to the scale adopted by the other eleven countries.

It must be said at once that this little anecdote suffers from some well known weaknesses of oral history. For one thing it is factually inaccurate. For example, the request that Britain be granted observer status in the provisional organisation was made the day Amaldi *arrived* in London, and not 'a few weeks later' as he says. For another (though this is an inference from the way this text is constructed) it exaggerates the role of the speaker in shaping events. From independent evidence we know that Cherwell had *already* agreed, however reluctantly, that Britain participate in CERN three weeks before he met Amaldi.† But there is more to it than that.

Within a week of arriving at CERN in October 1982, so another ten years on, I had the opportunity to have lunch with Amaldi. During the course of the meal he told me the story of his meeting with Cherwell, which I had never heard before. I did not have a tape recorder with me, but as soon as I returned to my office I wrote down what I could remember. Here is an extract from my notes made at the time:

> Amaldi claimed that Cherwell had come round in favour of UK involvement due to his influence. On the way to a celebration commemorating the opening of the Cosmotron, he had taken a detour via London for a couple of days. He had lunch with Lockspeiser and Cockcroft, and they went on to Cherwell's office in the War Ministry after that. Cherwell was very aggressive from the beginning, saying that this was a waste of money, just like UNESCO. Amaldi was riled. He said he didn't know why he had asked to come if this was Cherwell's

† The evidence for these claims is given in § 13.5.1 of Hermann *et al* (1987).

attitude, that if the UK didn't want to join, Europe would get ahead with the project without them [...]. Ten days later he [Amaldi] received a letter from Sir Ben saying that the UK would not join but would make a financial contribution as a gift. The amount corresponded exactly to what she would have had to pay as a member state.

The interest in giving this second long quotation is that it brings out the extent to which Amaldi's interpretation of this event has become embedded in his consciousness.† Is he describing the events that occurred, or is he, as one of Borges' characters suggests, simply repeating his previous account of such events ('The years pass and I've told this story so many times I no longer know whether I remember it as it was or whether it's only my words I'm remembering')? (Borges 1986) Accounts such as these, accounts repeated again and again by eminent physicists, by people who have participated in the launching and development of CERN, accounts which, for that very reason, are embodied in the publicity provided by the organisation—it is this already given, taken-for-granted collage of ideas, anecdotes and interpretations of key events which together constitute what we call the standard views of CERN's history.

The historian confronted with these standard views has little option but painstakingly to construct critical alternatives to them, in the (probably forlorn) hope of reopening discussion on certain important questions. To do that one has, of course, to try to set the factual record straight using appropriate documents. On the other hand, factual inaccuracy is not the only, and is certainly not the most interesting, characteristic of the standard views. They have other more elusive features, features indicative of a particular way of seeing the world, and of making sense of it. We have touched on one—the tendency to exaggerate one's own role. Let me now more carefully explore another, namely, the tendency to reconstruct rationally the past in the light of later events.

Kowarski's way of describing the decision to equip the envisaged European laboratory with a powerful accelerator nicely illustrates this tendency.‡ A crucial factor leading to the launching of CERN, writes Kowarski, was the recognition by scientists of 'the need both for new nuclear physics equipment and for an international effort'. 'The first need', Kowarski goes on, 'became *obvious* as early as 1947, when the synchro-cyclotron built in California [...] began to yield its first crop of

---

† Amaldi also told Margaret Gowing and Lew Kowarski essentially the same story when they interviewed him on 22 January 1974. The interview is in the CERN archives.
‡ What follows, and the quotations from Kowarski, are from Kowarski (1961) pp 1–4.

discoveries'. An awareness of the second was made manifest at a meeting of the European Cultural Conference in Lausanne in December 1949. Here a message from Nobel prize-winner Louis de Broglie was read out in which he proposed that a large number of European states collaborate to set up a scientific laboratory or institution endowed with more resources than any single participant could afford on its own. '*Curiously enough*', writes Kowarski, 'nuclear physics was not mentioned at all in this message'. Six months later another Nobel laureate, Isidor I Rabi, successfully put a similar resolution to a UNESCO Conference in Florence, asking the Director-General 'to assist and encourage the formation and organisation of regional research centres and laboratories [...]'. '*Once more*', comments Kowarski, '[...] the domain of research to be pursued [was not specified]', though 'meson (or high-energy) physics was the *most obvious and attractive candidate*'. Finally, in December 1950, at a meeting effectively organised by Pierre Auger, UNESCO's Director for Exact and Natural Sciences, 'the desirability of building a large particle accelerator for high-energy research was *explicitly* proclaimed' (my emphasis throughout).

How has Kowarski built this account? By *starting* at the *outcome* of a complex historical process and by reconstructing a chain of events which he sees as connecting it back to its first beginnings. He takes it for granted that that outcome was implicit in the minds of scientists in 1947, and he regards it as surprising, curious, that some aspect of it was not at least alluded to in the intermediate stages he identifies. Kowarski's unconscious model is that of a gestation process in which the end result is prefigured in the first seed, a seed which in growing moves ineluctably and inevitably towards its preordained destiny.

It is impossible for me now to explore at length the limitations of this way of writing history. Two observations will have to suffice. Firstly, at the factual level it is systematically misleading. For example, while it is true that *one* of the fields Rabi thought suitable for a regional laboratory was high-energy physics, it was not the only one—he also mentioned expensive biology and computing; what he seems to have had in mind was not CERN, but a multipurpose laboratory like that at Brookhaven of which he was a founder.† Secondly, methodologically speaking, this kind of approach leaves no room for uncertainty and confusion in the minds of human agents, it ignores as irrelevant criticism and opposition to a project unless the 'final outcome' was significantly altered by it, and it allows no role for chance and coincidence in history. It is difficult to admit this, to admit that the

---

† For an extensive analysis of Rabi's intervention at Florence see Dominique Pestre, §2.5 Hermann *et al* (1987).

historical process is complex, messy and unpredictable. But admit it we must if we are not to fall—as we all do so frequently—into the trap of imposing a spurious rationality on the past.

To conclude this chapter I want to make one more observation not entirely disconnected from the other material in this section. It is often difficult for the historian dealing with the work of important and influential scientists to remember that their subjects are not disembodied beings surveying the world from a detached and objective position, but humans like themselves working in a particular context, subject to a particular set of pressures and having particular aims in mind. Scientists too approach the world armed with preconceptions, they too have axes to grind, they too display some of the less attractive features of human behaviour. Marie Jahoda made the point forcefully a few years ago when she wrote (Jahoda 1983)

> There is no harmonious scientific community interested only in the advancement of knowledge and its benefits for mankind. Yes, there is passionate commitment to driving forward their breath-taking discoveries; but there is also ambition, jealousy, lack of foresight, moral ambiguity and arrogance in these scientists, and there is absolutely no reason to assume that they are any better in these matters than we of a lesser breed.

To recognise that fact, to bear it in mind when one is interviewing scientists or studying the documents produced by a scientific organisation like CERN, is to open to the historian a whole range of questions which would otherwise not arise. It is to take an important methodological step forward.

References

Amaldi E 1977 Personal notes on neutron work in Rome in the 30s and post-war European collaboration in high energy physics in *History of Twentieth Century Physics*, Proceedings of the International School of Physics 'Enrico Fermi', Course LVII, Varenna 1972 ed. C. Weiner (New York: Academic Press) pp 294–351 (at p. 339)

Borges J L 1986 *The Book of Sand* (Harmondsworth: Penguin) p. 52

Hermann A, Krige J, Mersits U and Pestre D 1987 *History of CERN, Vol. I, Launching the European Organization for Nuclear Research* (Amsterdam: North Holland)

Jahoda M 1983 Review of the DNA story (1982) (quoted by B Easlea 1983 in *Fathering the Unthinkable* (London: Pluto Press) p. 2)

Kowarski L 1961 *An Account of the Origin and Beginnings of CERN* (Geneva: CERN)

Kundera M 1985 Prague: a disappearing poem *Granta* **17** 85–103 (at p. 87)

5

Reflections of an Amateur 'Historian of Science'†

N Kurti

Inverted commas often serve as a simple shorthand for 'pretending to be' or 'so-called' and are used in this sense in the title of this chapter. When I talk to historians of science, or whenever I give a talk which has to do with the history of science, I always begin with the very old story about the professor of gynaecology who rings up his colleague in the history department and says 'I am due to retire in a few months time and I thought that I would write a history of obstetrics and gynaecology; could you spare an hour or so to give me a few good tips how to go about it'. The historian replies, 'Do come along. I too want to ask you some advice. I am also about to retire and have decided to take up midwifery. I should be most grateful if you could spare an hour or so and give me some tips about what to do when I attend my first delivery'. This illustrates the attitude of scientists and medical people who are apt to think that they can become historians of their subjects without first acquiring the skills and the methods of the historian. So what I propose to do here is to give examples of why I occasionally engage in something which might be called history of science; the reader will see that they are all very amateurish things and of no great importance.

Why does a physicist who has been doing physics all his life get interested in history? The main incentive is usually curiosity: why did

† Based on a presentation given at the conference on the History of Physics for the Physicist, 2–4 July 1986, Oxford.

something happen, why didn't something else happen? To look into this is very often amusing and very often demanding. So, since I am based in Oxford, let me begin with the well established fact that, for half a century after 1870, while the Cavendish Laboratory in Cambridge became the world's most important centre of physics under J C Maxwell (1831–79), Lord Rayleigh (1842–1919) and J J Thomson (1856–1940) Oxford physics was virtually unknown. The explanation is painfully simple. The Professor of Experimental Philosophy in Oxford during that period was R B Clifton (1836–1921) who was elected on the basis of one original paper of which he was a junior author and of the reputation of the excellent lectures he gave as Professor of Physics at Owen's College, the forerunner of the University of Manchester. He had no interest in research (he regarded the desire to do research as a sign of the restlessness of the spirit) and, during his professorship, there was practically no research at the Clarendon. The choice of the Electors to the Chair is puzzling and becomes worrying if one believes the widely accepted view that he was chosen in preference to Hermann von Helmholtz (1821–94). When I mentioned this to one of my Oxford friends he gave the following fairly plausible explanation. In the 1860s there were heated arguments about the future of the university. One faction, led by B Jowett (1817–93), the Master of Balliol, regarded the chief, almost the sole, aim of the University to be the teaching of undergraduates. The opposing faction, led by M Pattison (1831–84), Fellow of Lincoln College, wanted Oxford to follow the example of the great European universities and become also strong in research. One of Pattison's chief supporters was the distinguished philologist of German origin Professor Max Müller, at that time Professor of Modern Languages in Oxford. So, when Helmholtz's candidature became known, it was felt by many that it was bad enough to have one powerful German professor trying to reform Oxford; to have two of the same ilk would be asking for trouble and the Electors chose Clifton.

I accepted this explanation, which shows that I did not behave like a historian; I did not check the facts. But for a fortunate accident I would still believe, and so would many others, that in 1865 the prestigious University of Oxford preferred Robert Bellamy Clifton to Hermann von Helmholtz.

But a few years after I had been given this explanation there was a Conference in Oxford under the auspices of The Institute of Physics and the Physical Society and one of the Pro-Vice-Chancellors of the University, Dr Lucy Sutherland (1903–80), then Principal of Lady Margaret Hall, was invited as a guest of honour to the conference dinner. She was asked to make a brief speech and I promised to provide some anecdotal material.

The story of Clifton's election seemed to me a suitable subject but, since Dr Sutherland was a distinguished historian, I felt that I owed it to her to check the correctness of the story. The first thing was to find out whether Helmhöltz really did apply. Alas, there were no files about elections to Chairs in the University Archives and the only reference I could find in the Minutes of the Hebdomadal Council was the statement that Robert Bellamy Clifton, the newly elected Professor of Experimental Philosophy was made an MA by decree. So I decided to consult the biography of Helmholtz by Koenigsberger (1903). Unfortunately this otherwise excellent book had no table of contents and no index and was not even in strict chronological order. So I had to read it with some care and I came across some very fascinating things. Thus I found that Helmholtz had visited Oxford in March 1864, the year before Clifton's election to the Chair (Koenigsberger 1903 pp 50–1). A letter written to his wife from Oxford gives his impressions and, since what he says is relevant to the story of the Oxford Chair, I quote the letter in full, as it appeared in Frances Welby's English translation of the Koenigsberger biography (Koenigsberger 1906 p. 223).

> Yesterday morning I went to Oxford, and am staying with Max Müller. He is a clever young man of the world, whose like I have never yet seen in a professor of philology, and grasps everything, even the scientific matters with which he is less familiar, with extraordinary rapidity. His wife is an English lady, who is also most attractive, well-informed, and pretty, so that I spent two very pleasant days there. Oxford is probably unique of its kind in the world; its many old, and characteristically beautiful, and well-preserved buildings, with trim grass lawns and handsome trees, are all stately to a degree, and very magnificient. It is quite impossible to picture it at home until one has seen it, and I now understand the devotion of an Englishman to his University. The system works admirably for the education of 'gentlemen' but it cannot lead to much in science, and it needs an extraordinary interest in science to prevent a Fellow from sinking into indolence.

My appetite thus whetted I went on reading in the hope of finding some reference to Helmholtz's supposed application. There was nothing in 1865 but a letter Helmholtz wrote to his wife during a visit to Paris in April 1866, i.e. five months *after* Clifton's election, provided the explanation for this bizarre episode (Koenigsberger 1906 p. 232)

> ...At eleven o'clock I had to be back for a breakfast with M. Hermite and the mathematician Prof. Smith from Oxford. It was said in course of conversation that there had been some notion of inviting me to go to Oxford as Professor of Physics. However, they could not offer more than £700 salary, which of course is more than we get in Heidelberg, but

hardly enough to live comfortably in England. ... So I think Prof. Max Müller was right to say he could tell them decidedly that I should not accept it...

To put it bluntly Max Müller, without asking Helmholtz, who had been his guest the previous year, assured the electors that Helmholtz would not accept the position on that salary and that there was no point in asking him. I can think of only one explanation for this high-handed attitude, namely professional jealousy (Kurti 1977). Max Müller was undoubtedly a brilliant young scholar; Helmholtz himself thought highly of him. In fact the sentence which, according to another publication, followed the last sentence of the Koenigsberger version of Helmholtz's letter from Oxford, runs: '... from sinking into indolence. *Max Müller is at present perhaps the only man who does any work here*'. At that time Max Müller was generally regarded as perhaps the most distinguished and internationally best known scholar in Oxford and he was apprehensive that Helmholtz would outshine him. So, to forestall this possibility he virtually vetoed the invitation to Helmholtz.

Would Helmholtz have accepted the invitation? His remark 'Max Müller was right to say that I should not accept it' sounds firm, but the German is less definitive. Figures 5.1 and 5.2 give two versions of the German text, one as quoted by Koenigsberger (1903 p. 73) the other quoted in Ellen von Siemens-Helmholtz's (1929) biography of her mother *Anna von Helmholtz—Ein Lebensbild in Briefen*. The passages which are different in the two versions or occur only in one of them are underlined. The correct translation of 'Ich finde es also ganz in der Ordnung dass Herr Professor Max Müller erklärt hat' (figure 5.1) is not 'I think Professor Max Müller was right to say' but 'I find it quite in order for Professor Max Müller to say'. Similarly, in referring to the £700 stipend, Helmholtz did not say that it was hardly enough 'to live comfortably in England' but 'to live as comfortably in England (as in Heidelberg)'. Incidentally, the stipend of £700 that would have been offered was not all that low. When, in 1871, Sir William Thomson (Lord Kelvin) invited Helmholtz to become the first Cavendish Professor in Cambridge the stipend he could offer was only slightly more (Kurti 1985).

It was obvious that to clear up the discrepancies between the two versions of the letter I had to see the original, so I went to East Berlin to the Archives of the Akademie des Wissenschaften der DDR where the Helmholtz papers are kept. I saw quite a lot of fascinating stuff, including Thomson's letter to Helmholtz offering him the Cavendish Chair, but was told that the whole of the personal Helmholtz correspondence remained with the family. I then got in touch with one of Helmholtz's direct descendents Dr von Siemens in Munich, who

„... Um 11 Uhr musste ich zurück sein für ein Frühstück mit Mr. Hermite und dem Mathematiker Professor Smith aus Oxford. Dabei kam zur Sprache, dass man mich habe nach Oxford berufen wollen als Professor der Physik. Sie haben aber nicht mehr Gehalt zusammenbringen können als 700 £, was zwar mehr ist, als wir in Heidelberg haben, aber kaum genug, um in England mit derselben Behaglichkeit zu leben... Ich finde es also ganz in der Ordnung, dass Herr Professor Max Müller erklärt hat, er könne ihnen bestimmt sagen, ich würde daraufhin nicht kommen... Mr. Hermite war gegen mich sehr schmeichelhaft und introducirte auch noch Mr. Grandeau, um mich zu begrüssen und nach der Ecole normale zu führen.

Figure 5.1 Letter from Helmholtz to his wife, 11 April 1866, as in Koenigsberger (1903 p. 73).

Heute früh begab ich mich in den Louvre zu den Italienischen Bildern, an denen man freilich lange zu sehen hat. Um Elf einhalb mußte ich zurück sein für ein Frühstück mit Mr. Hermite und Professor Smith, Mathematiker aus Oxford. Dabei kam zur Sprache, daß man mich habe nach Oxford berufen wollen als Professor der Physik, daß ihnen Professor Max Müller aber bestimmt erklärt habe, ich würde das Anerbieten nicht annehmen. Deine Tante ist sehr unzufrieden, daß sie es mir nicht angeboten haben, obgleich sie zugibt, es würde höchst unweise gehandelt gewesen sein, es anzunehmen, und übrigens scheint ihr diese Möglichkeit sehr imponiert zu haben.

Mr. Hermite war gegen mich sehr schmeichelhaft und introduzierte auch noch Mr. Grandeau, um mich zu begrüßen und nach der École Normale zu führen.

Figure 5.2 Letter from Helmholtz to his wife, 11 April 1866, as in Siemens-Helmholtz (1929 p. 131).

gave me the following information. All the family papers, which were housed in a green safe, were deposited in the Academy Archives in the 1930s. The safe survived the bombing attacks in 1945 and was still in place in the late 1940s. Then some unauthorised person opened the safe and removed the contents, and no one knows what happened to them. All that one can hope is that some time in the future they will turn up in some old tea chest—or at an auction sale!

Let me tell you now about a case where, by consulting the original of a letter, it was possible to clear up a misconception about the announcement of the first liquefaction of oxygen. At the meeting of the Académie des Sciences in Paris on 24 December 1877 (*Comptes Rendus* 1877) a communication about the condensation of oxygen and carbon monoxide by the French chemist L P Cailletet (1832–1913) was read first and this was followed by reporting the receipt of a telegram from the Swiss engineer R P Pictet (1846–1929) announcing the liquefaction of oxygen at 8 PM on 22 December. To dispel any doubts about priority J B Dumas (1800–84) the 'Secrétaire Perpétuel', then read out a letter dated 2 December from Cailletet (1877) to his mentor, the chemist H E Sainte-Claire Deville (1818–81), which the latter had deposited at the Academy on 3 December in a sealed envelope: 'I must tell you, and you first, without losing a moment, that I have liquefied this very day carbon monoxide and oxygen... .' So begins Cailletet's letter which is printed in the *Comptes Rendus* (1877). Sainte-Claire Deville then explained that the reason for not publishing this discovery earlier was that Cailletet, a candidate for a corresponding membership of the Academy, did not want his work to figure in the discussion of his claim for membership before the results had been confirmed by competent judges. Cailletet's cautiousness was greatly embellished by Georges Claude (1870–1960) in his book *Liquid Air, Oxygen, Nitrogen* (Claude 1913) where he writes:

> The experiment, however, was not made public either at the sitting of the Academy of December 3rd or at that of the 10th, or at that of the 17th. What did this reserve mean, corresponding so little with the intense satisfaction which the happy experimenter could not help feeling?
> The explanation is to be found in a circumstance wholly to his honour. Cailletet, whose brilliant work, we have said, had already attracted the attention of the scientific world, had become a candidate for a corresponding membership of the Academy; the election was to take place on the 17th December, and he did not wish to appear to have influenced the voting by such a sensational communication. He was elected, but this exaggerated modesty nearly cost him dear. It is the rule, in fact, in France, that questions of priority should be settled by the dates of presentation to the Academy of Sciences. Now the last evening but one before the 24th December, the day on which Dumas was to have presented to the Academy the discovery of their new correspondent, a dispatch reached the Institute, it announced that on that very day, the 22nd of December, oxygen had been liquefied, and it was signed, Raoul Pictet!
> Happily for Cailletet, Sainte-Claire Deville, an old habitué of the Institute, had mistrusted possible indiscretions; he had taken on himself to lodge in a sealed envelope with the Academy the historical letter, the text of which we have quoted, and by his foresight he had definitely

assured for his protége the priority which had almost been taken from him.

This goes much further than Sainte-Claire Deville's statement but what is the origin of this speculation? There is nothing in Cailletet's letter to Sainte-Claire Deville published, seemingly in full, in the *Comptes Rendus* to indicate that he did not want his experiment of the 2 December to become public knowledge. Was there perhaps another letter from Cailletet to Sainte-Claire Deville giving such instructions? So I consulted the archives of the Académie des Sciences and, having found nothing, I looked up the original of Cailletet's letter. And, to my surprise and dismay, found that the printed version omitted, at the Secrétaire Perpétuel's scribbled instructions, three short paragraphs without any indication that it was not a true and full copy (Kurti 1978). Here are the three missing paragraphs in my translation:

> This is what I should like to do. If the election to the Academy is to take place presently, i.e. before the end of December, I will go to Paris and present then the complete results at the Academy. If, however, the election is not due to take place till later, I should be grateful if you could let me know and, if you think it useful, I'll go to be present at next Monday's (10 December) meeting.
> In any case oblige me by not talking about these results which I am happy to acquaint you with and *which could perhaps also be used for my candidature.* [my italics]
> Talking abut it would mean letting out the secret.
> ('En on parlant ce serait éventer l'affaire'.)

It is clear that Cailletet hoped that his discovery could be helpful in his election and was not against members of the Academy knowing about it. He simply did not want undue publicity at that moment because he may have been afraid that once the experiment became known someone might steal a march on him and improve the method before he had a chance of doing so. I believe that if the three missing paragraphs had appeared in the *Comptes Rendus* Georges Calude's speculations and the myth developed by succeeding historians of cryogenics would never have occurred.

Let me now turn to another area of the history of science, namely plagiarism. This is a rich hunting ground for an historian but I want to discuss one particular brand of plagiarism—I would call it innocent or subconscious plagiarism. About 40 years ago I nearly committed such an act and ever since I hesitate to pass judgment on people accused of plagiarism without looking at all the circumstances. In the late 1940s there was a lot of discussion in the Clarendon Laboratory and in many other low-temperature laboratories about orientating radioactive nuclei. It was a well established and theoretically understood fact that the emission of γ-rays during the radioactive decay of a nucleus is anisotropic, i.e. that the chance of a γ-quantum being emitted in the direction of the nuclear spin is not the same as perpendicularly to it. However, in a piece of radioactive substance the nuclear spins are orientated at random and the total radiation is isotropic. If all spins could be made to point in the same direction then the total radiation emitted by the sample would have the same intensity distribution as that emitted from a single nucleus, and by studying this *macroscopic* anisotropy one could learn about the radioactive decay process.

The conceptually simplest method to orient the nuclei is to place the sample in a magnetic field which would tend to turn the spins in the same direction. But because of the smallness of the nuclear magnetic moment, very strong magnetic fields and very low temperatures would be needed for the *ordering* tendency of the field to overcome the disordering tendency of the thermal agitation. Then C J Gorter (1948) and M E Rose (1949) independently had the idea that, instead of an *external* magnetic field, one could make use of the much stronger magnetic field that an unpaired electron spin in a paramagnetic ion exerts on the nuclear spin—termed magnetic hyperfine coupling. Therefore, if all the electron spins point in the same direction—this can be achieved by relatively small magnetic fields at low temperatures —nuclear spins would follow suit.

Then came R V Pound's (1949) ingenious idea to make use of the interaction between the electrical quadrupole moment associated with a nucleus and the crystalline electric field. This 'quadrupole hyperfine coupling' results in the nuclei being preferentially orientated with respect to a crystal axis but there is one important difference between this type of nuclear orientation and the two mentioned previously. In the electric quadrupole coupling the nuclei are 'aligned' along the same *direction* but they point in equal numbers in two senses, whereas in magnetic hyperfine coupling the nuclei are 'polarised'. Since, however, the anisotropy of electromagnetic radiations depends on the *direction* but not on the *sense* of the emitter both types of nuclear orientation are equally suitable for γ-ray studies.

In the course of the discussions on the 'Pound' method I had an idea—or rather I thought I had an idea. It was well known that the electric crystal field has an effect on the electron spins and, as a result, in certain paramagnetic crystals at low temperatures the electron spins are 'aligned' along a crystal axis. Therefore the nuclear spins will find themselves in local magnetic fields parallel to the same axis and will thus be aligned. This seemed to me a simple attractive method of aligning nuclei. I mentioned this idea to Bleaney who listened with mounting surprise and, when I had finished, said 'But that is what I proposed in our discussion a week ago' (Bleaney 1951a,b). And the explanation dawned on me. Bleaney proposed that in analogy to Pound's anisotropic electric quadrupole hyperfine coupling, one could use anisotropic *magnetic* hyperfine coupling for aligning nuclei but I apparently did not understand it. Nevertheless the idea germinated in my subconscious mind and finally, by using my rather naive picture of magnetic hyperfine coupling, I eventually understood it and believed it to be my original idea!

Another type of equally innocent plagiarism is the result of ignorance—of not knowing earlier work in the same field. However, ignorance leading to the *re*-discovery of something can be fruitful because, although the result may be the same, the thought-processes leading to it are often different and may shed new light on old knowledge.

Thanks to the proliferation of computerised abstracting, the use of key words and easily accessible databases there is in principle no longer any excuse for plagiarism through ignorance. But I feel that too conscientious a search of the literature to avoid even a remote possibility of being found guilty of plagiarism can be detrimental to originality. As I said before re-discovery need not be an act of sterile copying; it could be a creative act and the result might even appear in a new guise.

I do not suggest that the boundary between plagiarism and originality in science is as ill-defined as it is in music. But we should remember how Haydn's genius transformed a somewhat banal sequence of five tones of the major scale—probably used many times before him—into one of the most beautiful national anthems 'Gott erhalte Franz den Kaiser' and how Verdi created out of another five successive tones of the scale Violetta's heart-rending 'Dite alla giovine' in *La Traviata*. And we should be grateful that in Verdi's time there was no comprehensive database of all existing tunes. Fear of plagiarism might have prevented him from giving us his most famous 'hit' because the first two bars of 'La donna è mobile' appeared—some 80 years earlier—in the first movement of Mozart's F-major piano sonata (K 332).

References

Bleaney B 1951a On the spatial alignment of nuclei *Proc. Phys. Soc.* A **64** 315–16
—— 1951b Hyperfine structure in paramagnetic salts and nuclear alignment *Phil. Mag.* **42** 441–58
Cailletet L P 1877 *Comptes Rendus* **85** 1212–19

Claude G 1913 *Liquid Air, Oxygen, Nitrogen* transl. H Cottrell (London: J and A Churchill) pp 59–60
Gorter C J 1948 A new suggestion for aligning certain atomic nuclei *Physica* **14** 504
Koenigsberger L 1903 *Hermann von Helmholtz* vol. 2 (Braunschweig: Vieweg) pp 50–1
—— 1906 *Hermann von Helmholtz* transl. F Welby (Oxford: Clarendon)
Kurti N 1977 Vom Nernstschen Warmesatz zum Kernferromagnetismus *Abhandl. Akad. Wiss. DDR* **7N** 133
—— 1978 From Cailletet and Pictet to microkelvin *Cryogenics* **8** 451–8
—— 1984 Opportunity lost in 1865? *Nature* **308** 313–14
—— 1985 Helmholtz's choice *Nature* **314** 499
Pound R V 1949 On the spatial alignment of nuclei *Phys. Rev.* **76** 1410–11
Rose M E 1949 Scattering and absorption of neutrons by polarized nuclei *Phys. Rev.* **75** 213–14
Siemens-Helmholtz E von (ed.) 1929 *Anna von Helmholtz—Ein Lebensbild in Briefen* (Berlin: Verlag fuer Kultur Politik) p. 131

6

Could We Repeat It?†

Ronald G Stansfield

Autobiographical introduction

How far would it be possible to repeat an experiment from the published material describing it? This question arises right across the spectrum of science. My interest in it derives from the 'nitty gritty' of what early in life I was brought up to believe as a scientist, namely that the crucial safeguard against error in science, and specifically in physics, is that it is open to anyone to check by repetition the observations of the natural world, including the results of experiment, which a scientist reports and upon which his or her conclusions are based. I am not here concerned with the wider issues of principle which in recent years have increasingly interested philosophers and sociologists of science, for example that experiments can never really be repeated because the world has moved on in the meantime; or that the scientist cannot repeat the state of mind of the original experimenter or observer. These issues, as well as the practical constraints of ability to command the necessary resources to do something again, are certainly important. Here I am concerned with more pedestrian questions: if we wanted to repeat an experiment, would we know enough about what was originally done to achieve adequate replication? Would we be able to command the necessary manipulative and design skills to carry it out? Would the raw materials be available? In a

† Based on a presentation given in the conference on Experimenters and Instruments, 20 February 1985, IOP Headquarters, 47 Belgrave Square, London.

chapter entitled 'The Limits of Replication', Broad and Wade (1983) have considered the main issues systematically. The present chapter may be of interest and some value in complementing and illustrating their analysis; it seeks to record various instances which have directly concerned me.

My father Herbert Stansfield (1872–1960) was an experimental physicist (Wood 1960), so I have had a relatively long acquaintance with physics, direct or at one remove. In 1895 my father became a Fellow of the Physical Society of London. Just after he died in 1960, I became a Life Fellow of the Society. During this connection of over 90 years, not only has the Physical Society been absorbed into The Institute of Physics, but physics itself, its nature and practice, has greatly changed.

When I was a boy, my father talked much to me about physics and about the things in the world around us which interested him as a physicist. My first active involvement in research was in 1927, when I was almost 12 years old. The War Office was test-firing big guns on the artillery range at Shoeburyness, on the north side of the Thames Estuary, and took the opportunity to study the anomalous propagation of sound by refraction in the upper atmosphere. At Bristol University and at University College, Southampton, the Physics Departments were brought in as observing stations. Southampton, where my father was Professor, was expected to be within the zone of silence skipped by the sound, so we had the special challenge of showing a negative. To detect the sound, a Tucker hot-wire microphone was used, with an Einthoven string galvanometer recording on a moving strip of photographic paper (Stansfield 1927). To reduce wind noise, the microphone was placed in the middle of a bramble-patch near one of the department's ex-army wooden huts; my assignment was to sit with a stopwatch on the steps of the hut and record and time any noises, e.g. from a slammed door or a distant train, which might appear on the recorded trace.

A rarely noticed natural phenomenon pointed out to me by my father when I was a boy was the Reynolds line on slowly moving water. It appears to have been first recorded by Osborne Reynolds and often to have been independently rediscovered; Scriven and Sternling (1960) give much information. Different observers' descriptions vary; Schmidt (1936) said it is 'like a very fine thread or hair floating on the surface'. It seems to mark a boundary between areas of surface differing in contamination; Stansfield (1936) quoted my own observations: 'I often see the line on the river here [the upper Cam]; in any place where water is eddying up from below there may be a patch of clean surface marked by the bright line of boundary'. The line often appears to form loops, of no particular shape, lying sluggishly on the surface of the water; however, it behaves as if there were line-tension, so that if it forms a curved loop, the loop appears to contract increasingly rapidly as the curvature increases, ending by 'snapping together' with circular ripples spreading out from the point where it disappeared; this provides a quite characteristic identifying feature. I noted in 1975 (Stansfield 1975) that I was seeing the line less often than in the past; I thought then that this was because I was living and working in different places and saw different rivers and streams, for 'the Thames in London is too dirty to show the line'. Writing in 1988, I record that it is some years since I last observed the line, despite on various occasions deliberately looking for it in places where former experience made me expect to see it. My hypothesis is that at the present time the rivers and small streams where I have looked for the line all contain so much detergent residue that the surface conditions which produce the line never occur. Long ago my father observed that a few drops of soapy water added to a water-trough destroyed the line (Stansfield 1936). Does this mean that I will never be able to point out the line for the benefit of my grandchildren, as my father pointed it out to me? This is an unexpected instance of the problem of replication in science.

As well as giving me the habit of actively observing the world around me, my father taught me that scientists make hypotheses to explain what they have observed; that it is their duty to test each hypothesis by being always on the alert to notice anything inconsistent with the hypothesis, and also to do their best to think of ways to look for something which might be inconsistent with the hypothesis. If the hypothesis were upset, then the challenge was to think of a better one, one consistent with all that was observed and known. A hypothesis started as a tentative explanation; the more challenges and tests it survived, the more firmly it became established and moved from being 'hypothesis' to 'accepted theory'. But theories were never to be taken as final: the duty always remained to be on the watch for an observation, an experiment, inconsistent with the theory which would show it to be inadequate. I do not know how old I was by the time this basic principle had been instilled into me; I am sure it was firmly there by the time I was 15, i.e. by the year 1930. Moreover, I am sure that it was not something my father had recently acquired, but was something he regarded as long accepted. It may be of some importance for the history of the philosophy of science that this would make my absorbing it not later than four years before Popper (4 BP!), that is four years before the appearance of Popper's *Logik der Forschung* in 1934. With regard to philosophy, I lived a very sheltered life at school and then in the Cavendish Laboratory; I never heard of the 'hypothetico-deductive method' until after World War II. During the war I was engaged in operational research, using my natural science training and experience to study applied problems of behaviour within organisations. In this research we needed to find out more about regularities of behaviour of assemblages of diverse sorts; on the whole, the problems concerning assemblages of machines were less important than problems concerning assemblages of human beings. A consequence was that I became commensal with psychologists and sociologists, and found them speaking often of 'the hypothetico-deductive method' (Stansfield 1981).

The laboratory culture of the 1930s

Both the material and the social aspects of laboratory culture should be described here. Indeed, it is most important to keep in mind how intimately each aspect influences and helps shape the other. If in this chapter I pay attention mainly to the problems of understanding and re-creating artefacts and other items of material culture, rather than to customs and other elements in social culture, it is because it seems that current history of science under-plays attention to the material aspects compared with the ideas, knowledge, theories and other social aspects.

Also, it is hard to keep in perspective and in due balance the diverse aspects of material objects, of manual and other skills, of beliefs, ideas and knowledge, and of values and customs, which all interrelate and determine 'what goes on'. A major problem is that so much is too familiar to be noticed, except by special effort (if then) or under exceptional circumstances. To quote Hatch (1985), 'A large component of culture is below the level of conscious awareness'. Anthropologists are well aware that it is much easier to study other cultures than to study one's own.

Physicists refer often to a 'sealing wax and string' phase of laboratory work; trying to date this, I feel uncertain whether to notice especially that it first became conspicuous in the first or second decades of the present century, or that it was most salient in physics in the 1920s. In the 1930s, the period which has made the clearest impression on me as one of which I am aware, 'sealing wax and string' (together with associated cultural features) continued to be important; but the most characteristic material features of the decade, or at least those which come first to my mind, are glass, Plasticine, rubber tubing and Avominor meters (multimeters measuring milliamperes, volts and ohms, of a very widely used sort).

The difficulty of recognising the basic features of one's own culture came home to me when I tried to reconstruct further the typical physics laboratory of the 1930s, starting with the basic supplies laid on at, or adjacent to, most work-benches. These were electricity, gas for heating and water. In the UK, electricity was 50 cycles per second alternating current, probably at 200 or 220 volts; rarely at the coming standard of 240 volts, though the transition made progress during the decade. (In North America, the frequency was 60 cycles per second; a few serious misunderstandings have occurred because of failure to realise what was taken for granted in accounts written on one or other side of the Atlantic.) In a few laboratories, direct current was also available, at 110 volts DC. Gas was always available, often at taps actually on the work-bench; it was 'coal gas' from the town supply, as used in homes and other buildings, as was cold water, probably available from a tap over a sink adjacent to the work-bench. There might, or might not, be a tap with hot water alongside the cold tap. This water contained a variable amount of calcium, often substantial ('hard water'), as bicarbonate or carbonate. The water would come from a storage tank several stories higher up in the building, or at a similar height, so the pressure was enough to operate the 'filter pumps' often used to provide a vacuum—down to near the vapour pressure of water—when suction was needed, e.g. to pull liquid through a paper filter. These pumps, cheap and fairly robust, were almost always made of glass; inside the pump, a fine jet of water from a nozzle entrained air. The vacuum produced was not really good enough to provide backing for a high-vacuum pump; this would be a vapour-diffusion pump using mercury (low-vapour-pressure oil was just coming in to replace mercury as the working liquid), and for the backing 'Hyvac' pumps were common. This was a robust make of rotary pump, driven by a belt from a small electric motor; they were to be heard clicking away on the floor under many laboratory work-benches.

Another important aspect of the culture was the chemicals in common use, routinely available from the laboratory stores. Solvents, not forgetting tap water, included distilled water (used in special cases when the impurities in tap water would matter); also meths (methylated spirits), benzine, acetone and (ethyl) ether. Pure ethanol (absolute alcohol) was very rarely used, and then for special cleaning or other purposes; it was too expensive, because of the high excise duty on 'potable alcohol'. Meths was made undrinkable by additives, notably methyl alcohol and fuel oil, and clearly recognisable by methyl violet dye. Benz*i*ne was not to be confused with benz*e*ne, the chemical compound C_6H_6, also available but much less used. Benzine, used for cleaning when grease was to be removed, was a petroleum distillate fraction of fairly low boiling point; a similar fraction, of still lower boiling point, was available under the name of 'aviation spirit', but fire risk discouraged its use. When a piece of apparatus had to be clean and dry, it was usual to rinse it with acetone, then with ether to remove the acetone and finally to dry it with warm air (cautiously; ether was very inflammable, a known risk). For drastic cleaning, 'cleaning mixture' (concentrated sulphuric acid with potassium bichromate crystals) could be used; rarely and with reluctance, because it was so corrosive. Concentrated sulphuric acid was a strong acid readily available, but the acid most commonly used was probably dilute sulphuric acid, often four parts of water to one part of acid. Hydrochloric acid (the old name, 'spirits of salt', had become rarely used) was also in common use, strong or (mostly) dilute; nitric acid, strong or dilute, was readily available but treated with more respect. Ammonia dissolved in water (strong, at a density of 0.880—'880 ammonia'—or more dilute) was much used, as was 'washing soda' (crystals of hydrated sodium carbonate). Liquid air may perhaps be included under 'chemicals'; it was used for cooling apparatus, notably for mercury-vapour traps in high-vacuum work (then, at pressures below the vapour pressure of mercury down to, say, 10^{-6} mm of mercury). It was readily available from a large metal Dewar vessel in the laboratory stores; it came from the central air liquefier plant which supplied all the laboratories in the adjacent area of Cambridge University. By the end of the week, because of evaporation, the liquid in the Dewar became much enriched in oxygen, so that a cigarette would burn strongly in its vapour; a swab of cotton wool soaked in the liquid would burn especially strongly. Vacuum flasks were usually used in apparatus to hold liquid air, and also by the user to collect the liquid from the laboratory stores; mostly they were bought in local chemists or other shops, and were of the kind used by travellers and others to keep tea hot, usually the one pint size, sometimes one quart.

Regarding electrical matters in the laboratory, perhaps the most characteristic feature of the 1930s was the DCC copper wire used to conduct electricity from one piece of equipment, or one instrument, to another or from place to place within apparatus. It was much used also just to tie one thing to another. DCC stood for 'double cotton covered'; the copper wire was insulated by two layers of cotton thread, wound around the wire in opposite senses. (DSC—double silk covered wire—was very rare indeed, used when cost was quite subsidiary to the need to wind wire in the least possible volume). DCC copper wire came in various thicknesses; perhaps 26 SWG (standard wire gauge) was the commonest, with 30 SWG next, and 18 SWG a frequent size of thicker wire. The enamelled copper wire which replaced thread-wrapped wire was only just appearing in the 1930s. DCC wire had the practical advantage that when bare copper was needed, it was quick and easy to remove the covering by pulling the wire through a folded piece of fine sandpaper or emery paper; sandpaper and emery paper were about equally common, though emery was taking over. Any list of characteristic items must include the resistors, very cheap and extensively used; the resistance element was contained inside a small ceramic tube with, at each end, a metal cap carrying a length of tinned copper wire to provide an electrical connection. The body of the tube carried markings, colour coded to denote the value of the resistance. Electrical power for local use in apparatus was conveniently available from batteries; these were either secondary cells (rechargeable) of the lead-acid sort, giving 2.2 volts, or non-rechargeable, primary batteries on the Leclanché principle giving 1.5 volts per cell. Two, or a small number, of cells connected in series were often contained in a single unit to give a correspondingly greater voltage. Leclanché-type batteries were usually called 'dry cells', and lead–acid batteries 'accumulators'; 'HT (high-tension) batteries' were formed of many small dry cells in one (rectangular) package, connected in series to give (most commonly) 120 volts, often with intermediate voltage tappings available. They were used especially to provide the anode voltage for the thermionic valves, triodes or more sophisticated, which dominated electronic apparatus. Semiconductors were ill-understood; rectifiers and other devices using them were rare and crude. Oscillographs and other instruments using CRTs (cathode ray tubes) appeared during the 1930s, and were a characteristic feature of physics laboratories by the end of the decade. Avometers were a standard make of electrical multimeter, providing a choice of ranges to measure DC and AC voltages and DC millamperes, also ohms; they came in a variety of sizes. The smallest and cheapest, the Avominors (already mentioned above), were ubiquitous and used for many purposes. In 1933 all research students doing experimental work in the Cavendish Laboratory had to have their own, price £5 less 25 per cent discount if bought through the laboratory, i.e. £3 15s. net (£3.75). In the first decade or two of the twentieth century it seems to have been generally thought that in electrical instruments greater size and greater accuracy went together; by 1930 the fallacy of this had been recognised, and ammeters, voltmeters and other galvanometers had become smaller. Another custom, dominant up to the 1920s, had also given way; it was that of enclosing electrical instruments and equipment of all kinds as far as possible in French-polished boxes of wood, usually mahogany or a cheaper substitute. Cases of sheet metal took over; at first they were finished in black japan, but in the 1930s lighter-coloured finishes came in.

Another aspect of the culture of a laboratory is the pattern of communication between the people and the material provisions for such communication. When in 1937 I became secretary of the Cavendish laboratory 'Bun Club' (Stansfield 1982), each afternoon at 4 PM two of the youngest laboratory assistants brought large pots of tea to the room next door to the laboratory library; during the next half hour, teaching and research members of the laboratory drifted in for tea and informal conversation. Besides tea, milk and sugar, cakes were provided. These cakes were 'fancies', i.e. small iced sponge cakes, fancy biscuits, etc such as would be served as part of a set afternoon tea in a restaurant. They were delivered each day to the Laboratory by Messrs Hawkins, bakers and pastrycooks, whose shop in the centre of Cambridge included the well-known Dorothy's café and restaurant, and came from the previous day's left-overs. By teatime, they were none too fresh, and I was advised that people were likely to prefer fresh scones with strawberry jam. Fortunately I was then sharing 'digs'—i.e. lodgings licensed for students' accommodation by the University's Lodging House Syndicate—with another research student; he had just been quartermaster for the Summer Camp of the Cambridge Mission to Fruitpickers (well and amusingly described by Vidler (1977)). From this quartermaster experience, he was able to tell me how to obtain from the local wholesale grocers, Messrs Hallack and Bond, 'Magpie' strawberry jam; this was the 'seconds' brand from Messrs Chivers, the local (and excellent) manufacturers of jams and conserves, supplied in 7 lb jars. He told me also how to provide much better-quality tea for no increased cost by buying Indian tea in a 28 lb tin and China tea in a 7 lb tin. Lump white sugar was very cheap when bought in a 1 cwt wooden case (the minimum size; more than a year's supply for the club). Providing a choice between Indian and China tea was an innovation; coffee in those days did not come into consideration. Later I gathered that Fred Lincoln (the Cavendish 'character' who was in charge of the laboratory workshop and stores) had reservations about the storage and security of these bulk supplies, but at least he was saved the 'hidden cost' of sending a junior member of his staff to a nearby grocer's shop, perhaps once a week, to buy tea and sugar. All this now seems very remote from the size, formality and general style of the cafeteria-type refectory in the present Cavendish Laboratory. One wonders how the flow of information and of ideas, and the fostering of the sense of common membership of a community, has changed.

Repeating experiments

To return to the specific issue of repeating an experiment, my first direct experience came when I started as a research student. My first assignment was to try to replicate the finding reported in a paper just out in *Comptes Rendus de l'Academie des Sciences* of Paris. The author claimed to have obtained an excellent source of hydrogen ions by diffusing hydrogen gas through heated palladium into a strong electric field. In 1936 the Cavendish Laboratory was naturally very interested in any possibility of a more convenient source of hydrogen ions; the finding had interest also because it appeared to go against the second law of thermodynamics, as applied through Saha's equation. I set up an apparatus which diffused hydrogen through the wall of a hot palladium tube, which was electrically earthed, into an evacuated space between the tube and a nickel cathode at a high negative voltage. I found no hydrogen ions whatsoever (Stansfield 1938a) although as a by-product I learnt much I had never expected about the vicissitudes of carrying out an experiment, and also about strong-field cold-emission of electrons. The *Comptes Rendus* paper had said hardly anything about how the experiment was done and the discrepancy with my findings was a complete mystery. For a year the mystery remained, until I received a reply to my letter from the author, which had pursued him around half a dozen European capitals; he mentioned in passing that he had used a voltage of 100 kV to extract the hydrogen ions, and also that he had heated his palladium tube by a hot tungsten wire wound round it. My immediate thought was that he had bombarded his palladium with electrons, or had had gas-discharge ionisation of the hydrogen—the published paper had given no indication of such possibilities. As a tailpiece to this story, in *Europhysics News* in 1973 a note of his death described him as the doyen of physicists in his country.

Another of my experiments as a research student ran into rather different problems of replication. I took over an apparatus to measure the range in air of slow alpha particles, built and used by two research workers acting as a team. Particles of desired energy were selected by a magnetic field of variable strength which deflected them along a circle defined by three slits, these slits being in a vacuum between the poles of a large electromagnet. The energy of the particles was measured by the strength of the field. At that time it was not easy to measure to 1 part in 5000 a magnetic field inside a confined, evacuated space. My predecessors had 'flipped' a search coil; the arrangements were such that two people were needed to do this. Working solo, I had to think again. A promising method seemed to be to time the oscillations of a suspended diamagnetic crystal; but no-one seemed previously to have used this for precise measurement. Would there be snags? In particular, would temperature changes affect the measurements? From the crystal structure of calcite, theory suggested that temperature variation of the magnetic anisotropy, and of the moment of inertia of the crystal, should be acceptably small. But a direct check seemed desirable, as the matter was central to my experiment. The only direct observation of the effect of temperature on the magnetic anisotropy which I could find in the literature dated from 1855; Michael Faraday's book recorded that he heated calcite to a red heat and observed no change in the magne-crystallic action—but there were no details of how he did it, nor of how sensitive was 'no change'. The best I could do was to parallel, if not exactly repeat, Faraday's observation by cooling a calcite crystal, using liquid air. (I too 'observed no change'). My published paper (Stansfield 1938b) gave some description, including a line diagram, of my apparatus; but I was well aware that I did not know how effective was the cooling of the crystal—well aware too that the effectiveness would be sensitive to details of construction and of operation of the apparatus which my description did not cover. So someone else would be unable to check my findings, much as I had been unable to check Faraday's.

An earlier, well known issue in physics which turned on the repetition and confirmation of experiment, one which affected me, if only obliquely, is the value to be accepted for e, the charge on the electron. Robert Millikan's classic 'oil drop' determination (Millikan 1917) led him to the value $4.774 \pm 0.005 \times 10^{-10}$ esu. During the next two decades, further evidence from various sources, especially from x-ray reflection by crystals, became increasingly hard to reconcile with this figure. The discrepancy between Millikan's figure and an apparent x-ray figure of around 4.80×10^{-10} esu was felt increasingly acutely until the x-ray value became generally accepted.

The uncertainty was at its peak at about 1935–6, the time of my final undergraduate year. Because of the great importance for physics of the value of e, my course paid much attention to the various ways by which it can be determined; in retrospect, for me one of the most educative items in the course was the essay my supervisor set me to write, to examine the pros and cons and likely reliabilities of the possible methods. This was well before physicists had come round to realising that the trouble lay with the value used by Millikan for the viscosity of air.

After publishing his value for e and also his book, Millikan lectured widely, making known his work and drawing attention to the accuracy and importance of the determination. His extrovert presentation seems to have struck UK physicists: I remember clearly a joking remark from Rutherford, I think during an undergraduate lecture for the Part II Tripos course, to the effect that when Millikan spoke in Cambridge and put forward the claim for the accuracy of the value of e, he had said to him that if the real value eventually turned out to be within his claimed limits of uncertainty, 'you should go down on your knees and give thanks'—the last words said with true Rutherford gusto. Already as a boy I had been told by my father that Millikan had 'stumped' the world telling people about his measurement of e. My father also said (it was fully in character for him to remember and savour this story, but would have been quite out of character for him to invent it) that the suggestion had been put forward (not very seriously!) that for the unit of scientific publicity the 'microcan' should be adopted—the 'millikan' would be too large for practical use. I observe that the suggestion has not been taken up, even if the need for such a unit still remains or has increased; I have always felt that it fell down on the requirement that an acceptable standard unit must be adequately reproducible.

Replicating laboratory conditions of the past

My experience of difficulty in visualising past situations, possibilities and problems was extended as a result of my wife's interest in English literature. She wanted to know more about Thomas Beddoes MD (1760–1808) who greatly influenced the poet, Samuel Taylor Coleridge (1772–1834). He proved to be a man of very diverse interests. *Inter alia* he contributed regularly to the *Monthly Review* and early drew attention in the UK to the philosophy of Emmanuel Kant. He set up in Bristol the Pneumatic Institute where his patients could live under the same roof as the laboratory and he could carry out clinical trials to see if the recently discovered gases could be of use in treating diseases. He even recruited the young Humphry Davy (1778–1829) to be his laboratory superintendent (Stansfield 1984). Trying to understand the nature and significance of the experiments at the Pneumatic Institute, I had to free myself from automatic assumptions, e.g. that rubber tubes would have been used to connect pieces of apparatus and convey gases and liquids, and remember that things essential in my own experience did not then exist (nowadays I find equal difficulty in remembering that polythene has replaced rubber and in allowing for the different resistance to attack by chemicals). I had little or no idea of the level of purity of the gases—oxygen, nitrogen, hydrogen, nitrous oxide, carbon dioxide, water gas—prepared by Beddoes and Davy and used in their experiments on themselves and on patients, important as such information seemed to be. What, and how pure, were the raw materials available to them? To prepare oxygen, they heated 'manganese' (native manganese dioxide, from various fairly local sources) and Beddoes commented that some batches were much better than others. From the point of view of replication of apparatus, surviving information becomes very much more adequate from the time when the engineer, James Watt (1736–1819) involved himself in what Beddoes was doing. He did so, seeking consolation after the death of his daughter Jessie from tuberculosis in 1794; the doctors caring for her had been Erasmus Darwin (1731–1802) and Beddoes. Watt addressed himself to the task of redesigning Beddoes' apparatus for producing gases and administering them to patients (Stansfield and Stansfield

1986). The content and style of Watt's published accounts reflect both his direct involvement in designing and using the apparatus and his personal enjoyment of so doing; it shows also his ability to command the resources of the drawing office and 'sales promotion staff' of Boulton and Watt, the firm which made and marketed the equipment. One notices Watt's attention to important practical details such as the ability of black japan lacquer to resist corrosive gases, and his care in specifying the best way to make and apply the appropriate lute to make a joint gas-tight.

My next instance relates to a period some four centuries earlier. I met it when The Canon's Yeoman's Tale by Chaucer (1340–1400) was made a set text for an A-level English examination, soon after a new edition had appeared (Chaucer 1965). The students and their teacher needed not only to know the modern equivalents of fourteenth century names of chemicals such as argoille, realgar but also to be able to visualise what was going on; I was asked to help, particularly with this latter. Chaucer's story, put into the mouth of a laboratory assistant, is an account of chemical manipulations carried out by an alchemist purporting to transmute mercury into silver, and in particular of the manoeuvres and sleight-of-hand used in order to appear to have achieved positive results, and thereby to obtain further financial support. I have discussed elsewhere (Stansfield 1977) the details of Chaucer's account and the reasons from internal evidence why we must conclude that he wrote it from first-hand experience of, substantially, such a situation. Therefore we should accept his account as a valid case study in the sociology of science; moreover a very early one. Fairly recently, Terry Jones (1980) has discussed Chaucer's use of actual situations. His conclusions parallel mine; his book is also interesting in regard to wider issues of distortion of pictures of the past. The relevance of my study to the subject of the present chapter is that the key evidence was Chaucer's description of the behaviour of the Canon-alchemist and his yeoman assistant, accurately describing the symptoms of mercury poisoning. I could only conjecture an order-of-magnitude estimate of the concentration of mercury vapour in the Canon's laboratory; what I lacked was information about the likely ventilation conditions, or even about what sort of building it would be in. Holmyard's well known book on alchemy (Holmyard 1957) contains a plate showing the Canon's yeoman; but certainly the elegant and spacious premises in the picture owe everything to artistic licence, not to Chaucer's account.

Except for the problem of laboratory premises and layout, it seems to me that we could now reconstruct the materials, apparatus and implements used, even the clothes worn by the people affecting what they could or could not do in their manipulations; we could reacquire their manipulative techniques and skills. From Chaucer's account and using evidence from artefacts surviving in museums and from other sources, we could reproduce and re-enact, with a probable close approximation, in deed as well as in appearance, to the original, what took place late in the fourteenth century. Or rather, we could, were such re-enactment not now ruled out by attention to health and safety. We can only welcome both the progress in scientific ethics regarding experiments involving human beings and parallel developments in health and safety legislation which prevent such replication. I am struck when I reflect on how recent are these changes, seen in historical perspective. I have recorded elsewhere (Stansfield 1977) my own experience of changing awareness of health risks in the laboratory, particularly but not only with mercury. The change in attitude and practice in academic places of scientific work has taken place in a way, and with a speed, resembling a Kuhn-type revolution. Attitudes to mercury have changed in other respects also. It would seem that Chaucer, like Beddoes four centuries after him, regarded mercury and mercury compounds as readily available, with cost no special problem compared with other substances. They knew the medicinal and toxic effects of mercury, but not the danger of breathing its vapour. My own first experience of mercury and its compounds was when I used them as a boy in physics and chemistry experiments at home, on the dining room table or in the kitchen sink; they seemed easy to get but rather expensive. When I was a research student, cost was a known deterrent to the use of mercury in experiments, though materially less so than now. Serious attention to the dangers in the laboratory from mercury vapour, as far as I am aware, dates only from after World War II.

Changes in composition, meaning and nomenclature

Insights into the problems of repeating and confirming experiments of any period, along with substantial contributions to the sociology of science, have come from the important work, including case studies (for example, of gravitational radiation) of Harry Collins and his associates and others (e.g. Collins 1975, 1985, Pinch 1986; Broad and Wade 1985). Collins (1975) contrasts two models of scientific knowledge and its transfer between scientists, the 'algorithmic' and the 'encultural'. The algorithmic model 'implied that there is a finite series of unambiguous instructions which can be formulated, transferred, and when correctly followed will enable a scientist to copy another's experiment exactly'. The encultural model recognises that 'the meaning of an "exact copy of the original" is itself problematical'. This distinction would seem to be blurred if one remembers that instructions which are clear and unambiguous when they are written may later be quite otherwise. 'Instructions' will include specifications, as of materials, apparatus, etc to be used; I have been increasingly struck by the problems for replication arising from the way in which what is meant by a name changes with time, and indeed changes deceptively fast.

Many examples could be given, such as caoutchouc, glass, sealing wax. Descriptions from the first half of the nineteenth century refer to the use of caoutchouc; I know in general terms that this was a form of unvulcanised rubber, but I have no idea of how its properties compared with those of the rubber with which I was myself familiar when used in experimental apparatus. Gutta-percha is another, somewhat similar and likewise perishable, material once important in laboratory use but now vanished. Again, there have been a series of changes in the composition of the glass in everyday use in the laboratory for making apparatus, so that one may well be seriously uncertain about the fragility and other physical properties, and the chemical reactivity, of the apparatus described in old accounts. I began laboratory work when soda-glass was normal; Pyrex was coming in, but we were discouraged from using it because of its greater cost and the cost of the oxygen needed to work it, compared with using soda-glass and air from a foot-bellows. Pyrex, because it expanded much less on heating than did soda-glass, called for less skill in glass blowing if the product were not to crack on cooling; also it allowed more complicated designs to be made. A minor disadvantage was that, in practice, joints in Pyrex were apt to be less well melted than in soda-glass; the result was to leave pin-holes which leaked air into evacuated apparatus. The blowpipe used town gas—piped to laboratory benches—burnt either with oxygen or the laboratory air. A change which had considerable impact on laboratory techniques came around 1965, when town gas became 'natural gas' from North Sea sources; the town gas which it replaced may not have been quite the same in composition as as the traditional 'coal gas', but the changes had not been of material consequence for glass-blowing techniques. In 1984 the glass-blower in the Cavendish Laboratory, Cambridge, Mr Arnold Bloor, described to me what had been the effect of natural gas; it would not give so good a flame in the blowpipe.

Sealing wax and string, also Plasticine, are other basic materials traditionally used in physics laboratories which by now offer replication problems. As a schoolboy, the red sealing wax which I bought in any stationer's shop was closely similar to that in pieces of apparatus made by my father around 1890; by the late 1930s, the sealing wax from such shops came in a range of colours, and often was perfumed mildly. By then, besides red sealing wax, for vacuum apparatus workers in the Cavendish Laboratory used also Bank of England wax. This differed in being white in colour, less brittle and when heated becoming soft and stringy rather than definitely melting; it could not be poured into a hole or mould. The two kinds could therefore not be used interchangeably. String was made from hemp; the alternatives based on synthetic fibres did not then exist, so string was an electrical insulator but not a very good one if it were damp. In the 1930s Plasticine was a most useful and ubiquitous material for laboratory use; it could be assumed to be Harbutt's Plasticine. It came in an extensive range of colours and was widely sold in stationer's shops and toy shops as modelling material, for use especially by children. It was reputed to be made from a grease like Vaseline (i.e. petroleum jelly) absorbed in kieselguhr or similar material, plus colouring matter and a mild perfume; I never knew its vapour pressure, though this was important because we used it extensively to lute joints in vacuum apparatus, but the vapour pressure of Vaseline (which we also used, e.g. to grease taps in low-vacuum apparatus) was said to be of the order of 10^{-3} mm of mercury. Certainly, Plasticine hardened when such joints were left exposed to the air for periods of months, and could be re-softened by rubbing Vaseline into it. An important laboratory skill was in designing joints in apparatus so that they could be luted, and made airtight, by rubbing Plasticine in where two surfaces met at right angles; also the manual skill of rubbing the lute smooth, possibly lubricated with Vaseline, so as to remove any small holes which would leak air into vacuum apparatus. There existed also, specially manufactured for high-vacuum work, what we called Apiezon Plasticine, made by Metropolitan Vickers, of Old Trafford, who stated its vapour pressure as being less than 10^{-6} mm of mercury. It was very much more expensive than Harbutt's, so could only be got, and then only in minute amount, by making a very special case to the man in charge of the laboratory workshop and stores, Fred Lincoln. The problem for replication now arises because Harbutt's Plasticine is no longer made. Who now knows what was its composition? Evaporation and other processes will have greatly changed the composition and properties of such specimens of Plasticine as have survived in old apparatus. 'Blu-tack', perhaps today the nearest equivalent, is not closely similar. Also, how practicable would it be to recover the practical skills involved in designing, making and 'rubbing' a sound Plasticine joint? Incidentally, an important piece of practical 'know-how' was that semi-permanent vacuum-tight joints should be made with sealing wax (a relatively slow job), and demountable joints be made airtight with Plasticine (relatively quick); the Plasticine was readily removed with benzine-soaked cotton wool, without damaging a sealing-wax joint, whereas the sealing wax could be removed (with little effect on the Plasticine) by gently warming it and wiping with methylated spirits.

Another example of the question of availability is provided by the famous Christmas Lectures for a juvenile audience at the Royal Institution in 1899 (Boys 1890). Boys' book fascinated me when I was a boy, partly because my father had become a physicist when he was a student in Professor Arthur Rücker's physics department at the Royal College of Science, South Kensington. C V Boys was Associate Professor there. My father's great admiration for his skill, indeed genius, as an experimenter developed into lifelong friendship. My father had himself done research on soap films (Stansfield 1906). I repeated at home many of the experiments described by Boys (1890). Fortunately I was allowed to use the pipe for blowing soap bubbles which my father had made (and which I still have). It follows the design sketched by Boys in Fig. 68 of his book. (When I was a research student and had half a dozen lessons from the laboratory glass-blower, I made a similar pipe from soda-glass; it was much less well made than my father's, was inadequately annealed in cooling, and has not survived.) The immediate relevance to replicability arises because soap bubbles and soap films tend to thin and burst. As Boys wrote (1890 p. 143), 'Bubbles blown with soap and water alone do not last long enough for many of the experiments described, though they may sometimes be made to succeed'. He prescribed 'A Good Mixture for Soap-Bubbles', saying: 'Common yellow soap is far better than most of the fancy soaps, which generally contain a little soap and a lot of rubbish. Castille soap is very good, and this may be obtained from any chemist. ... Plateau added glycerine, which greatly improves the lasting quality. The glycerine should be pure, common glycerine is not good, but Price's answers perfectly.' In August 1988, in my nearest branch of Boots The Chemist, I enquired about Castille soap and Price's glycerine. Neither was available; a pharmacist remembered vaguely the latter. He called upon a colleague; she remembered it quite well from the past. Neither remembered Castille soap. The foregoing will explain why, a few years ago, I attended an Institute of Physics lecture on soap bubbles, planned particularly for younger students, and was interested to notice two bottles of 'Fairy' washing-up liquid (not soap-based) placed conspicuously on the demonstration bench on which some of Boys' experiments were repeated.

Repeating electrical experiments

Another area in which recent experience can be a misleading guide to understanding old experiments is that of electrostatics, specifically as regards what behaved as a conductor of electricity and what as insulator. My father enlivened some Christmastime science lectures for children by showing them how to make a simple gold-leaf electroscope, based on a square pickle jar with a large cork bung; they put a thick copper wire, insulated by a coating of sealing wax, through a hole made in the bung, turned the bottom end of the wire to be parallel with the bottom of the bung, and hammered it flat. At the next lecture in the series my father put a pair of strips of gold leaf, one hanging from each side of the flat on the wire, on each instrument brought to him by its maker. What is here relevant is that to prevent the gold leaves sticking to the glass when the electroscope was charged and the leaves diverged, the child who made the instrument put a strip of ordinary writing paper inside the jar, on the bottom and up two sides. When the leaves touched this, they were immediately discharged and fell together.

Until about forty years ago, one could safely assume that paper was normally a conductor in the context of electrostatics (though an insulator when in the context of current electricity). This applied to newspaper as well as to writing paper; amusing experiments based on rubbing pieces of newspaper to electrify them so that, for example, they would stick to the fireplace or two strips would spread apart like the leaves of an electroscope, required that the paper be specially dried, perhaps by holding it in front of the open coal fire. This assumption that paper conducted—parochial to the damp climate of the United Kingdom!—unobtrusively ceased to be valid when homes and other buildings became centrally heated, or at least kept normally $5\,°C$ or more warmer than was earlier customary.

The same change caused electrification of clothes, and consequent sticking, to appear as a significant problem; though here changes in fibre composition (especially the arrival of nylon) complicated the story. This is an area of the sociology of physics as yet largely unexplored. Modern readers of the history of the Cavendish Laboratory may well fail to appreciate the significance of the provision in the design of the new building opened in 1874 of a room in connection with experiments in electrostatics, a room in which the atmospheric humidity was to be reduced by a moving endless band of flannel (Moralee 1874). The ingenuity of the means for heating the flannel in one section of its path, so as to reduce the water content of the wool, makes one curious about how far it worked in practice.

It is interesting to notice that in his valuable manual *On Laboratory Arts* Threlfall (1898) devotes at least 30 pages (pp 240–70) to information about matters of electrical insulation; this is a reminder of the extent to which a coating of shellac was used, in the days when shellac (usually dissolved in methylated spirit) was a routine laboratory material, to make glass non-conducting of electricity. Threlfall says much also about another insulating material extensively used in the past, ebonite; his comment (p. 253) that 'This exceedingly useful substance can be bought of a perfectly useless quality' which 'for instance, deteriorates so rapidly when exposed to the air that it requires to have its surface renewed every few weeks', reminded me of the length of time I spent on various past occasions scraping ebonite to make it non-conducting; the most memorable was when, as a sixth former, I succeeded in making my school's Wimshurst machine (with ebonite plates), which for years had stood unused on a shelf in the physics laboratory, work again.

As a play on words, it is tempting to reply: 'Yes. Too often!' to the question, 'Can we repeat it?' in respect of the widely used quotation, 'when you can measure what you are speaking about, and express it in numbers you know something about it; but when you cannot measure it, when you cannot express it in numbers, your knowledge is of a meagre and unsatisfactory kind' (Thomson 1889). These words (often shortened, as here, from the original or misquoted) are almost always used by someone unaware, as is the recipient, of the context in which Thomson spoke, a context which much affected the meaning. It was, in fact, a fine example of the 'joking relationship' well know to social anthropologists. The future Lord Kelvin, in his capacity as a leading electrical engineer, was speaking as an important invited lecturer; he used his privileged position to 'take a rise' out of his civil engineer hosts. He drew it to their attention that during the previous decade, the accuracy of electrical measurements had improved by orders of magnitude, (which was not the case in civil engineering). This meant, to use modern terminology, that makers of instruments could exercise far better quality control over the purity of the copper they used and over replicating the performance of their instruments. He did not specifically mention the introduction of the Wheatstone bridge as a cause of the improvement.

The need for research into technological change

The foregoing has inevitably concentrated upon practical (including manual) skills, and technology. This underlines the importance of Rachel Laudan's comments on the reluctance of scholars to study technological change as knowledge changes, 'and particularly so if it were thought of as analogous to scientific change' (Laudan 1984 p. 6). In her shrewed analysis of the reasons for this, she points out that technological knowledge is quintessentially tacit, rarely articulated and even then in visual rather than verbal or mathematical form. 'Historians are well aware that technological knowledge can easily be lost. The pool of technological knowledge frequently shifts rather than expands with the appearance of new technologies, and is thus not always cumulative. There are losses as well as gains. If practitioners cease using a particular technology the knowledge of how to use it commonly dies with them. The mute presence of the remaining artifacts does not speak for itself, if there is no explicit statement separate from the artifacts' (p. 7). The present account illustrates extensively her points.

The implications of all of this indicate also the problems, and challenge, for museums whose job it is to record and make available the material evidence, perishable as well as imperishable, needed as basis for such an account. It is also needed to enable the readers of the account to understand it adequately. When, for example, I look at the tall glass jar closed with a cork bung, which I still have and which is full of my father's stock of flake shellac, or think of his mentions of Chatterton's Compound (widely used early this century; but I know little of what it was like or how it was useful), and then think of the Bostik period which has come and gone during my own lifetime, or of the present 'superglues' whose performance I know better than I understand their composition, I realise how much I need a guide, perhaps 'Laboratory materials through the ages', to help me understand the changing practicalities of experimental physics.

References

Boys C V 1890 *Soap-bubbles and the Forces which Mould Them* (Society for Promoting Christian Knowledge) [The Heinemann (1960) reprint gives an incorrect date of original publication and also of the 1912 'new and enlarged' edition *Soap-bubbles: their colours and the forces which mould them*]

Broad W and Wade N 1985 *Betrayers of the Truth* (London: Oxford University Press)

Chaucer G 1965 *The Canon's Yeoman Prologue and Tale* (from the *Canterbury Tales*) ed. M Hussey (Cambridge: Cambridge University Press)

Collins H M 1975 The seven sexes: a study in the sociology of a phenomenon, or the replication of experiments in physics *Sociology* **9** 205–24

—— 1985 *Changing Order: Replication and Induction in Scientific Practice* (London: Sage)

Hatch E 1985 Culture in *The Social Science Encyclopedia* ed. A Kuper and J Kuper (London: Routledge and Kegan Paul)

Holmyard E J 1957 *Alchemy* (Penguin: Harmondsworth)

Jones T 1980 *Chaucer's Knight: The Portrait of a Medieval Mercenary* (London: Weidenfeld and Nicolson)

Laudan R (ed.) 1984 *The Nature of Technological Knowledge. Are Models of Scientific Change Relevant?* (Dordrecht: Reidel)

Millikan R A 1917 *The Electron* (Chicago, IL: University of Chicago Press)
Moralee D 1874 The first ten years in Godby R *et al* 1980 *A Hundred Years of Cambridge Physics* 2nd edn (Cambridge: Cambridge University Physics Society) p. 14 (reproduced from *Nature* 25 June 1874 p. 141)
Pinch T 1986 *Confronting Nature: The Sociology of Solar-Neutrino Detection* (Dordrecht: Reidel)
Schmidt W 1936 Cause of 'oil patches' on water surfaces *Nature* **137** 777
Scriven L E and Sternling C V 1960 The Marangoni effects *Nature* **187** 186–8
Stansfield D A 1984 *Thomas Beddoes M.D. 1760–1808; Chemist, Physician, Democrat* (Dordrecht: Reidel)
Stansfield D A and Stansfield R G 1986 Dr Thomas Beddoes and James Watt: preparatory work 1794–96 for the Bristol Pneumatic Institute *Med. Hist.* **30** 276–302
Stansfield H 1906 Observations and photographs of black and grey soap films *Proc. R. Soc.* A **77** 314–23
—— 1927 Report on the experiment and samples of the recordings *University of Southampton, H Stansfield Archive in the University Library*
—— 1936 Line on the surface of water *Nature* **137** 1073–4. See also editorial notes on The Osborne Reynolds ridge *Nature* **138** 20, 612
Stansfield R G 1938a A search for ionization of hydrogen by diffusion through palladium *Proc. Camb. Phil. Soc.* **34** 120–3
—— 1938b Suitability of a diamagnetic crystal for the measurement of magnetic fields *Proc. Camb. Phil. Soc.* **34** 625–33
—— 1975 *The New Theology? The Case of the Dripping Tap (or, Student's Descriptions and the Forces Which Mould Them)* Paper presented 2 Sept. 1975 to Section N, Sociology of the British Association for the Advancement of Science (British Library Document Supply Centre shelf no Wq4-6987)
—— 1977 *Continuity of Funds for Research—The Canon's Yeoman's Tale* Read 5 Sept. 1977 to Section N, Sociology, of the British Association for the Advancement of Science (British Library Document Supply Centre shelf no Wq4-6986)
—— 1981 Operational research and sociology: a case-study of cross-fertilizations in the growth of useful science *Sci. Public Policy* **8** 262–80
—— 1982 Cavendish society *Br. Soc. Hist. Sci.* no 9 17–20
—— 1985 Ergonomics in *The Social Science Encyclopedia* ed. A Kuper and J Kuper (London: Routledge and Kegan Paul)
Thomson Sir William 1884 Electrical units of measurement *The Practical Applications of Electricity* (A Series of lectures delivered at The Institution of Civil Engineers) (London: The Institution of Civil Engineers) 149–74. (The words quoted appear, though the context is obscured, in Thomson Sir William 1889 *Popular Lectures and Addresses* vol. I (London: Macmillan) p. 73)
Threlfall R A 1898 *On Laboratory Arts* (London: Macmillan)
Vidler A R 1977 *Scenes from a Clerical Life: An Autobiography* (London: Collins) pp 78–80
Wood A B 1960 Herbert Stansfield *Year Book of the Physical Society 1960* (London: Physical Society) pp 81–2

Part 2

Applying the History of Physics

7

The Role of History in Physics Teaching†

A P French

Introduction

The other sections of this volume present us with a very rich feast of information and opinion about various areas of physics history. In this section various authors discuss the role or roles that such material may be able to play in our teaching of the subject. This is a question that has had a quite long history. As long ago as 1858 (and this may well not be a first case) Louis Pasteur, in his capacity as Director of Science Studies at the École Normale Supérieure in Paris, wrote a report on education (Pasteur 1858) with the title 'On the use of the historical method in science', and in it he pointed out what is still, I think, the fundamental question at issue:

> I know that most scientific discoveries can be described in a few words and demonstrated by a small number of decisive experiments. But if ones tries to understand the origin of these discoveries, if one follows carefully their development, one is struck by the slowness with which they have come into being. Two different methods can therefore be adopted to present a discovery. One method consists in stating the law and in demonstrating it in its present expression without bothering

† Based on a presentation given at the conference on the History of Physics for the Physicist, 2–4 July 1986, Oxford.

about the way in which it has come to life. The other method, more historical, evokes the individual efforts of the most important inventors ...and tries to transport the audience mentally into the period when the discovery was made.

And, as Pasteur said, the first method 'hides from the young people the slow and progressive [perhaps not so progressive!] march of the human mind. It develops in them an expectation of sudden revolution in thought, and an unfounded admiration of certain men and certain actions.'

Such matters have been discussed many times, though perhaps not continuously, since then. For the purposes of the present chapter I should like to draw attention to a few relatively recent publications that explore various aspects of the topic (Cohen 1950, Brush 1969, 1974, Brush and King 1972, Whitaker 1979). These publications are all addressed primarily to the teacher of physics rather than to the research historian. They do not arrive at any definitive conclusions, but they illuminate the problem in valuable ways, and make it clear that there are different kinds of questions about the possibility and the appropriateness of bringing an historical component into our teaching of physics.

Is history needed?

This is, I think, a useful first question, because it has a bearing on the nature of science itself. I B Cohen in his paper 'A sense of history in science' (Cohen 1950) points out that science is a cumulative discipline. It 'embodies the discoveries of the past insofar as they are valid or relevant in the light of our present knowledge—but without reference to the conditions under which they were made'. Cohen cites, as one example, the fact that it is important for a physicist to know that e/m for electrons has a certain value, but that today's scientist need not know the details of J J Thomson's original experiments—in particular the lack of good vacuum technology, which in fact made the experiment very difficult to do.

This notion of the cumulative character of scientific knowledge is certainly a very important one, and has been commented on also by Sir Peter Medawar in a delightful essay called 'Lucky Jim' (Medawar 1972) in which Medawar reviews James Watson's famous book, *The Double Helix*. He points out that Watson makes very little reference to previous workers in this field—not, Medawar suggests, from indifference, but simply from the fact that what mattered to Watson was where we are today, what we know today, without particular reference to exactly how we got there. To quote Medawar's precise words:

It is not good enough to dismiss this [lack of interest in the history of science] as cultural barbarism, a coarse renunciation of one of the glories of humane learning. It points towards something distinctive about scientific learning, and instead of making faces about it we should try to find out why such an attitude is natural and understandable. A scientist's present thoughts and actions are of necessity shaped by what others have done and thought before him; they are the wavefront of a continuous secular process in which The Past does not have a dignified independent existence on its own. Scientific understanding is the integral of a curve of learning; science therefore in some sense comprehends its history within itself.

To be sure, the teaching of science is not the same as *doing* science, but it would be hard to escape the conclusion that the only defensible answer to the question 'Is history of physics needed in physics teaching?' is 'Probably not'. This, however, still leaves plenty of room for debate.

Is the use of history desirable?

This is certainly a fruitful question. Florian Cajori, whose name is familiar in connection with the Berkeley edition of Newton's *Principia*, once gave a talk (Cajori 1899) to a group of science teachers in Colorado on 'The pedagogic value of the history of physics', and he commented on various reasons why one might want to introduce the history of the subject into the teaching. One reason was for the edification of the teacher: 'A knowledge of the struggles which original investigators have undergone leads the teacher to a deeper appreciation of the difficulties which pupils encounter'. Moreover, such knowledge will demonstrate to the teacher 'the futility of the pedagogical theory according to which the pupils in the laboratory should be made to re-discover the laws of nature'. (Presumably the 'discovery method' was enjoying some kind of vogue, just as it did for a while more recently.) But, for Cajori, the most important aspect was 'history as a stimulant, as a means of exciting interest... . Introduce historical matter incidentally and skilfully, and you will find it to be the honey that renders the bread and butter more palatable.'

Many physics teachers would, I think, give their support to this proposition. However, one must admit that it is a tricky business, fraught with perils. One of my recommended references (Brush 1974) is entitled 'Should history of science be rated X?'—suitable only for mature adults. Why did Brush say this? First of all, because the true history would essentially destroy the conventional image of the scientist as a rational, methodical, emotionless person who proceeds systematically through a subject and comes to inescapable conclusions. This would shatter cherished, though false, beliefs, and would undermine public faith in science. But his much less frivolous objection, to which he devotes most of the article and which is relevant to us, is that, when the teacher of science introduces historical materials, he or she must do so in a very selective way, because usually, as Brush puts it, 'he can only take from the past that which seems to have significance in the present'. Then, of course, any historian of science is likely to become upset, because it means that any such presentation will probably distort the true historical record. The danger is always to rearrange the facts in what *would* have been the logical order *if* the progress of physics had indeed been logical and sequential—what Whitaker (1979) has called 'quasi-history'. Specific cases are instructive here.

One example concerns the birth of the quantum theory. The story that many of us were taught, and which still abounds in textbooks, is that what led Planck to his theory of the black-body spectrum was the failure of the Rayleigh–Jeans law—the famous 'ultraviolet catastrophe'. But Brush and Whitaker both point out that this is a falsification of chronology as well as of Planck's approach to the problem and that Planck may not even have known about Rayleigh's work when he first announced his formula in 1900.

Another popular myth is that the Franck–Hertz experiment was undertaken as a test of Bohr's theory of quantised energy levels in atoms. But James Franck himself, in a delightful epilogue to a filmed demonstration of the Franck–Hertz experiment (Youtz 1961), has reminisced that he and Hertz were embarked on their experiment before Bohr's paper appeared; they did not know about it, and even if they had they might not have bothered to read it, since most papers on atomic theory at that time were probably wrong!

The photon concept is another fine field for distortion of history—but also a fascinating topic in the true story of the development of physical ideas. Despite what many textbooks say, Planck himself did not postulate quantisation of radiation. Einstein introduced it as an heuristic device in his 1905 paper on the 'emission and transformation of light'—which contains, *inter alia*, his photoelectric equation (Arons and Peppard 1965). But Bohr in 1913 (Bohr 1913), even though he boldly postulated discrete energy levels in atoms, still held to a classical picture of the radiation field, and did not accept the reality of quanta until well after the discovery of the Compton effect in 1923. (He resisted it even then.)

I would suggest that a presentation of the *correct* history of these and other developments would be a valuable way of deepening a student's understanding of physics. *Why*, for example, was Bohr so resistant to the photon concept? He must have had deep physical reasons for his objections, and knowing about them could not fail to be instructive. But clearly the presentation of physics through its historical development needs to be handled very carefully if one is not to make grave mistakes and misrepresentations; and that takes me to my next question—can it really be done at all in the physics class as distinct from a course in the history of science?

Is physics teaching through history possible?

Even if one assumes that the inclusion of physics history in physics teaching is desirable, there remain many questions as to what form it ought to take, and indeed what one *means* by history in this context. It is here, I think, that the historians and the physicists or physics teachers tend either to disagree strongly or perhaps to talk past each other, and there is a need to foster the kind of interaction that may lead to an informed use of physics history in our teaching.

The first and most obvious point is that it just is not possible to use the physics class as a place to present a full and correct historical account of any particular development. There simply isn't time to do justice to it. In any case, if it is a physics class, the physics has to take first place, and history, insofar as it enters, must accept a secondary role. Yet many physics teachers do want to give their students some sense not only of what we know, but also when we came to know it and *how* we came to know it.

Here, however, we run into a significant problem. Physics teachers must depend almost completely on accepting what the historians tell them. It is not possible for us, in general, to engage in researches of our own. Much of the false history is the result of physicists trying to play the role of historians without having the requisite professionalism, and the physics teacher must beware of assuming that history is somehow less demanding than the content of physics itself. Indeed, in some important respects it is the other way round; the task of establishing an accurate picture of the unfolding of events and ideas can be forbiddingly difficult, as has been made clear in some of the examples that have been mentioned in other chapters of this volume.

One particular element of this problem is the unreliability of personal recollections—something with which historians are familiar, but before which the average physics teacher is powerless. I myself vividly remember an autobiographical talk that the late Samuel Goudsmit gave a number of years ago, in the course of which he flatly said 'We lie!'. Of course he didn't mean deliberately, but simply that when years have passed since you have done a piece of work and you have described and discussed it with many people, you may no longer really know the truth.

A well studied example of this phenomenon is Einstein's recollection of the Michelson–Morley experiment and the role that it may have played in guiding him towards the special theory of relativity. This topic also involves one of the most famous and widely disseminated pieces of quasi-history of the '*post hoc, ergo propter hoc*' variety. For a very long time the elementary textbooks all portrayed the Michelson–Morley experiment, done in 1887, as the chief experimental result involved in the creation of the theory. But then, towards the end of his life, Einstein was interviewed several times by Robert Shankland. At that time he said that he had not known about the experiment prior to 1905; at any rate, he said, this was very doubtful. What had really influenced him was the Fizeau experiment, together with the fact of stellar aberration (Shankland 1963). Gerald Holton, in a detailed study of all the evidence (Holton 1969) concluded that Einstein's recollection was substantially correct, and that the Michelson–Morley experiment played at best a minimal role in his considerations. It remained unclear whether he had specifically known about it before 1905. But then, in 1982, the journal *Physics Today* carried the translation of some notes taken by a Japanese physicist (Jun Ishawara) of a talk that Einstein gave in Kyoto in 1922, entitled 'How I created the theory of relativity' (Einstein/Ono 1982). In it, Einstein is reported as saying that he had hoped to do an ether-drift experiment himself; he did not get around to it, but during his student years he came to know of Michelson's null result, and it was an important step towards his development of the theory. What are we to believe? The year 1922 was much closer to 1905 than were Einstein's conversations with Shankland in the 1950s. Should we then accept the earlier account, which would validate the story-book version of the history? If the historical record itself is in doubt, the physics teacher is certainly in trouble.†

The Einstein story is a rather special case, but the true historical record is undoubtedly obscured routinely in a large fraction of all research papers published today. We can be almost sure that the tidy description one usually finds does not represent the way in which the

---

† Recently there has appeared some further discussion in a special issue of *Physics Today* (May 1987) commemorating the centennial of the Michelson–Morley experiment. It contains a letter (Miller 1987) sharply critical of the Einstein/Ishiwara/Ono report, but also an article by John Stachel (Stachel 1987) concluding that Einstein 'was almost certainly aware in a general way' of the Michelson–Morley experiment from late 1899 on. Both authors uphold the view that the experiment *per se* was not very influential in Einstein's thoughts.

work was actually done; it has all been organised and condensed and put into orderly form. It is only if one goes back, let us say, to the nineteenth century, and to such massive tomes as the *Philosophical Transactions* of the Royal Society, that one finds articles in which the authors have had the time, space and opportunity to give essentially a blow-by-blow account of what they did. It doesn't happen any more, or almost never, except perhaps in such publications as the *Nobel Lectures* (a wonderful source, I think, for the physics teacher).

The enrichment of teaching through history

In view of all the difficulties discussed above, I think that the presentation of history as historians understand it is close to being an impossibility in the physics class. So then the question arises: 'What *can* we do, as physics teachers, to incorporate some physics history into our teaching?' My own belief is that there is a great deal, even though it may seem modest in a professional historical sense. I believe also that it can be done in a way that need not be offensive to the professional historian—and, indeed, must depend on reputable historical sources. But it will tend to take the form of little vignettes of anecdotal and biographical material incorporated into our teaching, through which we can give our students the idea that physics does *have* a history—something that they are prone to overlook.

Let me begin at the very lowest level—a recognition that physics is actually done by humans. This is something that students, if they are just working from a traditional textbook, may have little chance to realise. All they see are the bare names, so that Ohm is just a law, Coriolis is just an acceleration and poor old Avogadro is a mere number. There is often no suggestion that these people had first names, lived in a particular period in history, and so forth. And I think that it is very easy for anyone who has read about the history of science to any extent to forget that many students today come to us with almost total ignorance of such things, so that one has to go out of one's way to draw their attention to them. Samuel Goudsmit, if I may venture to quote him again, was giving a talk in 1976 to commemorate the fiftieth anniversary of the discovery of electron spin (Goudsmit 1976); this was also the two hundredth anniversary of the birth of the USA. Goudsmit, noting that the USA was only four times as old as spin was at that time, commented that it was therefore 'not surprising that most young physicists do not know that spin had to be introduced. They think that it was revealed in Genesis or perhaps postulated by Sir Isaac Newton, which young physicists consider to be about simultaneous'—a little savage, perhaps, but nevertheless I think he had a point. And on this matter of the impersonality with which physics is presented—this is perhaps a little frivolous, but I'd like to mention it briefly—someone once wrote a detailed biographical note about 'Claude Emile Jean-Baptiste Litre', who was alleged to have lived in eighteenth-century France, the scion of several generations of makers of glass bottles, and after whom the litre was supposedly named. This spoof was published, apparently in all seriousness, in the *International Newsletter of Chemical Education* (Weber 1982).

But removing the impersonality is not enough. It has been remarked that the textbook seems to have an existence independent of the past. I would agree; not only is it impersonal, but it seems to be nowhere in time really, except perhaps in the present. And one of the nice things we can do to counteract this is to introduce into our teaching some details of a personal or historical kind that may not only enliven a bare account of the physics but may also teach some physics in the process. For example, when Joseph Henry, at the same time as Faraday, was exploring electromagnetism, he insulated the wire of one of his electromagnets by wrapping it with strips of silk torn from a petticoat of his wife (Coulson 1950). This tells us that in those days you could not just go out and buy insulated wire; you had to get the copper and insulate it yourself, which students of today would not even imagine to be necessary. An even nicer example, pedagogically, is to be found in Bohr's first paper on his atomic theory. A very important thing for him was to recognise, and to be fairly sure, that the hydrogen atom had only one electron. How did he know this? Because (Bohr 1913) he had seen J J Thomson's positive-ray parabolas, and whereas all other materials showed two or more parabolas, hydrogen had only one, and that gave Bohr confidence that he was dealing with a one-electron system.

I am sure that any physics teacher has favourite examples of this kind, and I consider it very unfortunate that so little of this sort of thing gets into our teaching. It requires relatively little of our precious classroom time, and can be a notable source of enrichment.

The use of original data

Another way of combining good history with good physics—one that I personally find very delightful—is to go to the original literature for actual physical data. Many textbooks of physics look as if they had been composed completely in a closed room. There is text, there are equations and there are line drawings—nothing real, nothing that puts the reader into direct contact with the physics as it was originally done.

The use of real data can provide an antidote to this, and I should like to give one or two examples from my own teaching experience.

Let me introduce my first example with something that doesn't seem to have anything to do with physics, something that I came across a number of years ago and found quite interesting. I was writing a little piece about statistics, and I found out how a blood count is done. A small drop of blood is diluted and then spread out on a haemocytometer slide, which is etched with an array of small squares. The numbers of blood cells per square are small enough to be easily counted, and their average is converted into the blood count in cells per mm^3 of the original blood. But from time to time, to check on the reliability of the procedure, a plot is made of the numbers per square, to see if they fit, as they should, a Poisson distribution curve. Figure 7.1(a) shows such a plot, which I constructed from a set of photomicrographs of haemocytometer slides (Todd and Sanford 1948). The points scatter considerably, because the small numbers have relatively large statistical fluctuations, but the general picture is acceptable.

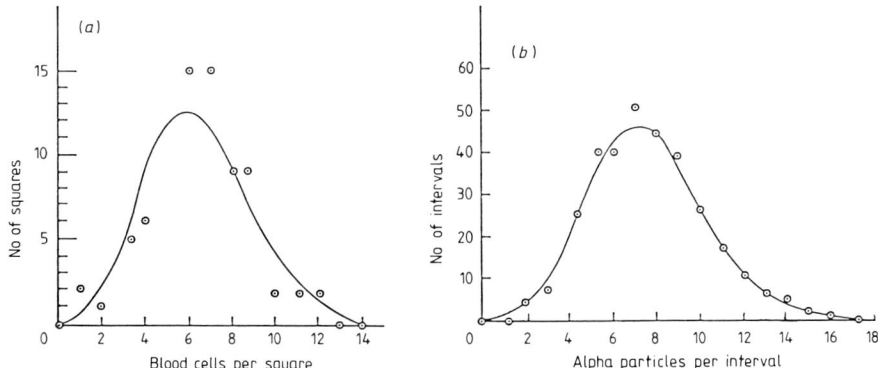

Figure 7.1 (a) Statistical distribution of numbers of blood cells on squares of a haemocytometer slide, from photographs in Todd and Sanford (1948). (b) Statistical distribution of numbers of alpha particles recorded in equal time intervals from a weak radioactive source, from data of Rutherford and Geiger (1910).

But what does this have to do with physics teaching? The answer is that a very interesting analogy to this exists in the early history of radioactivity. It must be remembered that radioactivity was a great shock to the world of physics. Radioactive decay was a process in which, for the first time in physics, the phenomenon appeared to follow no causal laws but, instead, to exhibit purely random behaviour. Was this indeed the case, or did the atoms in a radioactive sample influence one another in some way? It was a real physical question. To answer it, Rutherford and Geiger took a very weak radioactive source (of long life) and counted the numbers of alpha particles emitted in equal successive time intervals (Rutherford and Geiger 1910). The results (figure 7.1(b)) conformed nicely to a Poisson distribution; it was indeed a random process, and this was perhaps the first direct proof of the fact.

Another nice example (of course the range of possibilities is almost limitless if one digs into the original literature) comes from 1957, when the first Sputnik went up and its 20 MHz signals were being recorded over the world. In one of its early orbits it passed over MIT (Massachusetts Institute of Technology) and the frequency of the signal was recorded as a function of time (Brown *et al* 1957). The results are shown in figure 7.2. It is a beautiful example of the Doppler effect, and from the over-all change of frequency between approach and recession, one can immediately find the speed. But one can do more than this, for from the detailed shape of the curve, the precise way in which the frequency varies with time, it is possible to deduce the altitude of the satellite. It makes a very manageable but interesting exercise for students.

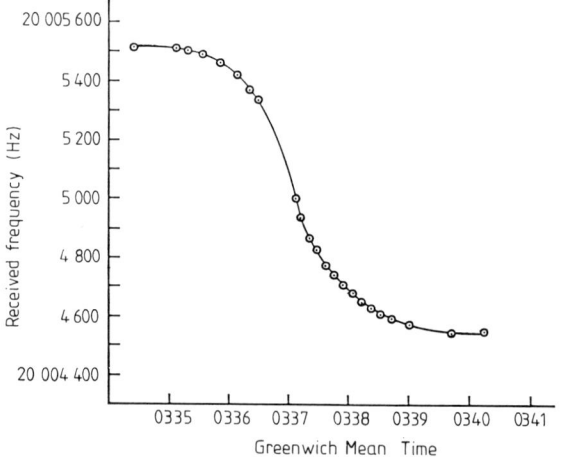

Figure 7.2 The Doppler effect for the radiofrequency signal emitted by Sputnik I, from data of Brown *et al* (1957).

Roemer: a cautionary tale

Despite my enthusiasm for bringing history in various ways into the teaching of physics, I am painfully aware of the hazards of accepting and transmitting false information from secondary or tertiary sources (or worse). Probably I have been guilty of this many times, but one undeniable instance is particularly embarrassing. It concerns the famous legend that Ole Roemer, in 1676, deduced a value for the speed of light from his observations on the apparent departures from exact periodicity of the eclipses of Jupiter's moons. The basic idea is that a satellite going around Jupiter is like a clock ticking regularly, but as observed from the earth at different places around its orbit, the time-lag for light to reach the earth is involved, and hence there is a measurable accumulated time-lag corresponding to the time needed for light to cross the diameter of the earth's orbit, a lag that Roemer reported to be about 22 minutes (Roemer 1676). And then nearly all the textbooks say that, from this, Roemer deduced the speed of light to be about 200 000 km s^{-1}. But it would appear that he did no such thing. He undoubtedly measured the time-lag, but it was Huygens, a few years later, who actually took this information and combined it with information about the diameter of the earth's orbit to arrive at a definite numerical value for the speed of light (Huygens 1690). Unfortunately I am one of many textbook authors, nearly fifty of us, who have helped to perpetuate the wrong history and have been publicly identified in a very interesting discussion of the matter (*De Mora Luminis*) by Andrzej Wroblewski (Wroblewski 1985). What makes this error particularly inexcusable is that, as long ago as 1940, I B Cohen published a full account of Roemer and his work (Cohen 1940). The trouble is that physics textbook authors tend not to read journals of science history!

I bring up this case, not just from an urge to confess, but also because it involves some genuinely interesting history. Why *didn't* Roemer proceed from his observations to a calculation of the speed of light? Was it because he did not know the earth's orbit radius? No, because, by a remarkable coincidence of dates, Jean Richer and Giovanni Cassini had, during 1672–3, made a credible determination (within about 10 per cent of the currently accepted value) of the distance of the earth from the sun. They conducted triangulation measurements on Mars (Cassini in Paris, Richer at Cayenne, near the equator) when Mars was at its closest approach to the earth, and this, used in conjunction with the Copernican model, established the linear scale of the solar system as a whole (Richer 1679, Cassini 1693). Moreover, since Roemer was working at the time in Cassini's observatory in Paris, he could hardly have failed to know about this result. Why, then, did he not take what we today would regard as an obvious final step? The answer may be that the question of real interest at that time was not the numerical value of c, but simply whether the speed of light was finite or infinite. Wroblewski points out that

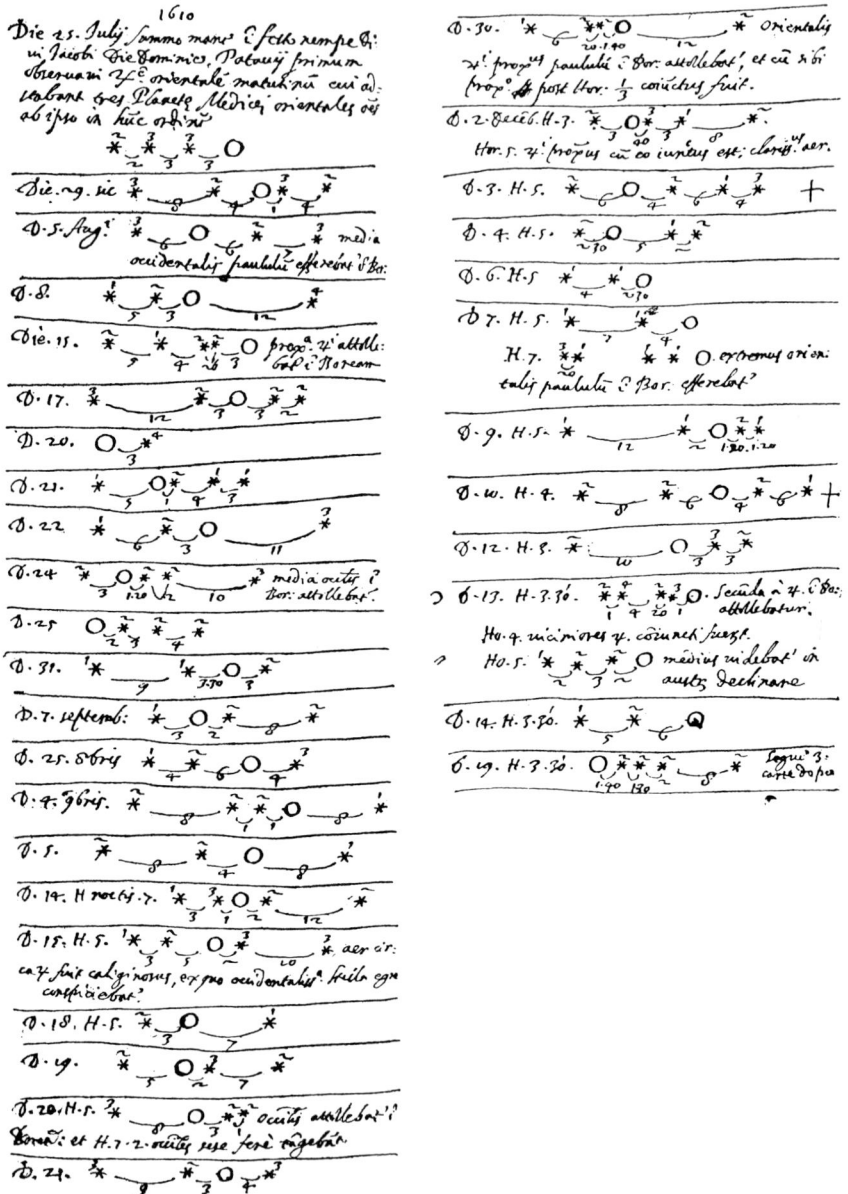

Figure 7.3 A page of Galileo's observations on the moons of Jupiter, from Galileo (1610).

instantaneous propagation had been the dominant hypothesis from Aristotle to Descartes; the prime importance of Roemer's work was that it made this assumption untenable. But many textbooks continue to credit Roemer with the quantitative calculation.

The moons of Jupiter and harmonic motion

I should like to end with one other example where I did go to original sources, or as close as I could get, and where the result was, I thought, interesting both historically and pedagogically. The subject, once again, was Jupiter's moons—or at any rate the four of them that Galileo discovered in 1610. Figure 7.3 is a page of Galileo's notebook, showing the positions of these moons relative to Jupiter night after night, and giving the quantitative values for their distances from Jupiter as a function of time (Galileo 1610). How he did these observations has not, I think, been completely established,† but it was a wonderful piece of observation, as I think I can demonstrate, because, since the orbits of the moons are more or less in the plane of the rest of the solar system, and since the orbits are more or less circular, one has here a case of motion which, as viewed from the earth, is circular motion seen in its own plane. We all teach our students that this is simple harmonic motion, and figure 7.4 shows it from some of Galileo's own data for Callisto, the outermost of the

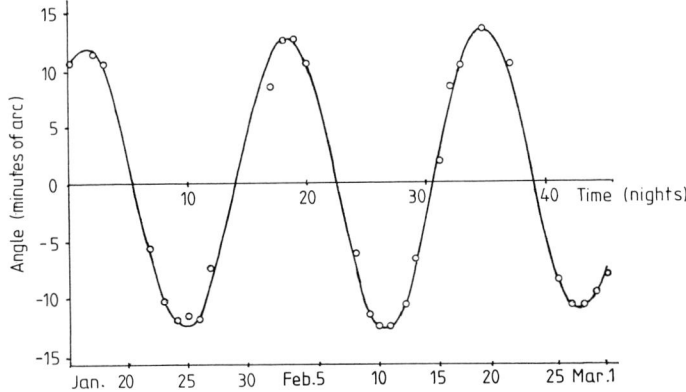

Figure 7.4 A graph of the angular position of Callisto with respect to Jupiter as a function of time, constructed from some of Galileo's observations during 1610, immediately following his discovery of the 'Medicean planets'.

† See, however, Drake and Kowal (1980).

'Medicean planets'. I found it exciting to think that, to the best of my knowledge, this nugget had remained buried and unrecognised for three and a half centuries; it will, I hope, encourage other physics teachers to go digging in the same sort of way for really rewarding pieces of physics history that might be used in their teaching.

Concluding remarks

My intention in this chapter has been to try to take a realistic view of the extent to which the history of physics can be incorporated into physics teaching, with some examples of the ways in which it might be done. I am conscious that it may not satisfy some professional historians of science—may even offend them in certain respects—but this activity, like politics, is the art of the possible, and I feel that it is better to do something than to do nothing at all. Above all, I should like to plead for a real meeting of minds between the historians and the physics teachers. Each group can, I think, learn from the other, and the health of both disciplines will be improved through interaction between them.

References

Arons A and Peppard M 1965 Einstein's proposal of the photon concept—a translation of the *Annalen der Physik* paper of 1805 *Am. J. Phys.* **33** 367–74
Bohr N 1913 On the constitution of atoms and molecules *Phil. Mag.* **26** 2–25 (in p. 8)
Brown R R, Green P E Jr, Howland B, Lerner R M, Manasse R and Pettengill G H 1957 Radio observations of the Russian satellites *Proc. Inst. Radio Eng.* **45** 1552–3
Brush S G 1969 *Phys. Teach. (USA)* **7** 271–80
—— 1974 Should the history of science be rated X? *Science* **183** 1164–72
Brush S G and King A L (eds) 1972 *History in the Teaching of Physics* (Hanover, NH: University Press of New England)
Cajori F 1899 The pedagogic value of the history of physics *School Rev.* **7** 278–85
Cassini G 1693 *Mém. Acad. R. Sci. 1666–99* **5** part 1 (Paris 1733) pp 168–75, 181–2, 212–16, 283–90, 309–10, 331, 418
Cohen I B 1940 Roemer and the first determination of the velocity of light *Isis* **31** 327–79
—— 1950 A sense of history in science *Am. J. Phys.* **18** 343–59
Coulson T 1950 *Joseph Henry: His Life and Work* (Princeton: Princeton University Press) p. 51
Drake S and Kowal C 1980 Galileo's sighting of Neptune *Sci. Am.* **243** (6) 74–81

Einstein A/Ono Y A 1982 How I created the theory of relativity (notes by J Ishawara 1922 tr. Y A Ono 1982) *Phys. Today* **35** (8) 45–7

Galileo G 1610 *Opere* vol. III part 2 (Florence: Barbera, 1931) pp 425–53. See also Galileo's *Sidereus Nuncius* (1610), reproduced in *Opere* vol. III part 1 pp 7–96.

Goudsmit S A 1976 Fifty years of spin: it might as well be spin *Phys. Today* **29** (6) 40–8

Holton G 1969 Einstein, Michelson and the 'crucial' experiment *Isis* **60** 133–97

Huygens C 1690 *Treatise on Light* (New York: Dover, 1962) pp 7–10

Medawar P 1972 *The Hope of Progress* (London: Methuen) pp 101–9

Miller A I 1987 Einstein and Michelson–Morley *Phys. Today* **40**(5) 9–13

Pasteur L 1858 *Oeuvres* vol. VII (Paris: Masson, 1939) pp 160–3

Richer J 1679 Observations astronomiques et physiques *Mem. Acad. R. Sci. 1666–99* **7** part 1 (Paris 1729) pp 231–326

Roemer O 1676 Démonstration touchant le mouvement de la lumière trouve par M. Romer Journal des sçavans *J. Sçavans* (Dec. 7) 233–6

Rutherford E and Geiger H 1910 *Phil. Mag.* **20** 698–704

Shankland R S 1963 Conversations with Albert Einstein *Am. J. Phys.* **31** 47–57

Stachel J 1987 Einstein and ether drift experiments *Phys. Today* **40**(5) 45–7

Todd J C and Sanford A H 1948 *Clinical Diagnosis by Laboratory Methods* 11th edn (Philadelphia Saunders) p. 201

Weber R L 1982 *More Random Walks in Science* (Bristol: The Institute of Physics) pp 155–6

Whitaker M A B 1979 History and quasi-history in physics education *Phys. Ed.* **14** 108–12, 239–42

Wroblewski A 1985 De Mora Luminis: a spectacle in two acts with a prologue and epilogue *Am. J. Phys.* **53** 620–30

Youtz B 1961 *Franck–Hertz Experiment* (Watertown, MA: Educational Services Inc.)

8

Historians and Physics Teachers: Partnerships for Packages†

Brian Davies

For several years science teachers have been encouraged by professional bodies such as the Royal Society, The Institute of Physics and the Association for Science Education to introduce and use history in their lessons. During this time the teachers were somehow coping with all manner of change in curricula, teaching methods, assessment procedures and new kinds of examinations. At a time when so many demands were being made on them, teachers with no expertise in the discipline were understandably uncertain about the value of and the best ways of introducing historical aspects, and reluctant to spend time on history in already full timetables. Now, however, when historically and socially oriented attainment targets are part of the National Curriculum for Science in England and Wales, we have an opportunity to help teachers transform their uncertainty and reluctance into enthusiasm. The transformation can be best effected only if teachers come to appreciate the possible benefits to be gained from using historical aspects of science with their pupils, and if they are provided with curriculum-related and *instantly usable* packages of pupil-oriented

† Based on a presentation given at the conference on the History of Physics for the Physicist, 2–4 July 1986, Oxford.

materials. In my article 'Changing the image' (Davies 1986) I gave several examples of the benefits stemming from the use of history in science teaching, so there is no need to go over the same ground here. Instead let us look straight away at the ways in which teachers and historians could collaborate, and at some history of science packages which could quickly be produced by small groups and which would, I think, be welcomed by many science teachers.

As it is impossible to imagine what is happening in our classrooms nowadays by remembering how things were in our own schooldays more than a decade or so ago, the teacher members of working groups would need to spend time explaining current classroom practices and educational thinking to the historians, and to make clear what they hoped to achieve with their pupils when working in some area of the syllabus. The historians could then advise on appropriate historical sources and possible classroom approaches, and from then on the partners could share their different kinds of expertise to develop curriculum materials in a reasonable time. (Should they ever run out of ideas for things to do they could always glance at John Roche's paper 'Suggested historical projects in physics' (unpublished) where they will find more than 230 suggestions to inspire them.)

The nature of the history packages

Many history of science packages will feature biographical data, scientists' original descriptions of their own discoveries, contemporary reactions, analyses of the significance and descriptions of the historic and scientific contexts of those discoveries. Such packages are more likely to be successful—that is, willingly *used* by science teachers in general—if they

   (i) are entertaining as well as instructive;
   (ii) are pupil-oriented, requiring the pupils to take a very active part in investigative/thought-provoking/problem-solving/role-playing exercises;
   (iii) contain a lot of pictorial and visually attractive materials;
   (iv) take up no more than a double period (about 70 minutes) in school;
   (v) are accompanied by good pupil and teacher notes;
   (vi) are capable of being used in more than one way, with those ways explained in the accompanying notes;
   (vii) are *immediately* usable (that is, no extra preparation time will need to be spent by the teacher, in chasing up sources, making transparency masters, etc, before the materials can be used in class);

EINSTEIN ON HIS THEORY.

By Dr. Albert Einstein.

I respond with pleasure to your Correspondent's request that I should write something for The *Times* on the Theory of Relativity.

...Since the time of the ancient Greeks it has been well known that in describing the motion of a body we must refer to another body. The motion of a railway train is described with reference to the ground, of a planet with reference to the total assemblage of visible fixed stars. In physics the bodies to which motions are spatially referred are termed systems of co-ordinates. The laws of mechanics of Galileo and Newton can be formulated only by using a system of co-ordinates.

The state of motion of a system of co-ordinates cannot be chosen arbitrarily if the laws of mechanics are to hold good (it must be free from twisting and from acceleration). The system of co-ordinates employed in mechanics is called an inertia-system. The state of motion of an inertia-system, so far as mechanics are concerned, is not restricted by nature to one condition. The condition in the following proposition suffices: a system of co-ordinates moving in the same direction and at the same rate as a system of inertia is itself a system of inertia. The special relativity theory is therefore the application of the following proposition to any natural process:—" Every law of nature which holds good with respect to a co-ordinate system K must also hold good for any other system K′ provided that K and K′ are in uniform movement of translation.

The second principle on which the special relativity theory rests is that of the constancy of the velocity of light in a vacuum. Light in a vacuum has a definite and constant velocity, independent of the velocity of its source. Physicists owe their confidence in this proposition to the Maxwell-Lorentz theory of electro-dynamics.

The two principles which I have mentioned have received strong experimental confirmation, but do not seem to be logically compatible. The special relativity theory achieved their logical reconciliation by making a change in kinematics, that is to say, in the doctrine of the physical laws of space and time. It became evident that a statement of the coincidence of two events could have a meaning only in connexion with a system of co-ordinates, that the mass of bodies and the rate of movement of clocks must depend on their state of motion with regard to the co-ordinates.

THE OLDER PHYSICS.

But the older physics, including the laws of motion of Galileo and Newton, clashed with the relativistic kinematics that I have indicated. The latter gave origin to certain generalized mathematical conditions with which the laws of nature would have to confirm if the two fundamental principles were compatible. Physics had to be modified. The most notable change was a new law of motion for (very rapidly) moving mass-points, and this soon came to be verified in the case of electrically-laden particles. The most important result of the special relativity system concerned the inert mass of a material system. It became evident that the inertia of such a system must depend on its energy-con-

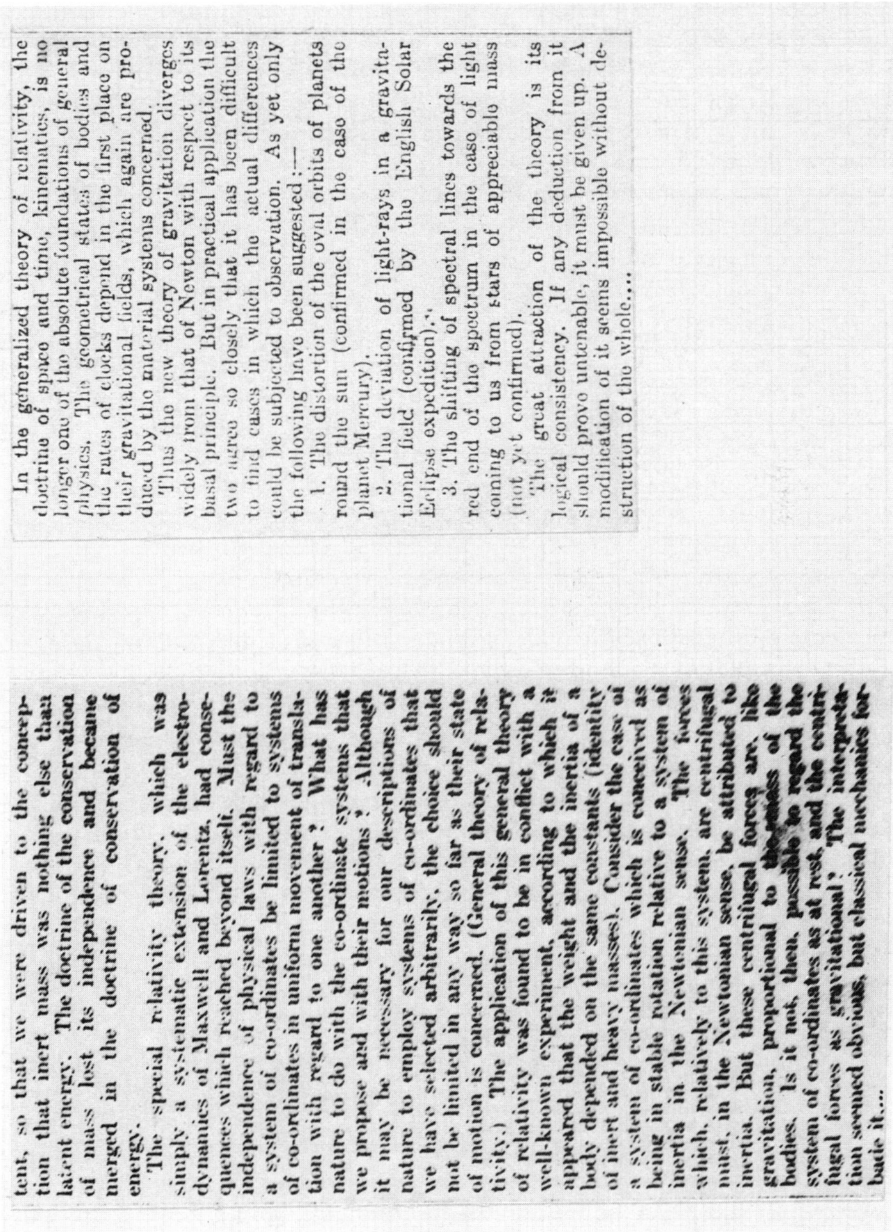

In the generalized theory of relativity, the doctrine of space and time, kinematics, is no longer one of the absolute foundations of general physics. The geometrical states of bodies and the rates of clocks depend in the first place on their gravitational fields, which again are produced by the material systems concerned.

Thus the new theory of gravitation diverges widely from that of Newton with respect to its basal principle. But in practical application the two agree so closely that it has been difficult to find cases in which the actual differences could be subjected to observation. As yet only the following have been suggested:—

1. The distortion of the oval orbits of planets round the sun (confirmed in the case of the planet Mercury).
2. The deviation of light-rays in a gravitational field (confirmed by the English Solar Eclipse expedition).
3. The shifting of spectral lines towards the red end of the spectrum in the case of light coming to us from stars of appreciable mass (not yet confirmed).

The great attraction of the theory is its logical consistency. If any deduction from it should prove untenable, it must be given up. A modification of it seems impossible without destruction of the whole....

...tent, so that we were driven to the conception that inert mass was nothing else than latent energy. The doctrine of the conservation of mass lost its independence and became merged in the doctrine of conservation of energy.

The special relativity theory, which was simply a systematic extension of the electrodynamics of Maxwell and Lorentz, had consequences which reached beyond itself. Must the independence of physical laws with regard to a system of co-ordinates be limited to systems of co-ordinates in uniform movement of translation with regard to one another? What has nature to do with the co-ordinate systems that we propose and with their motions? Although it may be necessary for our descriptions of nature to employ systems of co-ordinates that we have selected arbitrarily, the choice should not be limited in any way so far as their state of motion is concerned. (General theory of relativity.) The application of this general theory of relativity was found to be in conflict with a well-known experiment, according to which it appeared that the weight and the inertia of a body depended on the same constants (identity of inert and heavy masses). Consider the case of a system of co-ordinates which is conceived as being in stable rotation relative to a system of inertia in the Newtonian sense. The forces which, relatively to this system, are centrifugal must, in the Newtonian sense, be attributed to inertia. But these centrifugal forces are, like gravitation, proportional to the masses of the bodies. Is it not, then, possible to regard the system of co-ordinates as at rest, and the centrifugal forces as gravitational? The interpretation seemed obvious, but classical mechanics forbade it....

Figure 8.1 Einstein on his theory (Einstein 1919).

(viii) include sufficient supplementary material to help the teacher in any likely discussion sessions;
(ix) are free of copyright restrictions;
(x) are well printed and produced.

Possible packages—the basic contents

Here are some suggestions for the basic content of suitable history of science packages—illustrated by several examples used in my own teaching and lecturing—and a mention of some existing sources of useful materials which teachers can adapt for their own uses.

Old newspapers

Selections from old newspapers are always interesting to search out and great fun to use, but I know of no *collection* of pieces suitable for science history sessions in schools. Scientists' descriptions of their own work, contemporary reactions to discoveries (often with quite savage cartoons), scientific prejudices, contrasts between the views of scientific and non-scientific people to developments in science, foreigners' comments on the state of English science at various times—all of these are just examples of the kind of issue which can be readily illustrated. A short article by Einstein in *The Times* of 1919 (figure 8.1), proves that all readers of the paper in those days were consummate physicists while Liebig's letter quoted by Faraday in 1845 (figure 8.2), shows that English science of the nineteenth century was very much more instrumental than the pure academic work found in the UK 70 years later.

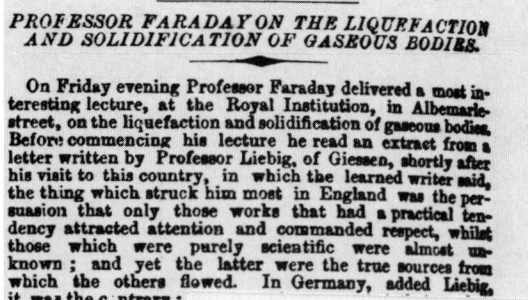

Figure 8.2 Faraday on Liebig (Faraday 1845).

A delightful example of science in action can be seen in extracts taken once again from *The Times* of 1845, and reproduced in *The Electric Telegraph*, one of a series of pictorial histories published by the Science Museum, London and BP Educational Services. Wheatstone and Cooke were setting up Britain's first commercial electric telegraph between London and Gosport with considerable help from the railway company which could see some commercial advantages in the rapid transmission of messages 'down the line' (figure 8.3).

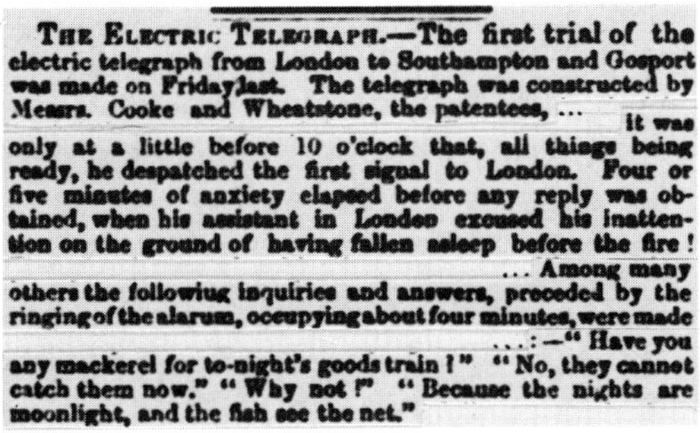

> THE ELECTRIC TELEGRAPH.—The first trial of the electric telegraph from London to Southampton and Gosport was made on Friday last. The telegraph was constructed by Messrs. Cooke and Wheatstone, the patentees, ... it was only at a little before 10 o'clock that, all things being ready, he despatched the first signal to London. Four or five minutes of anxiety elapsed before any reply was obtained, when his assistant in London excused his inattention on the ground of having fallen asleep before the fire! ... Among many others the following inquiries and answers, preceded by the ringing of the alarum, occupying about four minutes, were made ...:—"Have you any mackerel for to-night's goods train!" "No, they cannot catch them now." "Why not?" "Because the nights are moonlight, and the fish see the net."

Figure 8.3 A trial of the electric telegraph (*The Times* 1845).

Old encyclopaedias

Old editions of encyclopaedias contain 'state of the art' articles which were usually written by contemporary experts. The summaries are usually easy to understand, and give as good an account of the old ideas of science as any history of science text. In his will my uncle left me a set of *Chamber's Encyclopaedia* for 1786, and from them one can learn very quickly, for example, about static electricity, crystallography and the atmospheric sciences. A number of strategic searches and cullings from these old encyclopaedias—and here 'old' can mean anything over 25 years—could yield wonderful results. So too, could cullings from old textbooks and the journals of the leading philosophical and scientific societies throughout the UK and Ireland.

Popular scientific lectures

Science has always been exploited for profit, and often for entertainment. As the handbill (figure 8.4) from Bern Dibner's famous book (Dibner 1957 p. 25) *Early Electrical Machines* shows, hardly a Nuffield O-level teaching trick was missed when it came to using electricity for

> Newport, March 16. 1752.
>
> ## Notice is hereby given to the Curious,
>
> That at the COURT HOUSE, in the Council-Chamber, is now to be exhibited, and continued from Day to Day, for a Week or two;
>
> A COURSE of EXPERIMENTS, on the newly-discovered
>
> # Electrical FIRE:
>
> Containing, not only the most curious of those that have been made and published in *Europe*, but a considerable Number of new Ones lately made in *Philadelphia*; to be accompanied with methodical LECTURES on the Nature and Properties of that wonderful Element.
>
> By *Ebenezer Kinnersley*.
>
> **LECTURE I.**
>
> I. OF Electricity in General, giving some Account of the Discovery of it.
> II. That the Electric Fire is a real Element, and different from those heretofore known and named, and *collected* out of other Matter (not created) by the Friction of Glass, &c.
> III. That it is an extreamly subtile Fluid.
> IV. That it doth not take up any perceptible Time in passing thro' large Portions of Space.
> V. That it is intimately mixed with the Substance of all the other Fluids and Solids of our Globe.
> VI. That our Bodies at all Times contain enough of it to set a House on Fire.
> VII. That tho' it will fire inflammable Matters, itself has no sensible Heat.
> VIII. That it differs from common Matter, in this; its Parts do not mutually attract, but mutually repel each other.
> IX. That it is strongly attracted by all other Matter.
> X. An artificial Spider, animated by the Electric Fire, so as to act like a live One.
> XI. A Shower of Sand, which rises again as fast as it falls.
> XII. That common Matter in the Form of Points attracts this Fire more strongly than in any other Form.
> XIII. A Leaf of the most weighty of Metals suspended in the Air, as is said of *Mahomet's* Tomb.
> XIV. An Appearance like Fishes swimming in the Air.
> XV. That this Fire will live in Water, a River not being sufficient to quench the smallest Spark of it.
> XVI. A Representation of the Sensitive Plant.
> XVII. A Representation of the seven Planets, shewing a probable Cause of their keeping their due Distances from each other, and from the Sun in the Center.
> XVIII. The Salute repulsed by the Ladies Fire; or Fire darting from a Ladies Lips, so that she may defy any Person to salute her.
> XIX. Eight musical Bells rung by an electrified Phial of Water.
> XX. A Battery of eleven Guns discharged by Fire issuing out of a Person's Finger.
>
> **LECTURE II.**
>
> I. A Description and Explanation of Mr. *Muschenbroek's* wonderful Bottle.
> II. The amazing Force of the Electric Fire in passing thro' a Number of Bodies at the same Instant.
> III. An Electric Mine sprung.
> IV. Electrified Money, which scarce any Body will take when offer'd to them.
> V. A Piece of Money drawn out of a Person's Mouth in spite of his Teeth; yet without touching it, or offering him the least Violence.
> VI. Spirits kindled by Fire darting from a Lady's Eyes (without a Metaphor).
> VII. Various Representations of Lightning, the Cause and Effects of which will be explained by a more probable Hypothesis than has hitherto appeared, and some useful Instructions given, how to avoid the Danger of it: How to secure Houses, Ships, &c. from being hurt by its destructive Violence.
> VIII. The Force of the Electric Spark, making a fair Hole thro' a Quire of Paper.
> IX. Metal melted by it (tho' without any Heat) in less than a thousandth Part of a Minute.
> X. Animals killed by it instantaneously.
> XI. Air issuing out of a Bladder set on Fire by a Spark from a Person's Finger, and burning like a Volcano.
> XII. A few Drops of electrified cold Water let fall on a Person's Hand, supplying him with Fire sufficient to kindle a burning Flame with one of the Fingers of his other Hand.
> XIII. A Sulphurous Vapour kindled into Flame by Fire issuing out of a cold Apple.
> XIV. A curious Machine acting by means of the Electric Fire, and playing Variety of Tunes on eight musical Bells.
> XV. A Battery of eleven Guns discharged by a Spark, after it has passed through ten Foot of Water.
>
> *As the Knowledge of Nature tends to enlarge the human Mind, and give us more noble, more grand, and exalted Ideas of the AUTHOR of Nature, and if well pursu'd, seldom fails producing something useful to Mankind; 'tis hoped these Lectures may be tho't worthy of Regard & Encouragement.*
>
> ❖ Tickets to be had at the House of the Widow Allen, in Thames Street, next Door to Mr. John Tweedy's. Price Thirty Shillings each Lecture. The Lectures to begin each Day precisely at Three o'Clock in the Afternoon.
>
> Reproduced thru the courtesy of the Rosenbach Foundation

Figure 8.4 Early electrical entertainment (from Dibner 1957 p. 25).

public entertainment all those years ago. How much more enjoyable for pupils to be introduced to electricity through a mixture of this kind of material and related practical experiments (excluding perhaps the electrical annihilation and roasting of the turkey) than through the more usual introductions. I should like to see more examples of announcements of this kind, coupled with selections from texts and demonstrations from popular lectures, brought together for class use.

Book and textbook illustrations

The first illustration shown here (figure 8.5) is from Bern Dibner's book just mentioned. It illustrates the discovery of the electric capacitor. The other (figure 8.6) is taken from a late-eighteenth-century French science fiction thriller (de la Folie 1775) also cited in Dibner, and which is a very enjoyable read. (The hero is coming from another planet in an attempt to obtain special subterranean 'crystal', badly needed for parts of his electrostatic flying machine, to enable him to return home. The electro plane works on the principle that there is a

Figure 8.5 The (painful) discovery of the capacitor (from Dibner 1957 p. 27).

change of pressure from ambient whenever there is a strong electrostatic field present. By orientating the pressure differences the machine can be driven and steered.) Booklets of examples such as these, based around different themes, would be most useful to science teachers.

Figure 8.6 Eighteenth-century science fiction (from de la Folie 1775).

Old advertisements from magazines and newspapers

'Scientific' advertisements can be most instructive. A recent one from Hitachi (figure 8.7) clearly shows that the Japanese have an early lead in materials in which the speed of light exceeds that in a vacuum by a factor of about 1.6!

However, it is older kinds of advertisements I really had in mind when I wrote the heading immediately above, for they tend to have the sort of science in them which is more accessible to school pupils, and can therefore lead to greater participation in discussions and in writing exercises. For example, these two advertisements (the first (figure 8.8) taken from *Mad Old Ads* by Richard Sutphen (1968 p. 75) the second (figure 8.9) from a cover of *Physics Bulletin*) could be ... well, let's leave it to the reader to decide how he or she might like to use them with science classes!

Figure 8.7 A surprise from Hitachi.

Poetry

Brief selections of poetry and drama from Chaucer's time to the post-Newtonian poets can bring unexpected elements and insights into science lessons, and can show the development of popular scientific understanding and the changing attitudes, of a mainly classically educated elite, to science and scientists, especially after the time of Newton. It is important that all pupils understand something of the background to the present popular (or should I say 'unpopular'?) images and media presentations of science and scientists.

Paintings: slide sets and reproductions

The unusual colours seen in some sixteenth-century paintings due to the chemical changes in the pigments on the canvas; the artistic

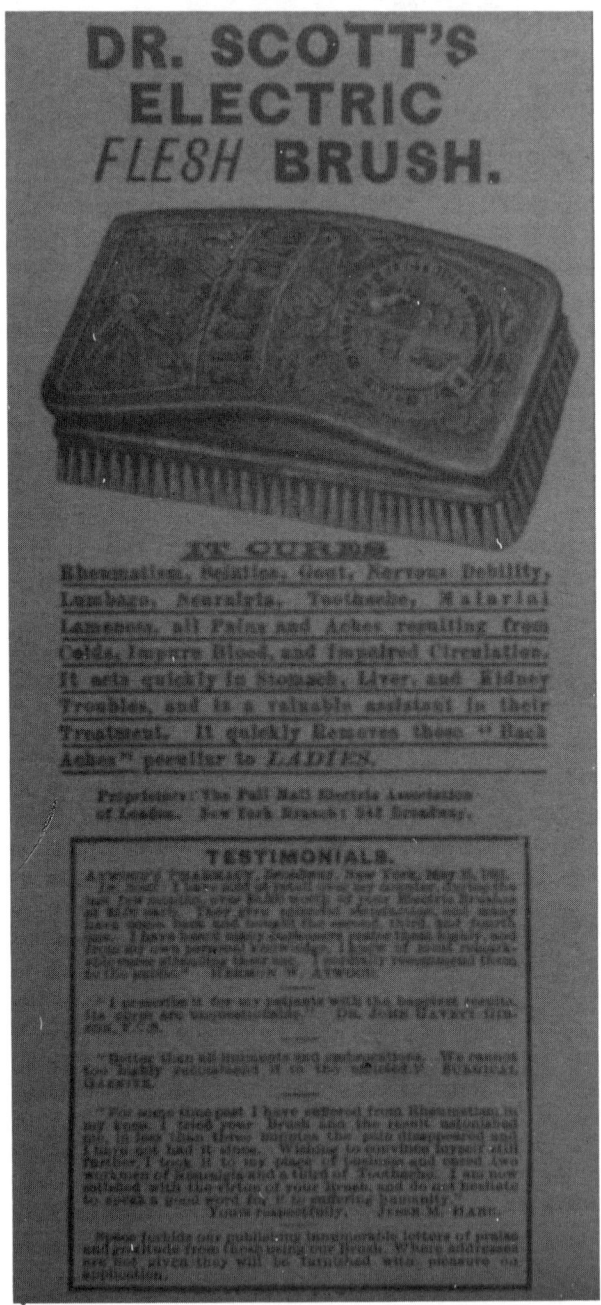

Figure 8.8 Contact electricity (from Sutphen 1968).

Historians and Physics Teachers

PHYSICS BULLETIN
Volume 33 Number 5 May 1982

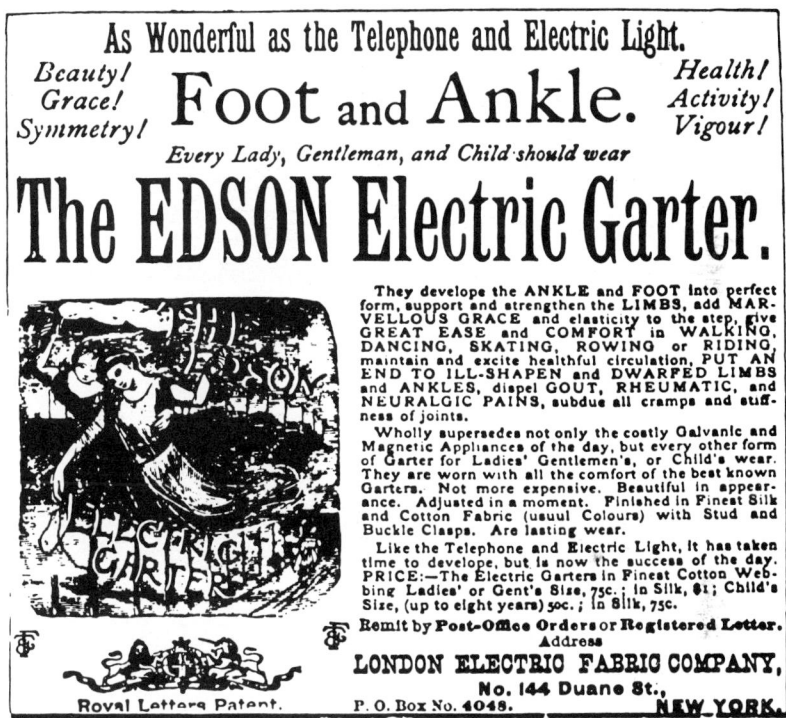

Figure 8.9

techniques used by Giorgiano or Rembrandt; the representations of rainbows or the use of light and shade; the use of colour fading to suggest atmospheric distances; the array of physics techniques in action in art history nowadays... . These subjects lend themselves to a broader education through science which I believe, more teachers should bring to their pupils.

Social aspects are easily brought into science lessons. For example, two slides of paintings from the London Science Museum's 'Energy in the Home' package, point up the contrast between a group of

Figure 8.10 Condensing lenses (from the Science Museum, London, and BP Educational Services).

Figure 8.11 Candle power (from the Science Museum, London, and BP Educational Services).

impoverished lace workers forced to share the light from a single candle (figure 8.10) (by using water-filled flasks as light focusing devices) while a Parisian ballroom of the same era glows with light (figure 8.11)—and a fair amount of heat, one imagines—from several hundred candles. Scope for science, scope for individual comment and argument, scope for personal thought and writing.

Sound collections

The voices of several great scientists have been preserved on wax discs, or by other early recording techniques, from at least the time of J J Thomson in the 1890s. The quote below was transcribed from a wax tape made by J J Thomson himself in *c*. 1897 (it was kindly lent by the Atomic Energy Establishment, Risley).

> Could anything at first sight seem more impractical than a body which could only exist in vessels from which all but a minute fraction of the air had been extracted, which is so small that its mass is an insignificant fraction of the mass of an atom of hydrogen, which itself is so small that a crowd of these atoms equal in number to the population of the whole world would be too small to have been detected by any means then known to science?

A collection of sound recordings, with notes, photographs and exercises on the life and work of the individuals, and accompanying slide sets showing what these physicists looked like could help remove those feelings of remoteness when pupils learn about the giants of modern science.

'Networks'

In his article in this volume on the vector potential, John Roche includes what he calls a 'pedigree' for the concept of vector potential. The pedigree is a flow chart which brings out the relationships between various pieces of research and many different scientific ideas, all of which funnelled down into the final complex concept signified by the word 'potential'. In the past I had used some similarly derived networks of ideas and research projects in teaching the evolution of theories and research programmes. But it was only when listening to John Roche's original lecture that the general usefulness of such network diagrams for science teachers suggested itself. So, why not produce a booklet of networks of two types, one type like the first one below (figure 8.12), showing the historical relationships between

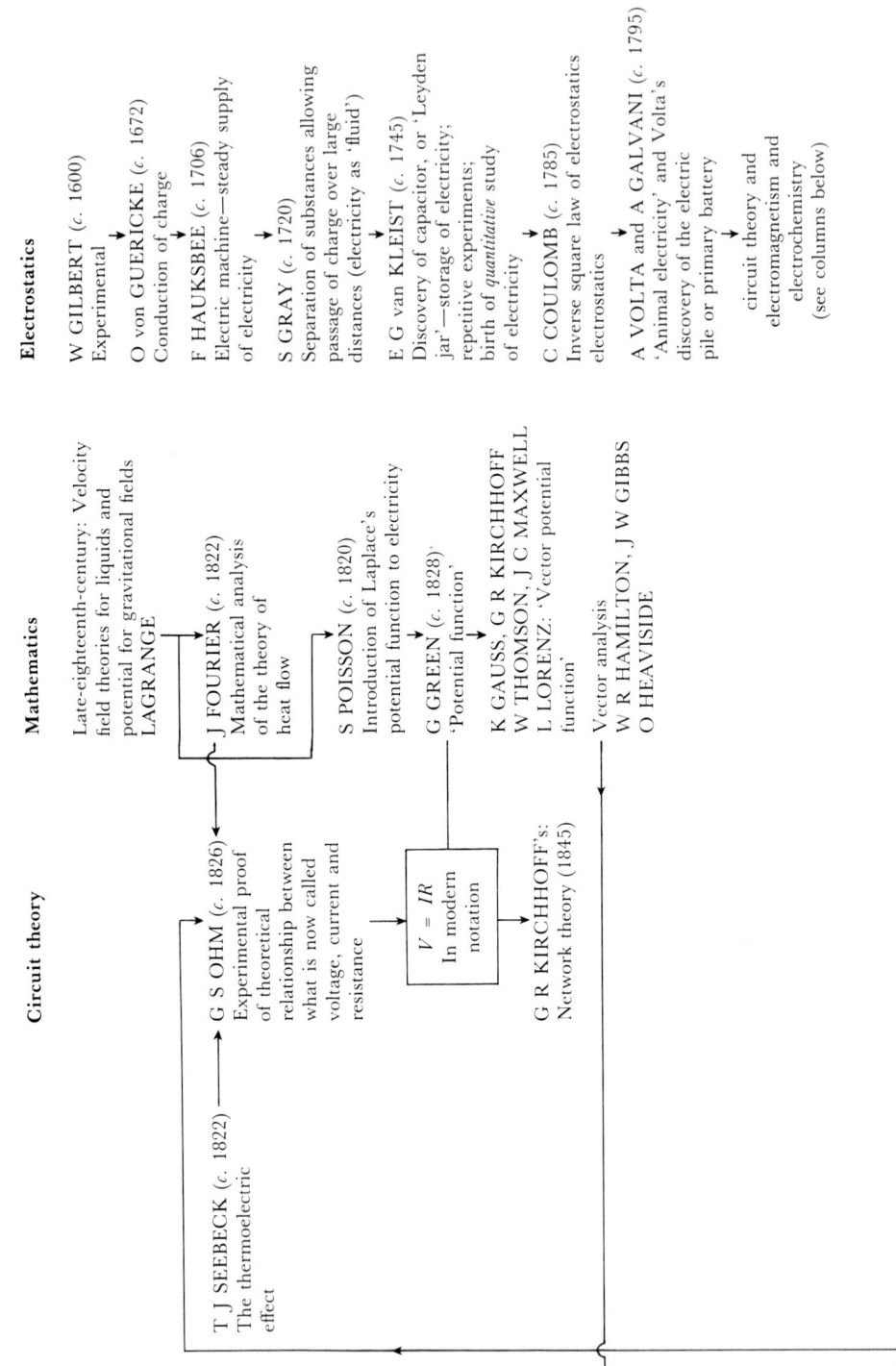

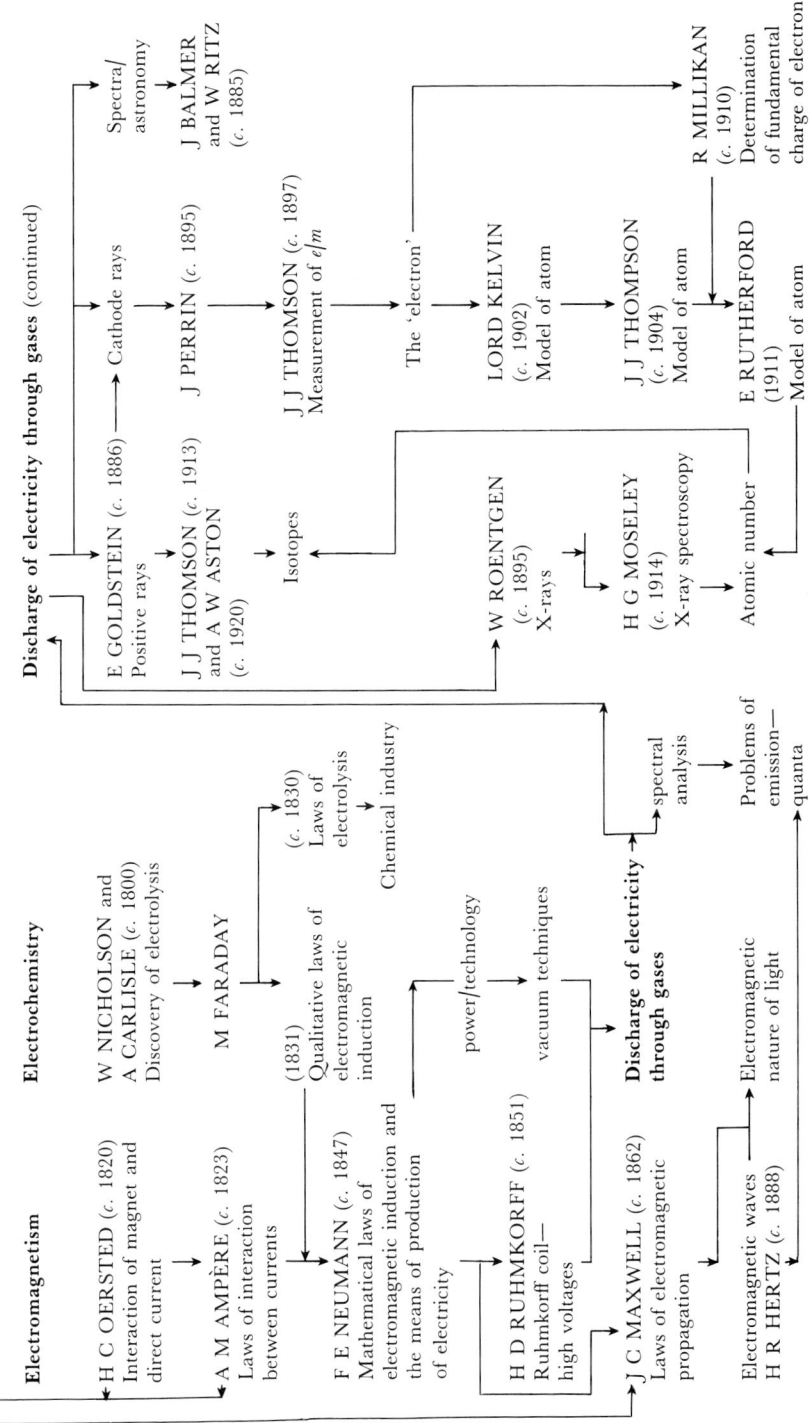

Figure 8.12 An electric current network.

different areas of research; and another type showing the pedigree, or funnelling of *ideas* needed to reach a simple-sounding concept such as 'the Bohr atom'?

Alongside these two main kinds of network, perhaps some 'definition mini-networks' could illustrate how many (tough) ideas are often built in to deceptively simple-looking scientific definitions. Take the example of 'voltage'. This is work per unit charge...: and (a) work = force × distance; force = rate of change of momentum; momentum = mass × velocity; velocity = vector change of distance with time; mass = tough; time = toughish; and (b) charge = ...*very* tough. Yet 'voltage' is one of those ideas we expect teenagers to throw around effortlessly, often at the same time as 'current' and 'resistance'.

Concluding remarks

The recommendations outlined only require low technology. They are people intensive, and not, I hope, too ambitious in scope or too costly. (With a *great deal* of money a history of science interactive videotape database could be produced, and some of us in The Institute of Physics have been looking at such a possibility as a long-term educational and commercial venture.)

A number of low-cost history of science packages are already available from sources such as the American Association of Physics Teachers, the Association for Science Education, The Institute of Physics, the Whipple Museum and the Science Museum (London). A compilation and review of what is currently available and of value to teachers needs to be produced as soon as possible.

Acknowledgments

I am grateful to the Burndy Library, Norwalk, Connecticut, USA, for permission to reproduce the figures from Bern Dibner's *Early Electrical Machines*. I also thank Elizabeth Doctor and Susan Gold on the library staff of the Wellcome Institute for Historical Research, London, for their expert help in tracking down copies of rare works such as de la Folie's.

References

Davies B 1986 Changing the image *The Times Educational Supplement* 18 April p. 61
Dibner B 1957 *Early Electrical Machines* (Norwalk, CT: Burndy Library)

Einstein A 1919 Dr Einstein's theory *The Times* 28 November p. 13
Faraday M 1845 Professor Faraday on the liquefaction and solidification of gaseous bodies *The Times* 4 February p. 5
Hitachi 1985 Advertising campaign summer/autumn 1985 (and *Snippets* **9** autumn 1985)
de la Folie C L 1775 *Le Philosophe sans Pretention, ou l'Homme Rare* ouvrage physique, chymique, politique et morale (Paris: Chez Clousier) facing p. vii
Sutphen R 1968 *Mad Old Ads* (London: W H Allen) p. 75
The Times 1845 The electric telegraph 4 February p. 5

9

A Critical Study of the Vector Potential†

John J Roche

History as an instrument of clarification and consolidation

A detailed study of the historical evolution of a theory in physics, by a professional physicist, can help to improve our understanding of it in various ways. History can suggest that certain technical terms, invented a century ago, perhaps, are now obsolete or misleading. 'The permeability of free space' or 'the coercive force' are obvious examples from magnetism. Historical study can reveal to the physics teacher that certain redundant explanations still linger on in modern textbooks. It can show that valid insights have been lost sight of, or that errors of interpretation introduced long ago into the foundations of a theory may survive unnoticed into the present. It may also show that a given modern theory which seems to be homogeneous is, in fact, an uneasy confederation of several competing traditions.

More positively, a trained physicist who studies carefully the foundational literature of a subject wins access to all of the explanations offered in the past for that subject and learns a rich repertoire of explanatory terms. This, coupled with the historically prompted recognition of flaws just mentioned, can lead to an improved reconstruction of the theory.

† Based on a presentation given at the conference on The History of Physics for the Physicist, 2–4 July 1986, Oxford.

As I suggested in the editor's introduction I believe that almost every topic in physics today need to be critically examined and carefully reconstructed using history as a guide. Classical physics would seem to be particularly ripe for such a treatment and an improved understanding of classical physics would surely pave the way for important clarifications in modern physics.

My own interests lie mainly in macroscopic or 'classical' electromagnetism and I hope to illustrate the value of critical history in the following study of the theory of the vector potential. The term 'vector potential' was apparently first introduced by Maxwell in 1871 in an unposted letter to his friend Peter Guthrie Tait (1831–1901) (Maxwell 1871, Bork 1967). However, without apology I shall use the term to refer to similar concepts which were introduced much earlier by Maxwell, and by others before him.

Faraday's theory of the electro-tonic state

It is important to distinguish clearly between the history of the physical interpretation and the history of the mathematical development of the vector potential (figure 9.1). For Faraday the former only was considered, without the latter. For Maxwell both developments went hand in hand. On the Continent the physical interpretation was virtually ignored.

The first kind of electromagnetic induction discovered by Faraday, in 1831, was the induction of a secondary current by a changing primary current (Faraday 1838, p. 3; 1932, pp 367–9). Influenced largely by Ampère he did not at first think of this as a magnetic phenomenon. He called it 'Volta-electric induction' and appears to have thought of it as analogous to electrostatic induction (Faraday 1838, pp 3–4, 7, 23). In a conductor experiencing electrostatic induction a brief displacement of charge takes place and the conductor, in the language of Volta, is said to be in a state of 'tension' (Volta 1779) (a term which still survives in modern electrical engineering). Analagously, for Faraday, when the primary current has reached its steady value and the secondary current has disappeared, the secondary conductor is in a state of tension of a new kind, which he called the 'electro-tonic state' (Faraday 1838, p. 16; 1932 p. 401). According to the Oxford English Dictionary, in nineteenth-century medical vocabulary a 'tonic state' is a state of tension. Faraday thought of the electro-tonic state as a condition of forced parallel alignment of certain particles in the conductor. A counter-current flowed while the electro-tonic state was being created and a forward current appeared while it was being destroyed (Faraday 1838, pp 3, 22). This led Faraday,

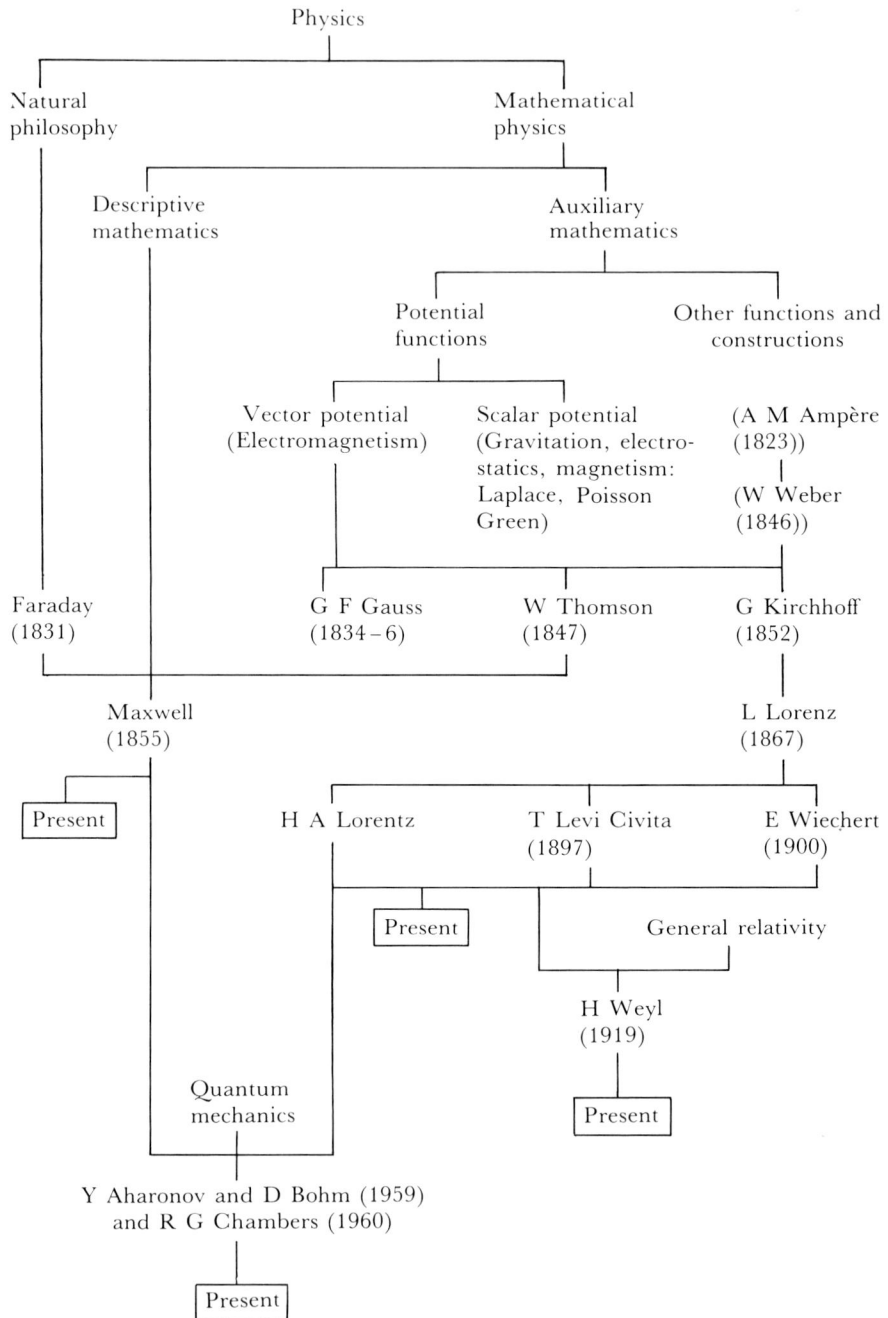

Figure 9.1 A short pedigree of the vector potential.

William Whewell (1719–1806) and others to suspect that some kind of electrical inertia was at work (Faraday 1971, pp 293–4) rather than a simple induction of like currents by like currents, as in the theory of Ampère (Ampère 1827, pp 372–3).

A few months after his electrical explanation Faraday hit upon his well known magnetic explanation of electromagnetic induction, in terms of lines of magnetic force cutting across conductors (Faraday 1838, pp 32–3, 66–9, 74; 1855, pp 344–5). This dominated his thinking for some years, during which time Faraday relegated his electro-tonic state to the status of an unsupported hypothesis (Faraday 1838, pp 67, 69). However, he felt a growing need to incorporate the electro-tonic state into his more developed theory of electromagnetism and electromagnetic induction.

Faraday's rather fragmentary and hesitant subsequent theory may, perhaps, be summarised as follows. He conceived the electro-tonic state as a directed quantity, perpendicular to the local magnetic field and parallel to neighbouring currents. It extended throughout the medium around the source currents and it existed within the currents themselves. It was created and destroyed when the magnetic field itself was created and destroyed. Faraday began to think of the electro-tonic state as a link in the chain of production of the magnetic field and as involved also in the creation of electromagnetic induction.† In these half-developed notions of Faraday we may recognise many of the qualitative features of the vector potential. The latter was introduced by Maxwell in 1855 with the expressed purpose of quantifying Faraday's electro-tonic state. However, before we consider Maxwell, we must examine the intervening twenty or so years.

Following Faraday's discovery of electromagnetic induction many mathematical physicists on the Continent attempted to deduce his results from Ampère's law of force between current elements. Ampère has two versions of this law, one of which is now known as Grassmann's law (Ampère 1827, pp 214, 232). By exploring and transforming the latter version of Ampère's law C F Gauss (1777–1855) discovered the vector potential function in 1835. He found that the induced EMF was given by the rate of change with time of this function.

Here are Gauss's own words, translated (Gauss 1863, vol. v, p. 609):‡

---

† Faraday (1838, pp 19–22, 340–2; 1855, pp 368, 551, 530; 1932, vol. II, pp 55, 332, 370, vol. III, p. 215, vol. IV, pp 54–5; 1971, vol. I, p. 342, vol. II, p. 332, vol. III, p. 315).

‡ I am grateful to Dr H G Schneider of St Cross College, Oxford, for his assistance with this translation

The law of induction (Found out 1835, January 23, at 7 a.m. before getting up)
1. The electricity producing power, which is caused in a point P by a current-element γ, at a distance from $P = r$, is during the time dt the difference in the values of γ/r corresponding to the moments t and $t + dt$, divided by dt, where γ is to be considered both with respect to size and direction. This can be expressed briefly and clearly by

$$-\frac{d(\gamma/r)}{dt}$$

Gauss also discovered that the curl algorithm applied to this function gave the local magnetic intensity (Gauss 1863, vol. v, p. 612). Maxwell was to rediscover both results independently in 1855. It seems likely that Gauss thought of his vector potential function as an auxiliary quantity only rather than as representing a physical state (Gauss 1863, vol. v, pp 627–8). Gauss's manuscript was not published until 1867 by which time a far more developed theory of the vector potential was in print, as we shall see next.

A vector potential function was independently discovered by G Kirchhoff (1824–87) in 1852 by a route quite different from that of Gauss. W Weber (1804–91), in 1846, by transforming Ampère's first and preferred version of his law of force between current elements, had arrived at the following expression for the EMF induced in a secondary circuit element of length α due to a changing current i in a primary element of length α':

$$\text{EMF} = -\frac{1}{2}\frac{\alpha\alpha'}{r} a \cos v \cos v' \frac{di}{dt}$$

where v, v' represent the angles between the circuit elements and the line joining them (Weber 1892–4, vol. III, p. 253). By writing this out in cartesian components Kirchhoff was led to the discovery that the induced EMF per unit length in the secondary conductor is equal to the rate of change with time of a certain vector function, U, V, W, which is closely related to the modern vector potential (Kirchhoff 1892). He also found that the resultant EMF per unit length, or in modern terminology, the resultant electric field intensity, is given by

$$u = -2k\left(\frac{\partial\Omega}{\partial x} + \frac{4}{c^2}\frac{\partial U}{\partial t}\right), \text{ etc}$$

where Ω is the scalar potential due to local unbalanced electric charges (Kirchhoff 1892). Kirchhoff also showed that

$$\frac{\partial U}{\partial x} + \frac{\partial V}{\partial y} + \frac{\partial W}{\partial z} = \frac{1}{2}\frac{\partial\Omega}{\partial t}$$

an equation which is closely related to that known today as the

'Lorentz condition' (Kirchhoff 1892). For Kirchhoff, however, it was not a condition but a deduction from his integral definition of the vector potential. Kirchhoff does not suggest that his vector potential function is other than a useful auxiliary construction. The introduction of such functions is as old as mathematical physics itself (Neugebauer 1962). The possibility cannot be excluded, however, that Kirchhoff may have hoped that his vector potential function would eventually acquire a physical meaning, since he did think of the corresponding scalar potential function as having a physical meaning (Kirchhoff 1892, pp 49–55; 1850, pp 463–8).

The next major Continental development of the vector potential function was due to Ludwig Lorenz (1829–91) of Copenhagen. In his study of the propagation of elastic disturbances through a solid medium Lorenz had introduced a retarded scalar potential function in 1861 (Lorenz 1861). In 1867 he applied a similar approach to electromagnetism, treating the whole of space as a weakly conducting medium. He based his analysis on Kirchhoff's theory of the total EMF present in a conductor (Lorenz 1867). Guided by his own theory of elastic disturbances Lorenz postulated retarded scalar and vector potential functions Ω, α, β, γ in place of Kirchhoff's present-value functions. He also transformed Kirchhoff's rather clumsy function into the modern retarded integral definition of the vector potential:

$$\alpha = \iiint \frac{dx'\, dy'\, dz'}{r} u'\left(t - \frac{r}{a}\right)$$

$$\beta = \iiint \frac{dx'\, dy'\, dz'}{r} v'\left(t - \frac{r}{a}\right)$$

$$\gamma = \iiint \frac{dx'\, dy'\, dz'}{r} w'\left(t - \frac{r}{a}\right)$$

where α, β, γ are the cartesian components of Lorenz's vector potential, u', v', w' are the components of the retarded current density and a is the velocity of propagation (Lorenz 1867, p. 291). Furthermore, in place of Kirchhoff's equation linking the scalar and vector potentials Lorentz proved analytically that

$$\frac{d\bar{\Omega}}{dt} = -2\left(\frac{d\alpha}{dx} + \frac{d\beta}{dy} + \frac{d\gamma}{dz}\right)$$

which I shall call the Lorenz equation (Lorenz 1867, p. 294). Apart from units it is formally identical with the modern Lorentz condition, although it was not thought of as such. It should be noted that neither Kirchhoff nor Lorenz, unlike Gauss, linked the vector potential to magnetism.

The subsequent development of the vector potential function on the Continent stems largely from Ludwig Lorenz. It is important to note, however, that the theories of Kirchhoff and Lorenz did not pretend to be general theories of electromagnetism but dealt with special topics only. It should also be noted that all of these Continental developments were quite independent of Maxwell.

We now return to England for a very different line of evolution of the vector potential. British mathematical physics in the mid-nineteenth century was heavily influenced by Scottish and Irish physicists and by a commitment to a mechanical explanation of electromagnetic phenomena. It was strongly influenced also by the tradition of J B Fourier (1768–1830) whose mathematical physics was severely descriptive and tightly controlled by standardised units, and was characterised by a reluctance to mix in auxiliary mathematics with descriptive mathematics. An important difference between Britain and the Germanic countries in particular, was that most British theorists accepted the electromagnetic theory of Biot and Laplace, which recognised a qualitative distinction between electricity and magnetism, rather than the electrodynamics of Ampère, which reduced magnetism to electricity (Whittaker 1962, vol. I, chs v–ix).

In 1847 William Thomson (1824–1907) (later Lord Kelvin) in Glasgow independently introduced a vector potential function to magnetism and electromagnetism in the course of his theoretical search for an elastic model which would display some of the formal properties of the magnetic field (Thomson 1882–1911, vol. I, pp 76–80). Thomson found vector potential functions whose curl gave the magnetic field intensity due to a magnet and a current element, respectively. His model also demanded that the divergence of the vector potential was zero. In a publication of 1851 Thomson made the vector potential an integral part of his theory of magnetised bodies, without relating it to any mechanical model (Thomson 1851). Since Thomson's vector potential applied to stationary magnets and steady currents only, the question of retardation did not arise.

For Thomson, therefore, the vector potential was not introduced in relation to electromagnetic induction, as on the Continent, but in relation to permanent and 'galvanic' magnetism. Given Thomson's general commitment to a mechanical theory of magnetism and electromagnetism and to descriptive mathematics, it seems possible that he hoped that his vector potential function would eventually find a mechanical meaning within such a framework. Thomson, however, was never satisfied with any of his mechanical models for electromagnetism. (Thompson 1910, pp 1062–6, 1072–3).

In 1855 James Clerk Maxwell (1831–79) found a mathematical representation for Faraday's hypothesis of an electro-tonic state (Maxwell 1965, vol. I, pp 187, 204–6). Much of the underlying qualitative framework of Maxwell's theory was, of course, derived from Faraday. His mathematical tools were partly derived from Thomson's magnetic vector potential (Maxwell 1965, vol. I, p. 209) and partly also, perhaps, from Kirchhoff's electrodynamic vector potential (Maxwell 1965, vol. I, p. 208).

Maxwell showed that the rate of change with time of Thomson's magnetic vector potential function always gave the intensity of the local induced electromagnetic 'force' or, in modern terminology, the local induced electric field intensity (Maxwell 1965, vol. I, pp 202, 204). Maxwell called his vector potential function the 'electro-tonic function', after Faraday, since it was related to the local magnetic field intensity, and to the local induced EMF, in almost perfect agreement with Faraday's concept of the electro-tonic state. Maxwell's function, therefore, was incorporated into a broader network of relationships than either Kirchhoff's or Thomson's vector potential.

Maxwell had another motivation for introducing the electro-tonic function beside his faithfulness to Faraday's ideas. He writes (Maxwell 1965, vol. I, p. 203)

> The electromotive force in a closed conductor is measured by the rate of change of the entire electro-tonic intensity around the circuit ...We have now obtained in the [electro-tonic function] the means of avoiding the consideration of the quantity of magnetic induction which *passes through* a circuit. Instead of this artificial method we have the natural one of considering the current with reference to quantities existing in the same space as the current itself.

In other words Maxwell felt uneasy about making the change in the number of lines of force passing through the circuit responsible for what happens at its periphery. Cause and effect were not in proper contact. The electro-tonic function, however, *was* located at the periphery of the circuit and Maxwell felt that it was a better candidate for the immediate cause of electromagnetic induction.

In 1855, Maxwell was not yet confident that his electro-tonic function did describe a real physical property of the electromagnetic field (Maxwell 1965, vol. I, p. 207). In the course of the next ten years, however, he did find, to his own satisfaction, just such a physical meaning for it.

Maxwell was influenced by William Thomson rather than by Faraday in his attempts to provide a mechanical foundation for all electromagnetic phenomena. In an extraordinarily ambitious mechanical model published in 1861, he visualised the electric currents and the electromagnetic field as an elaborate linked hydrodynamical system with line vortices corresponding to the magnetic lines of force, and with rotating particles separating these vortices (Maxwell 1965, vol. I, pp 451–513). Translations of these particles or 'idle wheels', as he called them, constituted the electric current.

Maxwell also found a mechanical interpretation for his electro-tonic function, or vector potential, in terms of this model. If the translating particles (or source currents) which are responsible for the system of vortices are suddenly brought to rest this also brings the vortices to rest and in so doing the source currents themselves experience an impulsive forward momentum. This reactive momentum, per unit charge, is equal to the local electro-tonic function, or vector potential, in magnitude and direction (Maxwell 1965, vol. I, pp 476–9). It is defined at every other point of the field as well, when the source currents are brought to rest. Maxwell called it the 'reduced momentum' at every point of the field. For Maxwell, therefore, the vector potential had become a latent momentum in the electromagnetic field. Furthermore, electromagnetic induction became an inertial effect, as Faraday and others had suspected.

In Maxwell's 1861 publication there is also the first appearance of what is now known as the 'Coulomb gauge'. This is a stipulated mathematical condition which Maxwell imposed on his definition of the vector potential function, namely

$$\frac{dF}{dx} + \frac{dG}{dy} + \frac{dH}{dz} = 0$$

where F, G, H are the cartesian components of the vector potential function (Maxwell 1965, vol. I, p. 476). This equation first appeared in Thomson's publication of 1847, as we have seen, and it is written here by Maxwell in Thomson's notation of 1851. Thomson, however, had deduced this equation from his mechanical model whereas Maxwell is following a paper by Stokes of 1849 (Stokes 1880–1905) in imposing it as an arbitrary mathematical condition which would lead to a determinate solution for his function (although he did not work out such a solution here). Maxwell was fully aware that the equations in which his electro-tonic function appeared, relating it to the magnetic and electric field intensities, did not specify it fully (Maxwell 1965, vol. I, pp 198–9). However, for Maxwell, the zero divergence of the vector potential was never an arbitrary mathematical choice. From his first encounter with the equation in Thomson's paper of 1847 Maxwell clung tenaciously to the belief that it was demanded by the underlying nature of electromagnetism, as we shall see.

Remarkably, in the same year, 1861, in which both Maxwell and Lorenz published papers incorporating the vector potential, B Riemann (1826–66) of Göttingen arrived independently at a vector potential function which he interpreted as a velocity of the aether

(Whittaker 1962, p. 291). Riemann's lecture on this subject was not published until 1875, however, and it appears to have had little influence on discussions of the physical interpretation of the vector potential.

By 1865 Maxwell had arrived at a physical definition of the vector potential (which he now calls the electromagnetic momentum) which was in principle non-hypothetical and was independent of any mechanical theory of electromagnetism (Maxwell 1965, vol. I, pp 536, 555). In slightly modernised terms the vector potential for Maxwell is measured at any point of the electromagnetic field by the impulsive momentum given to unit charge placed at that point, when local magnets or currents are quickly removed. Maxwell now thought of Faraday's electro-tonic state as a latent or hidden momentum defined at every point of the field.

In his *Treatise on Electricity and Magnetism* published in 1873 Maxwell gives the same physical definition of the vector potential as in 1865, but he now calls it the 'electro-kinetic momentum' (Maxwell 1873, vol. II, pp 176, 232). He states that its components are 'perfectly determinate for every point of space (Maxwell 1873, vol. II, p. 44). It is a property only of those charges placed in the neighbourhood of magnets or currents and it is zero-valued everywhere in space remote from such currents (Maxwell 1873, vol. II, p. 256).

Maxwell's next problem was to discover a calculable mathematical function which would relate the vector potential function to the local electric currents. Maxwell imposed the same condition as he had done in his 1861 paper, namely

$$\nabla \cdot \boldsymbol{A} = 0$$

and he thereby arrived at the expression (Maxwell 1873, vol. II, pp 256-7)

$$\boldsymbol{A} = \int \frac{c}{r} \, dx \, dy \, dz.$$

Maxwell's original gothic letters have been romanised here. c stands for the present value everywhere of Maxwell's total current density, that is, the conduction current added to the 'displacement current'.

Although Maxwell's function $\boldsymbol{A}$ is a perfectly valid *mathematical* solution to the field equations expressed in terms of $\boldsymbol{A}$, it has caused considerable perplexity up to the present (Brill and Goodman 1967). Why did Maxwell choose an explicitly present-value solution when Lorenz's retarded vector potential function was well known to him? Why did he make $\nabla \cdot \boldsymbol{A} = 0$ when Lorenz had effectively found that

$$\nabla \cdot \boldsymbol{A} + \frac{1}{c} \frac{\partial \Omega}{\partial t} = 0?$$

Was Maxwell aware that his condition $\nabla \cdot A = 0$ also led to a present-value and, therefore, a non-causal solution of the scalar potential function (Jackson 1975, p. 223)? If so, why was he not troubled by this lack of causality when he believed that the vector potential function was a physical state and also that electromagnetic waves were propagated at the speed of light and not instantaneously?

Such questions are possible only with hindsight but from Maxwell's perspective his choice was altogether rational and consistent with his general theory of electromagnetism. The choice $\nabla \cdot A = 0$ was indeed partly motivated by mathematical convenience (Maxwell 1873, vol. II, p. 32) and earlier custom, as we have seen, but Maxwell also felt it physically necessary to make $\nabla \cdot A = 0$ otherwise, he believed, non-transverse and therefore non-physical electromagnetic waves would result (Maxwell 1965, vol. I, pp 580–2, Cranefield 1954, Heimann 1970). It should be noted that the distinction between the pure radiation zone (where $\nabla \cdot A$ is always zero) and the near and intermediate zones (where $\nabla \cdot A$ need not be zero) was first introduced, after Maxwell, by Hertz in 1888 (Hertz 1893, p. 121, Jackson 1975, p. 392).

In Maxwell's theory both the displacement current and the conduction current contribute to the magnetic field and to the vector potential and it must have seemed a perfect vindication of his decision to make $\nabla \cdot A = 0$ that a solution thereby resulted which contained the total current density. G F Fitzgerald showed in 1882, however, that the 'displacement current' does not contribute to the magnetic field intensity (Fitzgerald 1882) nor does it contribute to the vector potential.

Finally, Maxwell must not be criticised for failing to introduce retarded potential functions. In Maxwell's understanding the static or quasi-static electric and magnetic fields are not propagated outwards from charges and currents at the speed of light. Only the radiation fields fit this picture. The view that all fields are propagated at the speed of light was introduced after Maxwell, again by Hertz (Hertz 1893). Furthermore, Maxwell had no reason to believe that Lorenz's retarded potential functions were a general solution to his own field equations. Indeed this was first proved, as we shall see, by Levi Civita in 1897.

An outline of the mathematical developments of the vector potential function after Maxwell

Heaviside in 1886 (Heaviside 1892) and Hertz in 1890 (Hertz 1893) attempted to banish the vector and scalar potential functions from electromagnetism as part of a programme to emphasise what they believed to be the true physical content of Maxwell's theory. Although this proscription contributed to the view that the vector potential does not represent a physical quantity, it failed entirely to eliminate its use as an auxiliary function, at least, in electromagnetism.

There were now in existence at least three competing vector potential functions of importance, the older present-value function of William Thomson which was still applied to stationary magnets and currents, the retarded function of Lorenz and Maxwell's function. Maxwell's function appears to have found little favour until the present century. It is not difficult to see why. Lorenz's integral definition was already well established when Maxwell published his own. Hertz's discovery of electromagnetic waves in 1888 and his arguments that electrostatic and magnetostatic fields are also propagated with the speed of light strongly favoured Lorenz's retarded functions, at least by implication. Furthermore, Lorenz's function was much simpler than Maxwell's since it integrated only over the conduction currents and not over all of space. It also introduced a pleasing symmetry between the scalar and vector potential functions which Maxwell's algorithms failed to achieve. We find that Lorenz's functions were indeed employed in the 1890s in the writings of Fitzgerald, Lorentz, Poincaré and others (Whittaker 1962, vol. I, p. 325, O'Rahilly 1938, Fitzgerald 1891).

The case for Lorenz's functions received further support in 1897 when T Levi Civita (1873–1941) demonstrated that Hertz's field intensity equations (and therefore Maxwell's also) can be deduced from Lorenz's retarded vector and scalar potential functions (Levi Civita 1897). In 1906 H A Lorentz (1853–1928) of Leiden completed the demonstration of the perfect compatibility between Maxwell's equations and the retarded potentials by proving the inverse result, namely, that Maxwell's equations lead to the retarded potential functions of Lorenz, provided the *condition*

$$\nabla \cdot A + \frac{1}{c} \frac{\partial \phi}{\partial t} = 0$$

is satisfied (Lorentz 1916). What had been a deduced consequence of a postulated function for Lorenz became a matter of mathematical convenience for Lorentz.

Earlier, in 1903, Lorentz had shown which notional alterations in the vector and scalar potential functions are still possible within the Lorentz 'gauge', in the sense that they do not alter the values of E and B (Lorentz 1904). As with almost all other Continental authors, the potential functions were auxiliary quantities only for Lorentz and were not tied down to any physical explanation or measuring definition.

Herman Weyl (1865–1955) invented the terms 'calibration' or

'gauge' transformation to describe such notional alterations in the electromagnetic potential functions, in a series of papers from 1919 onwards devoted to a unified theory of gravitation and electromagnetism (Weyl 1922, Pais 1986).

Other mathematical developments in the potential functions may be summarised briefly as follows. A Liénard (1869–1958) (Liénard 1898) and E Wiechert (1861–1928) (Wiechert 1900) worked out in 1898 and 1900, respectively, the values of the retarded potentials for a single moving charge. Earlier expressions had related the vector potential function, in particular, to current elements only.

It was recognised by various authors early in the present century that the vector potential plays the role of a momentum vector in the canonical formulation of electrodynamics (Thomson 1904). It features also in the electrokinetic potential function in that formulation (Schwarzschild 1903). It was also recognised, soon after the foundation of special relativity, that the Lorenz potentials are relativistically covariant (Pais 1982).

The development of the physical interpretation of the vector potential after Maxwell

As we have seen, a strong theme in all of Maxwell's writings on electromagnetism, including the *Treatise*, is his commitment to a physical meaning and measuring definition for the vector potential. Furthermore, Maxwell was backed up in this by the authority of Faraday. It is surprising to discover in the subsequent literature, therefore, that Maxwell's physical interpretation of the vector potential quickly became a very marginal tradition. On the Continent it was noted but largely ignored (Wiechert 1900, p. 533, O'Rahilly 1938, p. 185). Even in Britain its only champion of note was J J Thomson (1856–1940) who showed in 1903 that the more general electromagnetic field momentum vector, $(1/c)\,\boldsymbol{E} \times \boldsymbol{B}$, which he had introduced in 1893, reduced to Maxwell's vector potential (multiplied by the charge) in special cases.†

---

† Thomson J J (1893), p. 13; 1904, pp 348–9:

> Hence we see that the vector potential at any point is the momentum due to the magnetic force produced by the system giving the vector potential and the electric field due to unit charge at that point. We thus get a physical interpretation of the vector potential, and see that instead of being merely an analytical device, it represents a most important physical property of the system.

Why was Maxwell's physical interpretation of the vector potential allowed to lapse? It was surely not his arguments, since similar arguments are being used today in the modern revival of Maxwell's interpretation. I believe that a number of factors conspired. Maxwell's algorithm for calculating the vector potential was based on questionable physical presuppositions. Lorenz's auxiliary or non-physical interpretation of the vector potential was well established on the Continent when Maxwell published and Continental investigators took the lead in theoretical electromagnetism soon after Maxwell. The efforts by Hertz and others to simplify electromagnetism involved a new austerity about which field variables should be accredited the status of physical properties. Summing up, therefore, it would seem that circumstances were not favourable in the late nineteenth century, for Maxwell's physical interpretation of the vector potential. The subsequent covariant generalisation of electromagnetic energy and momentum may have reduced its possible physical significance still further.

I have carried out a fairly thorough but not an exhaustive search of electromagnetic literature this century and I can find no text after J J Thomson in 1903 and before 1949 which supports Maxwell's physical view of the vector potential. References became more frequent after this date and at present Maxwell's interpretation receives some support.† Nevertheless, it remains a marginal tradition only in modern macroscopic electromagnetism. Surprisingly, also, none of the post-war authors who regard the vector potential as a physical state point out that Faraday and Maxwell were the first to do so. Is it possible that they have rediscovered the momentum interpretation independently by pondering over electromagnetic theory?

At present the physical interpretation of the vector potential finds stronger support in quantum theory than in classical electromagnetism. The theoretical investigations of Aharonov and Bohm in 1959 and the experimental confirmation by R E Chambers of Bristol in 1960 have shown that the presence of a vector potential, even in the absence of a magnetic intensity B, is accompanied by a change in the gradient of the phase of the quantum wave of an electron (Aharonov and Bohm 1959, Chambers 1960).

While there now seems to be a willingness to accept that the vector potential may have some sort of physical meaning in quantum physics there appears to be a considerable reluctance to grant it such a status in classical electromagnetism, despite the growth of interest in Maxwell's interpretation. This question will now be examined carefully.

† Eherenberg and Siday (1949); Feynman (1967, vol. II 17-5 to 17-6; vol. III, 21-2 to 21-6), Calkin (1966), Bork (1967), Konopinsky (1978, 1981).

An examination of the case for a physical meaning of the vector potential

In the more advanced reaches of classical electromagnetism today both the vector potential function A and the scalar potential function ϕ are generally understood as mathematical tools only. These functions are defined in terms of the given electric and magnetic intensities by the relations

$$E = -\frac{\partial A}{\partial t} - \nabla \phi \quad \text{and} \quad B = \nabla \times A.$$

These equations alone do not specify A and ϕ completely and the freedom to choose or modify these functions, within the constraints imposed by the condition that the values of E and B should be unaltered by the process, is called the 'principle of gauge invariance'. The further specifications required to establish A and ϕ more precisely, or even exactly, are said to determine the 'gauge' of these functions (Jackson 1975, pp 219–23). Such specifications are selected for calculational or exploratory convenience only and suggest strongly, if not compellingly, that A and ϕ are no more than semi-arbitrary auxiliary or service functions for the intensity vectors E and B, and do not, therefore, have a physical meaning.

No attempt will be made here to argue the case for a physical interpretation of the potential functions as they have just been defined. Instead, a different approach will be adopted and different kinds of questions raised.

One such question concerns the status today of classical electromagnetism as a physical theory. There is, of course, a solid experimental foundation for the theory which is constantly being tested directly, indirectly and by implication in the myriad research and technological applications of electromagnetism. Nevertheless, some of the most basic laws are difficult to test directly, many electromagnetic concepts are poorly understood, and there are important unresolved difficulties concerning the mathematical formulation of the theory (Rohrlich 1965, Jackson 1975, pp 47, 240–1 and ch. 17). All of this suggests that there is a need for caution in proposing explanations in electromagnetism.

Another, more specialised, question concerns the promotion, or otherwise, of a mathematical quantity in physics to physical status. Throughout the history of physics deductive mathematical exploration, and the analysis of experimental data, have prompted physicists to introduce or invent auxiliary mathematical quantities as an aid to their investigations. Some of these quantities eventually acquire a physical interpretation, others do not, still others are granted such an interpretation in certain circumstances only but not in others and, finally, others remain with an uncertain status.

For example, the velocity potential in hydrodynamics and the Lagrangian have always been regarded as mathematical artefacts, only. The mutual potential energy functions of Laplace and Gauss have acquired a physical meaning in electrostatics as have the atomic ranking numbers of Mendeleev in modern atomic theory. Ampère's conventional current does represent the physical current in some circumstances but not in others. The vector and scalar potentials, the magnetic flux function, and the concept of 'action' all have an uncertain physical status.

This raises a general question concerning the criterion used by the physics community in deciding whether or not to grant physical status to a given mathematical quantity. Detailed case-studies are required to answer this question in a satisfactory manner. However, a reflective examination of well known cases suggests that, before an invented mathematical quantity can be promoted to physical status there must be a foundation in experimental evidence for doing so or a positive role must be found for it in physical theory. If it can be measured directly *and* given a definite physical meaning within accepted theory then it becomes a strong candidate for physical status. The vector and scalar potentials will be examined in terms of these criteria.

It will be clear from the study so far that any investigation of the physical status of the potential functions must be carried out within the general context of a theory of energy and momentum in electromagnetic systems. There is, of course, a highly developed general theory of electromagnetic energy and momentum. In simple regimes these are represented by

$$E = \int \left(\frac{\varepsilon_0}{2} E^2 + \frac{1}{2\mu_0} B^2 \right) dv \quad \text{and} \quad \boldsymbol{P} = \int \varepsilon_0 \, \boldsymbol{E} \times \boldsymbol{B} \, dv$$

respectively, where ε_0 and μ_0 represent the electric and magnetic interaction constants, respectively, and the volume of integration is chosen large enough to satisfy certain simplifying assumptions. Although powerful and apparently valid in its fully covariant form (Jackson 1975 pp 791–6) this presentation of the theory of electromagnetic energy and momentum is rather abstract and difficult to interpret. What follows can be seen as a tentative contribution to explanatory research in a rather small area of a large topic.

In order to reduce the possibilities of vagueness in the present investigation it is necessary to have precise operational definitions of energy and momentum which apply to electromagnetic systems. In mechanics the momentum of a moving body or system may be defined operationally as the force which it exerts while it is being brought uniformly to rest in one second or, more generally, as the impulse which it exerts during any arresting process. This leads to a physical explanation of momentum as the capacity of a moving body or system to exert a force or impulse while it is being arrested. The energy of a body or system may be defined in the traditional manner as its capacity to do work in an appropriate mode of action. These definitions, drawn originally from mechanics, will now be applied to electromagnetism.

Electrokinetic momentum

The momentum of a train of electromagnetic waves is not a difficult concept in the present context since it is the impulse which the train will exert while it is being absorbed. The 'bound' elecromagnetic momentum of a system is a far more subtle concept. There appear to be several kinds of bound momentum, that associated with moving charged bodies, that associated with moving, neutral, current-bearing conductors and that associated with the interaction between charges and neutral currents. The present study will be concerned mainly with the latter.

It is sometimes stated that the bound momentum is located in the bound fields. This may, indeed, be so but these fields are tied to charges and currents. In the present study the bound fields and their sources will be treated as a single system and the question of the location of momentum or energy will not be raised.

Consider the case in which a charge q is placed in the vicinity of a neutral current. If both charge and current are held fixed in space and the current is reduced to zero electromagnetic theory predicts that the charge will deliver (and experience) an impulse. The same theory also predicts that the circuit experiences no reactive impulse. According to the operational definition of momentum given above the system has, indeed, a momentum since it is capable of delivering a one-way impulse while all its charges are being arrested. Following Maxwell and recent usage this will be called the 'electrokinetic momentum' of the system, p_e.

Since the stopping impulse is channelled through q it seems appropriate in the above arrangement to think of the electrokinetic momentum as a property of q. The electrokinetic momentum of a charge is different from mechanical and radiation momentum in that its realisation as an impulse is conditioned on bodies other than itself being brought to rest. It is an interaction momentum, rather than a proper momentum, of q.

Electromagnetic theory shows that the electrokinetic momentum of a charge is directly proportional to the magnitude of the charge. It follows, therefore, that $p_e = qA_e$, where A_e represents the electrokinetic momentum per unit charge at a particular point. By analogy with the 'intensity of the electric force', E, which is the force per unit charge, A_e will be termed the 'intensity of the electrokinetic momentum'.

Perhaps it is appropriate here to outline the argument so far. It is being suggested that electrokinetic momentum is an authentic, measurable and, therefore, physical momentum of the special system being considered. It is also being suggested that the momentum per unit charge A_e, like the force per unit charge E, should be treated as a true physical property of the charge. These concepts will now be examined in more detail.

When the source current is steady A_e is constant. At points remote from the source current the value of A_e due to that current is zero since the neutralisation of the source will not deliver an impulse to distant charges. Guided by these boundary conditions and by more general physical considerations electromagnetic theory suggests that A_e may be uniquely related to the source current, in the simple regime which we have been considering, by the equations

$$A_e = \frac{\mu_0}{4\pi} \int \frac{j_{\text{ret}}}{r} dv \quad \text{and} \quad p_e = qA_e$$

where j_{ret} is the retarded current density at r. It should be emphasised that the purpose here is to define p_e and A_e, in principle at least, by experiment, and the above equations ideally should have the logical character not of a definition, nor of a solution to a differential equation, but of a testable experimental law.

Electromagnetic theory also predicts that when A_e changes at a given point it is associated with an electric intensity there, as expressed by the equation $E = -\partial A_e/\partial t$. It follows, therefore, that the force on the charge is related to the electrokinetic momentum in this simple regime by the equation $F = Eq = -dp_e/dt$.

Electrokinetic momentum has some remarkable invariant properties. The impulse delivered to the test charge is independent of how quickly the source currents are brought to rest. Invariance is not confined to the static case which we began with. The same impulse is delivered to the test charge, in magnitude and direction, if the source currents (or magnets) are slowly and steadily withdrawn, in an arbitrary direction, from its neighbourhood. Even when both source current and test charge are slowly and steadily withdrawn to a great distance from each other in an arbitrary direction the total mechanical impulse generated in the system, including the effects of both electrical and magnetic forces on the charge and circuit is again given by $p_e = qA_e$. During any stage of the above process the resultant of the magnetic and electrical forces on the system is given by

$F = -d\boldsymbol{p}_e/dt = -qd\boldsymbol{A}_e/dt$, where $d\boldsymbol{A}_e/dt$ is the convective derivative of $\boldsymbol{A}_e$ (Calkin 1966). Clearly, only when the source current is stationary can $\boldsymbol{p}_e = q\boldsymbol{A}_e$ be regarded as a property of the charge q alone.

As is well known, and as we have suggested already, theory predicts that a non-reactive mechanical impulse $\boldsymbol{p}_e = q\boldsymbol{A}_e$ will be experienced by a fixed charge when a current is switched on in a fixed circuit in its neighbourhood (Goldstein 1980). If the charge is moving slowly, and is attached to a body with sufficient inertia so that the convective change in $\boldsymbol{A}_e$ is negligible while the current is being brought to rest, then the charge will gain a momentum $\boldsymbol{p}_e = q\boldsymbol{A}_e$. However, if the current is switched on again this momentum will be destroyed. If the charge is allowed to leave the neighbourhood of the current the system will lose a net momentum equal to the original mechanical momentum generated. Although theory predicts that non-reactive mechanical momentum can be generated within an electromagnetic system, there appears to be no way in which it can be cyclically accumulated or withdrawn as such from the system. The sum of the mechanical and electrokinetic momentum is, of course always conserved, so there is no question of the violation of the principle of conservation of total momentum.

It would seem, therefore, that in the regime which we have been considering, the electric and magnetic forces are not the only properties of the charge. A more complete description of its properties should include its electrokinetic momentum. The intensity of the electrokinetic momentum $\boldsymbol{A}_e$ is related to the source currents by an expression which is, of course, a particular solution of the vector potential equations. In such a regime, therefore, one vector potential function would seem to have a privileged status in that it represents a physical property of the system.

When there are no charges in the neighbourhood of a current system there is, of course, no electrokinetic momentum. $\boldsymbol{A}_e$ at each point, nevertheless, represents the intensity of the electrokinetic momentum which the system would commit to that point, were a charge placed there. Every charge in the source circuit also has an electrokinetic momentum, $\boldsymbol{p}_e = q\boldsymbol{A}_e$, but the resultant momentum of the circuit is zero since it is electrically neutral.

As we have seen already, electrokinetic momentum, like mechanical momentum, is associated with a phase gradient of the electron wave. The present analysis suggests that, for a steady current $\boldsymbol{p}_e = h/2\pi \boldsymbol{\nabla} \theta = e\boldsymbol{A}_e$, where θ is the phase of the quantum wave and h is Planck's constant. As in the case of mechanical momentum, it does not seem correct to regard the electrokinetic momentum as the cause of the additional phase gradient. They are simply correlated properties of the charge. Furthermore, the presence of a steady current does not alter the wavelength of the electron, since this is determined by the mechanical momentum of the charge.

It is well known that the magnetic flux function is quantised in a superconducting ring (Feynman *et al* 1969, vol. III pp 21–7). It can be deduced from this, or demonstrated directly, that $\oint q\mathbf{A}_e \cdot d\mathbf{s} = nh$, where h again is Planck's constant. This states that the circulation of the electrokinetic momentum, and, therefore its 'action' around the ring, is quantised.

It is questionable whether these results confer physical status either on the magnetic flux or on the circulation of the electrokinetic momentum. One objection to granting physical status to these quantities is that 'action' seems to have no associated operational definition. Another objection is that useful mathematical artefacts and equations which mix pure mathematics and physics frequently emerge from the mathematical laws of physics as a kind of unexpected bonus, and without any supposition that they have a physical meaning. For example, the electric flux function around a static charge is directly proportional to the charge and is also 'quantised' because all charges are composed of elementary units. No one supposes from this, however, that electric flux is a true physical quantity.

Electromagnetic energy and more general regimes

The concept of the electrokinetic momentum of a charge has been introduced only for a quasi-stationary charge in the presence of a stationary current. Can the concept be generalised to collections of charged bodies and currents moving arbitrarily? Such a study would take us far beyond Maxwell's concept of electrokinetic momentum and needs to be linked, I believe, to an historical and interpretative study of the general theory of energy and momentum of electromagnetic systems. Such an investigation will not be attempted here. This study would be unbalanced, however, without some reflections upon a larger explanatory theory. I shall make some remarks, therefore, concerning the bound energy and momentum of various systems, generalising by stages.

The total electrostatic energy or mutual potential energy of a charged body may be defined operationally as the electrical work done by the system on each of its elementary charges when these are slowly withdrawn to a great distance from each other. This will be understood as a purely interactive energy between elementary charges. If each elementary charge has indeed an electrostatic self-energy this will be assumed, as is usual, to be included in its rest-mass energy.

The mutual potential energy of a system of static charges is not the only form of electrical energy which can be associated with the system. Each charge present also has an individual potential energy with respect to the system. This may be defined operationally as the electrical work which the system will do on that charge during its slow exit from the system, the other charges remaining fixed. An example of individual potential energy which does not quite meet these requirements is the work done by a Van de Graaff generator on an α particle released near its surface and expelled from its neighbourhood, or the ionisation energy of a photo-electron in an atom. Individual potential energy is just as important in physics as collective potential energy.

The individual electrostatic potential energy per unit charge, ϕ_e, is independent of the magnitude of the test charge. It follows, therefore, that $\Phi_e = q\phi_e$, where Φ_e represents the total (individual) electrostatic potential energy of the charge q. Again by analogy with the electric intensity E, ϕ_e will be described as the 'intensity of the (individual) potential energy' of a charge.

Both the individual potential energy, and the intensity of that energy ϕ_e will be zero for charges which are remote from the given source system of charges since the latter system can do no work on the test charge as it is withdrawn to an even more remote point from it (Roche 1989).

It is well known that in the static regime which we have been considering a function can be found which satisfies the physical requirements and boundary conditions of the intensity of the individual potential energy. For a discrete system of charges

$$\phi_e = \frac{1}{4\pi\varepsilon_0} \sum \frac{q_n}{\gamma_n}.$$

This can easily be generalised to a continuous system.

In electrostatic regimes, therefore, it would seem that a particular potential function can be given a physical meaning and has a privileged status and should be included in a full description of the properties of any test charge.

When an isolated charged body moves with a steady velocity v we might be inclined at first to assume that its only gain in energy is its mechanical kinetic energy, and its only gain in momentum is its mechanical momentum. However, at rest it has an electrostatic energy and when it moves a more careful consideration might suggest that the mass associated with this energy, and therefore the energy itself, should increase by the usual relativistic factor $\gamma = [1 - (v^2/c^2)]^{-1/2}$. A fully covariant theoretical treatment of electromagnetism predicts precisely this result. There is also a specifically electromagnetic momentum which is measured by the product of the enhanced electromagnetic mass into the velocity (Thomson 1904, Jackson 1975, pp 784–5). By analogy with relativistic mechanics the electrostatic energy may be thought of as a rest energy and the additional electromagnetic energy due to motion as an electrokinetic energy. These are, of course, exceedingly small quantities compared with their mechanical counterparts, but need to be included in the total energy and momentum budget of the system. When several moving charged bodies are present interaction will affect the total electromagnetic energies and momenta.

Interestingly, the energy of a stationary neutral electric current is purely electrokinetic since the electrostatic energy of such a system is zero. H A Lorentz (1916, p. 48) seems to have been the first to show that the electrokinetic energy of a current is greater than the mechanical kinetic energies of the drifting electrons. When a current bearing circuit moves, electromagnetic theory again predicts that its electromagnetic energy increases by the usual relativistic factor and it develops a specifically electromagnetic momentum due to the convection of this mass energy. When several such moving circuits are present their mutual interactions will again affect the electromagnetic energy and momentum of the system.

In a more general and isolated regime, consisting of arbitrarily moving charges and currents the total electromagnetic energy may be defined operationally as the electrokinetic energy released when all current-bearing conductors (and magnets) are brought to rest and neutralised, and all unbalanced charges are brought to rest, and all radiation absorbed. The remaining energy is elecrostatic and may be defined as above. Analagously, the total electromagnetic momentum of the system may be defined as the specifically electromagnetic impulse delivered by the system during the above arresting process.

Because of interactions and interchanges the total momentum and energy of such a system will be extraordinarily complex. This will need to be taken into account in any theory of a general role for an electrokinetic momentum, or for an electrokinetic vector potential.

Conclusions

The practice of gauge choice and transformation of the classical macroscopic vector and scalar potential functions remains unaffected by these considerations with respect to its mathematical validity and usefulness in calculating E and B, but it alters interpretations somewhat. It suggests that, in certain regimes at least, particular values of these functions have a physical interpretation in that they describe properties of charges. It also suggests that in such regimes a more complete description of the action of a system of currents and charged bodies on a given test charge is provided by the four quantities E, B, A_e and ϕ_e than by the quantities E and B alone. When the gauge is transformed from a descriptive to an auxiliary gauge, in these regimes, physical information may be lost about the system and only part of the information about the system is gauge invariant. This also emphasises that the potential functions in an auxiliary gauge are mathematical tools only, and the question of their physical status does not arise. In the general case, of course, the question of whether or not there is always a physical gauge has not been settled here.

A remark on nomenclature will conclude this chapter. Since the general vector potential function is related to both the electric intensity and the magnetic intensity by the equations $E = -(\partial A/\partial t) - \nabla \phi$ and $B = \nabla \times A$ and since A may exist where $B = 0$ it seems inappropriate to call A the 'magnetic' vector potential.

References

Aharonov Y and Bohm D 1959 Significance of the electromagnetic potentials in the quantum theory *Phys. Rev.* **115** 485–91

Ampère A M 1827 Memoire sur la théorie mathématique des phénomènes électrodynamique... *Mém. Acad. R. Sci. Ann. 1823 (Paris)* 175–388 (in 372–3)

Bork A M 1967 Maxwell and the vector potential *Isis* **58** 210–22 (in 216)

Brill O and Goodman B 1967 Causality and the Coulomb gauge *Am. J. Phys.* **35** 832–7

Calkin E J 1966 Linear momentum of quasistatic electromagnetic fields *Am. J. Phys.* **34** 921–5

Chambers R E 1960 *Phys. Rev. Lett.* **5** 3

Cranefield P F 1954 Clerk Maxwell's correction to the page proof of 'A Dynamical theory of the electromagnetic field' *Ann. Sci.* **10** 359–62

Ehrenberg W and Siday R 1949 The refractive index in electron optics and the principles of dynamics *Proc. Phys. Soc.* B **62** 8–21

Faraday M 1838 *Experimental Researches in Electricity* vol. I (London: Richard and John Edward Taylor)

—— 1844 *Experimental Researches in Electricity* vol. II (London: Richard and John Edward Taylor)

—— 1855 *Experimental Researches in Electricity* vol. III (London: Richard and John Edward Taylor)

—— 1932–6 *Faraday's Diary* ed. T Martin 7 vols (London: The Royal Institution)

—— 1971 *The Selected Correspondence of Michael Faraday* ed. P Williams 3 vols (Cambridge: Cambridge University Press)

Feynman R, Leighton R and Sands M 1969 *The Feynman Lectures on Physics* 3 vols (Reading, MA: Addison-Wesley)

Fitzgerald G F 1882 A note on Mr J J Thomson's investigation of the electromagnetic action of a moving sphere *Phil. Mag.* **13** 302–5
—— 1891 On an episode in the life of Hertz's solution of Maxwell's equations *Rep. Br. Assoc. 1890* 755–7
Gauss C F 1863–1929 *Werke* 12 vols (Leipzig: Akademie der Wissenschaften zu Göttingen)
Goldstein H 1980 *Classical Mechanics* (Reading, MA: Addison-Wesley) p. 56
Heaviside O 1892 *Electrical Papers* vol. II (London: Macmillan) pp 172–3
Heimann P M 1970 Maxwell and the modes of consistent representation *Arch. Hist. Exact Sci.* **6** 171–213 (in 203)
Hertz H 1893 *Electric Waves* (London: Macmillan)
Jackson J D 1975 *Classical Electrodynamics* (New York: Wiley)
Kirchhoff G 1850 On a deduction of Ohm's laws in connection with the theory of electrostatics *Phil. Mag.* **37** 463–8
—— 1892 *Gesammelte Abhandlungen* (Leipzig: Barth) pp 155–8
Konopinsky E 1978 What the electromagnetic vector potential describes *Am. J. Phys.* **46** 499–502
—— 1981 *Electromagnetic Fields and Relativistic Particles* (New York: McGraw-Hill) p. 158
Levi Civita T 1897 Sulla riducibilita della equazioni electrodinamiche di Helmholtz alla forma Hertziana *Nuovo Cimento* **7** 93–108 (in 104–6)
Liénard A 1898 Champ électrique et magnetique produit par une charge électrique concentrée en un point et animée d'un mouvement quelconque *L'Éclairage Électrique* **16** 5–14, 53–9, 106–12
Lorentz H A 1904 Weiterbildung der Maxwell's Theorie; Elektrontheorie in *Encyklopädie der Mathematischen Wissenschaften* vol. V (Leipzig: Teubner) pp 145–280 (in 157)
—— 1916 *The Theory of Electrons* (Leipzig: Teubner) pp 19, 239
Lorenz L 1861 Mémoire sur la théorie de l'élasticité des corps homogénes a l'élasticité constante *J. Reine Angew. Math.* **58** 329–51 (in 330, 334, 336)
—— 1867 On the identity of the vibrations of light with electrical currents *Phil. Mag.* **34** 287–301
Maxwell J C 1871 *Maxwell Papers* Cambridge Add. MS 7655/b/18 dated Jan 23 1871
—— 1873 *A Treatise of Electricity and Magnetism* 2 vols (Oxford; Clarendon)
—— 1965 *Scientific Papers* ed. W D Niven 2 vols (New York: Dover)
Neugebauer O 1962 *The Exact Sciences in Antiquity* (New York: Harper) chs v and vi, Appendix i
O'Rahilly A 1938 *Electromagnetics* (Cork: Cork University Press) pp 184, 517
Pais A 1982 *Subtle is the Lord: The Science and Life of Albert Einstein* (Oxford: Clarendon) pp 151–2
—— 1986 *Inward Bound* (Oxford: Clarendon) pp 344, 357 and nn 58, 59, 60
Roche J J 1989 Applying the history of electricity in the classroom: a reconstruction of the concept of potential in *Teaching the History of Science* ed. M Shortland and A Warwick (Oxford: Blackwell) pp 175–80
Rohrlich F 1965 *Classical Charged Particles* (Reading, MA: Addison-Wesley)

Schwartzschild K 1903 Zwei formen des princips der kleinsten action in electron theorie *Nachricht. Königlisch Ges. Wiss. Göttingen Math.-Phys. Klasse, 1903* 127

Stokes G G 1880–1905 *Mathematical and Physical Papers* ed. J Larmor 5 vols vol. II (Cambridge: Cambridge University Press) pp 254–5

Thompson S P 1910 *The Life of William Thomson, Baron Kelvin of Largs* vol. II (London: Macmillan)

Thomson J J 1893 *Notes on Recent Researches in Electricity and Magnetism* (Oxford: Clarendon)

—— 1904 On momentum in the electric field *Phil. Mag.* **8** 331–56 (in 354)

Thomson W 1851 A mathematical theory of magnetism *Phil. Trans. R. Soc.* **141** 283

—— 1882–1911 *Mathematical and Physical Papers* 6 vols vol. I (Cambridge: Cambridge University Press) pp 76–80

Volta A 1779 Observations sur la capacité des conducteurs électriques *J. Physique* **13** 252

Weber W 1892–4 *Werke* 6 vols vol. III (Berlin: Springer) p. 253

Weyl H 1922 *Space, Time, Matter* (London: Methuen) pp 285–6

Wiechert 1900 Electrodynamische elementargesetzen *Arch. Néerl. Sci. Exact. Nat.* **5** 549–73

Whittaker I 1962 *History of the Theories of Aether and Electricity* (London: Thomas Nelson)

10

The Photoelectric Effect—A Suitable Case for Surgery?†

Stuart Leadstone

This chapter is born out of a growing conviction that the time is ripe for a re-examination of many topics and concepts which are an established part of physics. The point of view from which I write is that of a physics teacher, concerned for the last 25 years with the preparation of A-level and Baccalaureate students for university entrance. The significance of this level of teaching is that it is the students' first encounter with the subject which demands a significant degree of formality and rigour. It is the stage during which the foundations of the subject are laid, and should therefore be characterised by both accuracy and depth. The difficulty of achieving a satisfactory blend of these two ingredients is daunting. The reflective teacher needs recourse, therefore, to informed specialist knowledge in order to refine the presentation of his or her material in order to do justice to the sophistication of the concepts involved while not going over the heads of the students.

Certain topics are remarkably intransigent when it comes to the ironing out of difficulties. Recent experiences in this connection have led me to a realisation that a 'consensus view' in physics is more of an ideal than an actuality. The dilemma for the teacher is obvious—how to arrive at an informed grasp of such topics so that they can be

† Based on a presentation given at the conference on The History of Physics for the Physicist, 2–4 July 1986, Oxford.

presented with the rigour and authority appropriate to the age and intellectual development of aspiring undergraduates? Insofar as present confusions and ambiguities in a given field are the result of its historical development it seems clear that a thorough re-examination of the history of the field is a necessary starting point. The general case for such a 'history as surgery' approach has been eloquently argued elsewhere (Roche 1984). In this article one candidate for historical surgery will be examined, namely the photoelectric effect, to which I now turn my attention.

The photoelectric effect

The photoelectric effect is an important topic in the atomic physics section of modern pre-university and undergraduate physics courses. Its importance lies in the evidence it gives for the quantisation of electromagnetic radiation. Most introductory texts, however, give treatments which ignore or gloss over important aspects of the phenomenon, and it is difficult to find material at a sufficiently non-specialist level which is authoritative and accurate. To be specific it is remarkably difficult to obtain clear answers to the following queries:

Q1 Which electrons are involved in the process—free, bound or both?

Q2 What is the significance and precise definition of the term 'work function'?

Q3 What is the reason for the spread of energies in the emitted electrons from zero up to a maximum value?

Q4 What is the significance of the 'stopping potential'? How is it related to the work functions of the photocathode and anode?

Each of these will now be examined in turn.

Which electrons?

A random survey of ten A-level and five undergraduate texts revealed that most authors do not commit themselves on this issue. Of those that did, three (Nelkon and Parker 1979, Bleaney and Bleaney 1965, Richtmeyer *et al* 1955) indicated that free electrons were ejected from the metal, whilst two (Noakes 1956, Open University 1971) stated or implied that bound electrons were the ones involved. Undergraduate texts usually, though not always, make it clear that it is the *free*, or to use a less ambiguous term, the *conduction* electrons which are expelled from the metal.

Work function

The work function of a metal is usually given a rather vague definition such as 'the energy required to liberate an electron from the metal'. In view of the wide range of energies possessed by the conduction electrons prior to liberation such definitions are meaningless. Consultation of dictionaries shows that the vagueness surrounding this concept is very widespread. Examples of seriously flawed dictionary definitions of the 'work function' are

> The minimum energy required by an electron for it to pass through a potential barrier (Collocott and Dobson 1974)

and

> The work to be done by the exciting [*sic*] electron (Isaacs 1985).

Those texts which give a clear indication that it is the conduction electrons which are ejected by the incident radiation do not always explain *why* work must be done for such an electron to escape. Even when the treatment is sufficiently advanced to make reference to the Fermi–Dirac energy distribution of the conduction electrons, further confusion can arise by failure to say whether the work function is measured with respect to the bottom or top of the distribution. Thus it comes as a surprise to discover that the quantity Ω in the following definition is *not* the work function of the metal (Richtmeyer *et al* 1955 pp 102–3)

> ... the work Ω that must be done by any electron against attractive forces in escaping from the surface of the metal ...

In the text cited here the situation is clarified by reference to figure 10.1 which shows the range of energies possessed by the conduction electrons in a metal at absolute zero. The quantity Ω is measured from the bottom of the Fermi–Dirac distribution to a level representing an electron with zero kinetic energy outside the metal surface. On the other hand the work function ϕ is measured from the top of the distribution, that is from the Fermi level E_F. The relation between the two quantities is thus

$$\phi = \Omega - E_F. \tag{10.1}$$

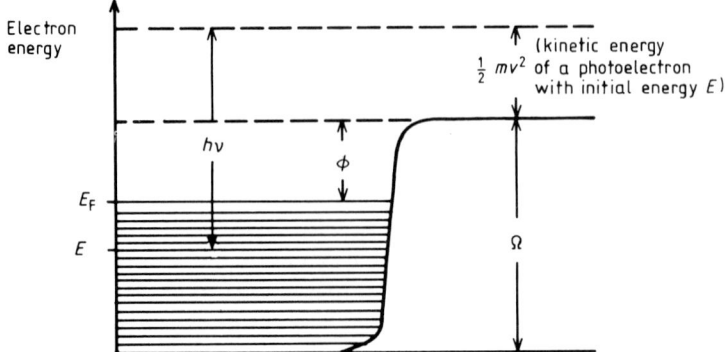

Figure 10.1 Energy level diagram for conduction electrons in a metal at absolute zero. (Adapted from Richtmeyer *et al* (1955), courtesy McGraw-Hill Book Company Inc.)

Spread of emitted energies

Explanations are broadly divided between the following:

(*a*) As an escaping electron makes its way to the surface it suffers collisions and hence loses some kinetic energy. Electrons starting from different depths within the metal will lose differing amounts of energy in this way; hence the range of energies with which they emerge.

(*b*) As indicated in figure 10.1 the electrons have different *initial* energies by virtue of the Fermi–Dirac distribution. An electron with initial energy E will therefore emerge from the metal with kinetic energy given by

$$\tfrac{1}{2} mv^2 = E + h\nu - \Omega \tag{10.2}$$

where ν is the frequency of the incident radiation and h is Planck's constant. In particular an electron at the Fermi level E_F will emerge with the maximum possible kinetic energy, given by

$$(\tfrac{1}{2} mv^2)_{\max} = E_F + h\nu - \Omega. \tag{10.3}$$

Using equation (10.1), equation (10.3) may be rewritten

$$(\tfrac{1}{2} mv^2)_{\max} = h\nu - \phi \tag{10.4}$$

which is the familiar form of Einstein's photoelectric equation.

Some authors admit the possibility of both causes (*a*) and (*b*). There is another simple explanation which appears currently not to be part of the consensus view. This is that '... not all of the quantum energy may be transferred to a single electron ...' (Beiser 1962). Why this is generally disregarded is not made apparent in most texts. No doubt its exclusion derives in part from its rejection by Einstein (1879–1955) in his seminal paper of 1905 (Arons and Peppard 1965). Lest this be lightly dismissed it *is* mentioned as a possibility by Martin Klein in his article 'Einstein and the development of quantum physics' contributed to *Einstein: A Centenary Volume* (French 1979).

It is appropriate at this point to say a little more about Einstein's 1905 paper, the third of the famous trio of papers which he submitted that year. The title is translated by A Arons and M Peppard (1965) 'Concerning an heuristic point of view toward the emission and transformation of light'. In it Einstein introduces the concept of light quanta to account for a number of phenomena associated with black-body radiation, fluorescence and 'the production of cathode rays by ultraviolet light'. The photoelectric effect is dealt with in one of nine sections. Let us see what Einstein actually says. In the translation referred to this is rendered:

> ... one can conceive of the ejection of electrons by light in the following way. Energy quanta penetrate into the surface layer of the body, and their energy is transformed, at least in part, into kinetic energy of electrons. The simplest way to imagine this is that a light quantum delivers its entire energy to a single electron; we shall assume that this is what happens. The possibility should not be excluded, however, that electrons might receive their energy only in part from the light quantum.
>
> An electron to which kinetic energy has been imparted in the interior of the body will have lost some of this energy by the time it reaches the surface.

The last sentence is a clear statement of explanation (*a*) above, whilst the penultimate sentence raises an intriguing possibility. Taken at face value it does seem to suggest the existence of some emission energies greater than that normally implied by $(\frac{1}{2} mv^2)_{max}$.

Contact potentials

The question of contact potentials is totally ignored in A-level texts, but their importance has been realised since at least 1912 (Compton 1912, Richardson and Compton 1912). Furthermore their measurement was an essential feature of the renowned investigations of R A Millikan (1868–1953) (Millikan 1914, 1916, Trigg 1971). In the 1914 abstract Millikan writes

The experiments herewith reported were undertaken for the sake of subjecting to rigorous experimental test the three assertions contained in Einstein's photoelectric equation...
These assertions are
1. That there is a linear relation between the frequency of the impressed light and the maximum energy of emission of the electrons ejected by it.
2. That the slope of the line representing the linear relation between PD [potential difference] and ν is h/e, i.e. that this slope times e is Planck's 'h'.
3. That the intercept of the PD line on the ν axis gives the frequency ν_0 at which the metal in question first begins to be photo-electrically active.

The second and third of these assertions have not heretofore been made the subject of accurate test nor can they be so made without simultaneous measurement *in vacuo* of both contact potentials and photo-potentials in the case of metals which are sensitive throughout a long range of frequencies.

The importance of contact potentials is thus clearly established by Millikan at the outset. The way in which the measured contact PDs were used by Millikan to correct the raw data is illustrated in figure 10.2. The vertical displacement of the experimental line to give the dotted line is clearly essential in order to determine the threshold frequency ν_0 (Millikan's third assertion). However, the *gradient* of the line is unaffected by the correction displacement (Hughes and Dubridge 1932), which conflicts with Millikan's statement that the measurement of contact potentials is essential to the testing of the second assertion.

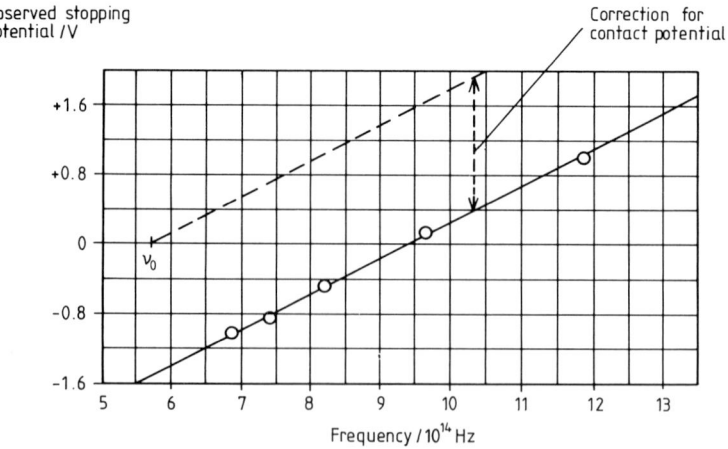

Figure 10.2 Millikan's data for lithium showing the correctness of Einstein's photoelectric equation. (After Millikan (1916).)

The Photoelectric Effect

That the issue of contact potential remains a matter of contention can be judged from more recent debate concerning the roles played by the work functions of the photocathode and anode (James 1973, Hodgson and Lambert 1975). Perhaps it is not surprising that textbook authors on the whole prefer silence to commitment. Some, however, do grasp the nettle (Bitter and Medicus 1973).

Other aspects of the photoelectric effect

If a thorough and rigorous account of the photoelectric effect is to be attempted then there are other issues to be raised. These are presented in the form of further queries and numbered consecutively with the original four listed earlier. (No claim for completeness is made here.)

Q5 What does the term 'photoelectric effect' encompass? Is it restricted to the optical/ultraviolet photoemissive effect in metals or does it have much wider implications?

Q6 What proportion of incident photons participates in the phenomenon? What happens to those not involved?

Q7 Comparing the photoelectric and Compton effects, why are momentum considerations usually ignored in treatments of the former but of paramount importance in the latter? (The possibilities for confusion occasioned by a superficial view of these two phenomena are indicated in figure 10.3).

Photoelectric effect	Compton effect
Commonly associated with metals	First observed in graphite (Compton 1923a,b)
↓	↓
'Free' electrons involved	'Bound' electrons involved
but	but
Momentum (and energy) conservation requires these to be *bound* if a photon is to give the whole of its energy to an electron (Richtmeyer *et al* 1955)	Momentum (and energy) conservation account for the Compton shift only if these are treated as *free*

Figure 10.3 Paradox resulting from a naïve view of phenomena

Q8 Does the radiation pressure exerted on the surface by the incident radiation show a sudden increase as its frequency increases through the photoelectric threshold? This question is prompted by the thought that as the photoelectric threshold is crossed a new process is initiated involving a proportion of the incident photons. In this process the metal surface must recoil because (i) the momentum of the incident photons is destroyed, and (ii) electrons are kicked out of the metal towards the side from which the photons have come. So the question effectively asks: is this recoil observable?

Q9 What is the current status of theories which attempt to explain the photoelectric effect without assuming quantisation of the electromagnetic field?

Is there a consensus view?

In an attempt to answer this question all the queries (excluding Q4 which was an afterthought) raised in this paper were submitted for scrutiny to four professional physicists with a special interest in the field (see acknowledgments). As might be expected their responses showed substantial though not complete agreement. An abbreviated summary of interesting points which this investigation produced is as follows.

Which electrons?

All three possible answers were given—*bound*, *free* and *both*. The discrepancies arose from different usages of the term 'free electron' and different views on the range of phenomena embraced by the term 'photoelectric effect'. All clearly indicated that the surface photoelectric effect in metals involved the conduction electrons.

Work function

All indicated that the work function is measured with respect to the Fermi level. Variations which remained related to the emphasis on or inclusion or omission of three details:

(i) specification of temperature, namely absolute zero;
(ii) specification of final position of electron, namely at infinity;
(iii) specification of final kinetic energy of electron, namely zero.

Spread of emitted energies

Broad agreement was displayed in that the range of emitted energies was attributed primarily to the range of energies present in the conduction band. Collision losses were admitted though it was not necessarily clear whether these were electron–electron collisions or electron–lattice collisions. Only one correspondent took up the suggestion that one photon might share its energy with several electrons, and ruled this out as a possibility.

The term 'photoelectric effect'

No consensus here. The use of the term varied from the most restricted sense (surface effect in metals) to the most general sense (any interaction between electromagnetic radiation and matter in any of its forms). Whilst this state of affairs is of no consequence to the experienced physicist it can disturb and hinder progress of understanding for the student. Use of language in physics is not uniformly precise, and it is submitted that this is pedagogically undesirable. Clearly, textbook authors carry a heavy responsibility in this connection.

Participating photons

Estimates of the proportion of incident photons which give rise to photoelectrons varied by three orders of magnitude—from 0.1 to 10 per cent. The fate of other photons could include reflection, electron excitation not leading to escape and Compton scattering. The relative probabilities of these processes will depend on several factors, notably the energies of the incident photons.

Momentum considerations

The importance of the crystal lattice as the 'third body' which can take up momentum in the photoelectric effect was unanimously acknowledged. Quantitative consequences can be ignored because the relatively enormous mass of the lattice means that a negligible amount of kinetic energy is transferred to it. In the Compton effect it is the very much higher energy of the photons, relative to the binding energy of the scattering electrons in the material, which makes it a valid approximation to treat the latter as 'free'.

Radiation pressure

There was agreement that the radiation pressure should increase, though opinions differed on the magnitude of the effect. One correspondent pointed out the misconception contained in the question implied by the expectation of a *discontinuity* in the radiation pressure. Since the quantum efficiency (i.e. the number of electrons emitted per incident photon) increases continuously from zero (ideally) at the threshold, any observed increase in radiation pressure should also be continuous. A further complication is signalled by the word *ideally* in parentheses. Because of the 'thermal tail' in the Fermi–Dirac energy distribution (figure 10.4) there exists, at all temperatures above absolute zero, a small proportion of the conduction electrons with energies $E > E_F$. The consequences of this are that neither the photoelectric threshold nor the kinetic energy maximum are the simple limits which elementary treatments assume. A realistic statement of the situation is given by Maurer (1967) in an authoritative account of the photoelectric effect:

> Since the photoelectric yield decreases rapidly as the frequency of the incident radiation is decreased but does not become zero, the apparent threshold frequency depends upon the sensitivity of the experimental apparatus. The apparent threshold also depends upon the temperature of the surface since the photoelectric yield increases with increasing temperature for frequencies near the apparent threshold...
>
> The relative number of fast electrons is small and the original investigators concluded that a maximum kinetic energy of emission existed. As in the case of the photoelectric threshold, the apparent

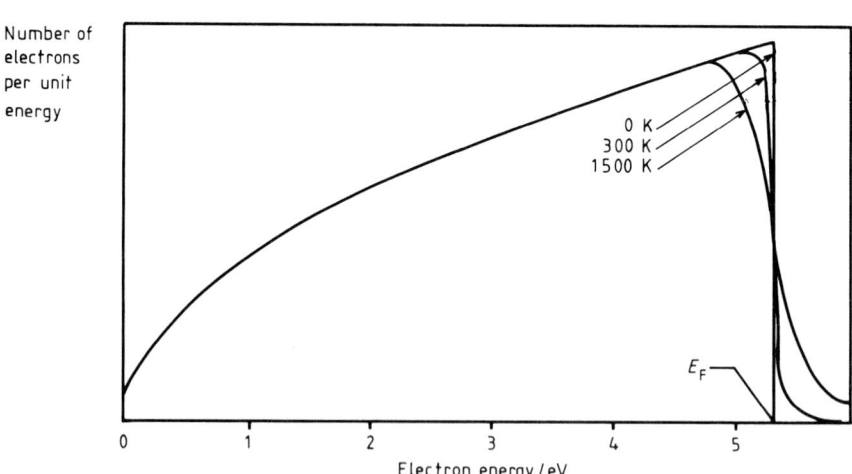

Figure 10.4 Fermi–Dirac energy distribution for conduction electrons in platinum at three different temperatures. (Adapted from Richtmeyer *et al* (1955), courtesy McGraw-Hill Book Company Inc.)

maximum emission energy is due to the finite sensitivity of experimental apparatus and the extremely rapid decrease in the number of fast electrons with increasing energy. A prime achievement of modern photoelectric research has been to give a definite and physical interpretation to the concepts of 'threshold frequency' and 'maximum kinetic energy of emission'.

Modern theories not assuming quantisation of the electromagnetic field

This is the deepest and most difficult of all the issues raised in this chapter. In view of the complexity it is inappropriate to attempt a full discussion here. Suffice it to say that the issue has a long pedigree, as may be judged from the following three quotations made from significantly different historical standpoints:

> Curiously enough, however, it has recently been shown that the Einstein equation *can* be derived on the basis of the wave theory of light—but only if we also adopt a wave theory of the electron! One may say, then, that the photoelectric effect, as well as its close relative the Compton effect, can be understood if both light and electrons are corpuscles, or if they are both waves, but not if one is a corpuscle while the other is a wave!
>
> (Hughes and Dubridge 1932)

> ... it is possible, even today, to account for the main features of the photoelectric effect along non-Einsteinian lines, that is, without assuming light quanta or photons to be incident on the atom.
>
> (Stuewer 1970)

> The status of theories, such as the semi-classical theory and the neo-classical theory... is now more parlous than it was in the mid-1970's.... But notice that the growing body of evidence for the necessity of field-quantisation
> (i) comes from experiments which involve the photoelectric effect, but are vastly more subtle than simple observations of the kind made by Millikan: all these 'strong' experiments involve a very rigorous analysis of correlations of the fields at different space–time points;
> (ii) these experiments, and their interpretations, do not validate the simple-minded picture of photons-in-flight. Quantisation manifests itself only when and where matter and radiation interact, and the question whether photons exist in the empty space between places where radiation and matter interact is really metaphysical rather than physical.
>
> (Sillito 1984)

I have endeavoured to show in this chapter that there are several 'grey areas' associated with the photoelectric effect on which light is not easily shed. Textbook inadequacies are the rule rather than the exception, and continue to be propagated with remarkable fidelity.

This is an appropriate place at which to take a fresh historical look at the development of the concepts inherent in the phenomenon, for its centenary year has just passed. By universal agreement the discovery of the effect is attributed to Heinrich Hertz (1857–94) in 1887 (Hertz 1893). (It is probable that the crucial entry in Hertz's diary was made on 3 December 1886 (Stuewer 1970).) By 1902 the investigations of Wilhelm Hallwachs (1859–1922), J P Elster (1854–1920) and F K Geitel (1855–1923), J J Thomson (1856–1940) and Philip von Lenard (1862–1947) had established all the salient features of the effect, save the linear relationship between maximum kinetic energy of emission and photon frequency. Taking in particular Lenard's results, Einstein made his famous conjecture in 1905 and predicted the said linearity. It is fascinating that Millikan approached his experimental research into the energy–frequency relationship in classic Popperian fashion. He was firmly convinced that he would falsify the prediction. In his *Nobel Lecture* of 23 May 1924 he says (Millikan 1965)

> After ten years of testing and changing and learning and sometimes blundering, all efforts being directed from the first toward the accurate experimental measurement of the energies of emission of photoelectrons, now as a function of temperature, now of wavelength, now of material (contact e.m.f. relations), this work resulted, contrary to my own expectation, in the first direct experimental proof in 1914 of the exact validity, within narrow limits of experimental error, of the Einstein equation, and the first direct photoelectric determination of Planck's h.

Even after his experimental accomplishment Millikan had distinct reservations regarding the implications of his work in relation to Einstein's theory. In 1917 he wrote (Millikan 1963)

> Despite then the apparently complete success of the Einstein equation, the physical theory of which it was designed to be the symbolic expression is found so untenable that Einstein himself, I believe, no longer holds to it, and we are in the position of having built a very perfect structure and then knocked out entirely the underpinning without causing the building to fall. It stands complete and apparently well tested, but without any visible means of support. These supports must obviously exist, and the most fascinating problem of modern physics is to find them. Experiment has outrun theory, or, better, guided by erroneous theory, it has discovered relationships which seem to be of the greatest interest and importance, but the reasons for them are as yet not at all understood.

The Photoelectric Effect

In the course of time, however, Millikan's experiments have come to be regarded as sound and precise confirmation of Einstein's hypothesis of light quanta as embodied in his photoelectric equation. This is certainly the official textbook view, but is it the consensus view? Strong dissent has recently been registered by R Keesing (1981, p. 148, 1987) who has challenged the validity of both Einstein's theory and Millikan's experimental results. Various considerations, notably the blurring effect of temperature on the Fermi–Dirac distribution (figure 10.4) led him to comment (Keesing 1981 p. 148)

> Acceptance of the strange results obtained by Millikan and others up until 1930 have had a number of unfortunate consequences, which include the following.
> (i) The use of Einstein's theory of photoemission from metals to determine Planck's constant, when in fact it is not applicable because of the unobservable nature of the stopping potential.
> (ii) For some time it was believed that photoelectrons and conduction electrons originated from different states in the metal. However this is now generally accepted not to be the case.
> (iii) Several generations of undergraduate text-book have made claims about the photoelectric effect which are not borne out by direct experiment and are incompatible with other branches of physics.

Later in the same article he concludes (Keesing 1981 p. 149)

> However there has been considerable diffidence in coming to the unavoidable conclusion that Einstein's equation for photoemission from metals is simply incorrect. The cardinal point of the theory, that is the postulate that $E = h\nu$ is without doubt correct: the irony of the situation is that the application of the postulate to photoemission from metals and the experimental verification were both in error.

The current crisis in physics teaching is well known. There is, I would suggest, a deeper crisis, that of the integrity with which the concepts and ideas of physics are communicated by specialists and authors through teachers to students. Perhaps we have been using sticking-plaster remedies long past the point where surgery is required. If the case presented here is valid, and the photoelectric effect is a candidate for the surgeon's knife, then one of its edges at least must be historically sharp.

Acknowledgments

I wish to record a special debt of gratitude to the four correspondents who responded so fully and so readily to my queries. They are

Professor P J Black, Chelsea College, University of London, Dr R G Keesing, Department of Physics, University of York, Dr R M Sillito, Department of Physics, University of Edinburgh and Dr J W Warren, Department of Physics, Brunel University. The views expressed in this chapter, however, remain exclusively the responsibility of the author.

In addition I would like to acknowledge the part played by many colleagues and students over the years who have stimulated and challenged my comprehension of physics, and especially my own teacher Mr Harry Lodge.

References

Arons A B and Peppard M B 1965 Einstein's proposal of the photon concept —a translation of the *Annalen der Physik* paper of 1905 *Am.J. Phys.* **33** 367–74

Beiser A 1962 *The Mainstream of Physics* (Reading, MA: Addison-Wesley) p. 333

Bitter F and Medicus H A 1973 *Fields and Particles* (New York: Elsevier) p. 149

Bleaney B I and Bleaney B 1965 *Electricity and Magnetism* (Oxford: Oxford University Press) pp 95–7

Collocott T C and Dobson A B (eds) 1974 *Chambers Dictionary of Science and Technology* (Edinburgh: Chambers) p. 1278

Compton A H 1923a Wave-length measurements of scattered X-rays *Phys. Rev.* **21** 715

—— 1923b The spectrum of scattered X-rays *Phys. Rev.* **22** 409–13

Compton K T 1912 The influence of contact difference of potential between the plates emitting and receiving electrons liberated by UV light on the measurement of the velocities of these electrons *Phil. Mag.* **23** 579–93

French A P (ed.) 1979 *Einstein: A Centenary Volume* (London: Heinemann) p. 139

Hertz H 1893 *Electric Waves* trans. D E Jones (London: Macmillan) pp 63–79

Hodgson E R and Lambert R K 1975 Letter in response to note 19 *Phys. Ed.* **10** 123–4

Hughes A L and DuBridge L A 1932 *Photoelectric Phenomena* (New York: McGraw-Hill) p. 21

Isaacs A (ed.) 1985 *Concise Dictionary of Physics* (Oxford: Oxford University Press) p. 289

James A N 1973 Photoelectric effect, a common fundamental error *Phys. Ed.* **8** 382–4

Keesing R G 1981 The measurement of Planck's constant using the visible photoelectric effect *Eur. J. Phys.* **2** 139–49

—— 1987 Photoemission from metals: mistakes arising from Millikan's erroneous experiment *School Sci. Rev.* **68** 447–55

Maurer R J 1967 *Handbook of Physics* ed. E U Condon and H Odishaw (New York: McGraw-Hill) pp 8-67–8-75

Millikan R A 1914 A direct determination of 'h' *Phys. Rev.* **4** 73–5
—— 1916 A direct photoelectric determination of Planck's 'h' *Phys. Rev.* **7** 355–88
—— 1963 *The Electron* (Chicago: University of Chicago Press) p. 230
—— 1965 *Nobel Lectures: Physics 1922–1941* (Amsterdam: Elsevier) pp 61–2
Nelkon M and Parker P 1979 *Advanced Level Physics* 4th edn (London: Heinemann) pp 918–19
Noakes G R 1956 *A Text Book of Electricity and Magnetism* 3rd edn (London: Macmillan) p. 385
Open University 1971 *Science Foundation Course* S100 (London: Oxley Press) Unit 29 p. 26
Richardson O W and Compton K T 1912 The photoelectric effect *Phil. Mag.* **24** 575–94
Richtmeyer F K, Kennard E H and Lauritsen T 1955 *Introduction to Modern Physics* 5th edn (New York: McGraw-Hill) pp 97–104
Roche J 1984 History as surgery *Phys. Bull.* **35** 414–15
Sillito R M 1984 Private communication
Stuewer R H 1970 Non-Einsteinian interpretations of the photoelectric effect *Minnesota Stud. Phil. Sci.* **5** 246–63
—— 1971 Hertz's discovery of the photoelectric effect *Actes XIIIe Congr. Int. Hist. Sci.* **6** 35–43
Trigg G L 1971 *Crucial Experiments in Modern Physics* (New York: Van Nostrand Reinhold) pp 76–87

11

Physics and Society†

Charles Boyle

> ... there is a tendency to forget that all science is bound up with human culture in general, and that scientific findings, even those which at the moment appear the most advanced and esoteric and difficult to grasp, are meaningless outside their cultural context.
>
> Erwin Schrödinger (1887–1961)

Introduction

It is perhaps to be expected that the main activities of historians of physics will centre round the analysis and discussion of significant specific developments in physics in definite periods in the past—the emergence of a new physical concept, for example, or the experimental investigations of a newly discovered phenomenon. But it may be useful to take a broad overview of the growth of physics in the past three or four hundred years, emphasising the societal context in which this growth took place and identifying a few salient features of the evolving relationship between physics and society. This chapter is written not from the viewpoint of the specialist historian but rather from the viewpoint of someone primarily interested in the many and varied interactions of physics with society in the present who would like some light thrown on the historical background to these interactions. And so, instead of providing answers (except tentatively in a few cases) this

† Based on a presentation given at a conference on Projects in the History of Physics 10 July 1985, Oxford.

chapter puts forward a series of questions for the historian to consider and suggests possible lines of research.

Why bother to look at the relationship of physics to society at all? Previous generations of historians of science felt that perfectly adequate accounts of the development of physics could be given without much reference to the cultural, political or economic environment. The assumption of these historians seemed to be, that in their scientific work, physicists (and, earlier, natural philosophers), sealed in ivory-tower laboratories, were immune to pressures and influences from the outside world; or, if they were not, it was of no concern to the historian of physics who had no interest in the ultimate origins of physical concepts, but only in their refinement and use. This is to say that physics was to be described as growing mysteriously like some self-sufficient plant that requires no inputs from its surroundings, but bears within itself all it needs for its own development. This approach has the unfortunate consequence of reinforcing the general public's view of scientists as a race apart, aloof, uninterested in other people's concerns, secretive and perhaps not to be trusted.

But we know that many of the greatest physicists were passionately interested in contemporary philosophical and religious questions that went far beyond the confines of their science, and they were unavoidably prey to the beliefs or prejudices (as we may judge them later to be) of their time. And we know too that physics, in anything like its modern form, has not existed in all cultures at all times. It required a special type of tolerant and relatively open intellectual climate in European society in the seventeenth century for the first great physical theories to emerge. From humble origins a giant has grown. The advanced industrial nations now spend one per cent or more of their gross national products (GNP) on physical sciences and on research into and development of physics-related technologies.

Thus physics has been profoundly influenced by society at large both in its content and in its mode of development. Ideas and beliefs from the wider society, no doubt much modified, find their way into the physicists' conceptual schemes (from which, again transformed, they may flow back into society). Society's material support for the scientific enterprise is also essential; it is differential support, some fields being well funded or otherwise encouraged while others are constrained almost to extinction.

Quite detailed and interesting internalist histories have of course been written on the work of individual physicists or on famous episodes, paying scant attention to the social or economic context. But history is more than the reconstruction of past events and the telling of stories, however detailed, about how this or that discovery was made. It is to do with interpretation, and (like physics itself) with testable

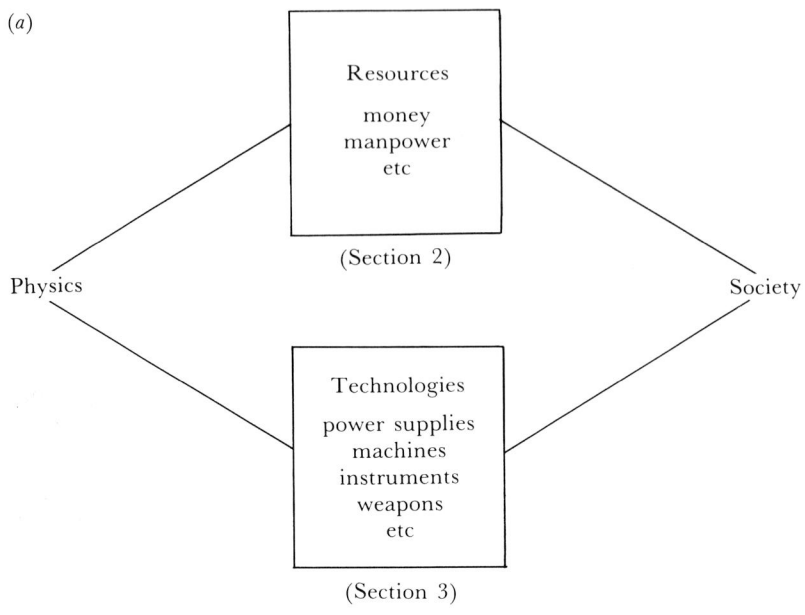
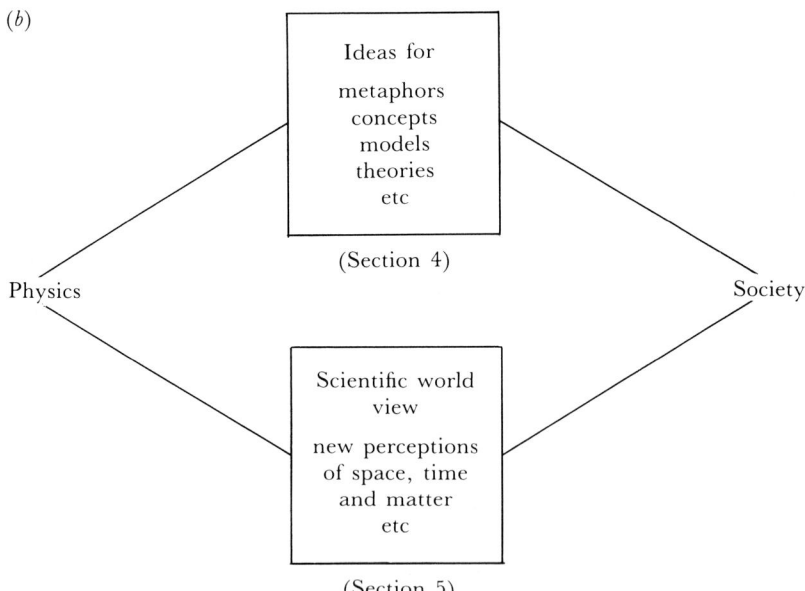

Figure 11.1 Interactions of physics and society. (*a*) Material level, (*b*) ideological level.

theories involving abstract concepts to identify and describe patterns of movement and change and to link phenomena which are at first sight disparate. The real challenge for the historian then is to provide an analysis that locates both the content and the resources of the natural philosophy or physics of a particular period firmly within the cultural context of that period in all its philosophical, social, political and economic dimensions. This is not an easy task.

Figures 11.1(*a*) and (*b*) illustrate a simple model of the interactions of physics and society which may provide a convenient starting point. The question then is to ask how this model has changed over time. Interactions at two different levels are considered. At the material level we have the flow of financial and other resources from various funding agencies to the community of physicists. This 'input' includes young researchers who have passed through the educational system. The 'output' is a wide variety of physics-related technologies ranging from television sets to nuclear weapons, though of course these end products incorporate much more than just physics. Equally important, however, are the interactions at the level of ideas—the concepts taken up in some mysterious ways by individual physicists from the intellectual currency of contemporary society and hammered out by the scientific community into the basic theories of physics, to return again eventually to the wider culture in a completely new form. To complete the model interactions between the two levels should also be taken into account.

Resources for physics

The most obvious way in which society affects physics is by funding it. Until the nineteenth century these funds came largely from the pockets of a few individuals, usually the scientists themselves or wealthy patrons. Royal grants were significant in France in the eighteenth century in supporting the work of the Académie des Sciences. During this period in England the Royal Society declined as the influence of wealthy Fellows with little serious interest in natural philosophy increased. English scientists were usually amateurs whose scientific work had to be done in free time left over from other activities.

The professionalisation of scientists, including physicists, occurred slowly in the nineteenth century and at different rates in different countries. It was linked to developments in education which gave science a more prominent place in the curriculum both in secondary and higher education and which generated a need for teachers and researchers. These developments were linked in turn to the growing need by industry for scientifically qualified staff. In the early phase of the Industrial Revolution science played an indirect role only, and it was not until the latter part of the nineteenth century that science-based, rather than craft-based, innovations became important, the work of physicists being particularly significant for the infant electrical engineering industries.

The idea of the modern research laboratory with teams of researchers working together systematically on projects using the best equipment available, can be traced back to the *New Atlantis* of Francis Bacon (1561–1626), published in 1624. It was in nineteenth-century Germany, however, that research laboratories in something approaching their modern form were established, both in universities and industry. Germany inherited world leadership in science from France during this period and this organisational innovation, later widely copied in other countries, may partly explain German supremacy. Not only is the level of resources important for the successful growth of physics, but also the organisation of these resources. Popular histories of physics are usually centred round a few outstanding and charismatic individuals. Almost as important, however, are the many less-well-known members of the scientific community who in their experiments and discussions modify, refine and apply the concepts of the great innovatory geniuses.

Even towards the end of the nineteenth century spending on physics research was insignificant by modern standards, though it had increased dramatically. The work of Price in the 1960s drew attention to the rapidity of the growth of science since the seventeenth century. On the basis of studies of numbers of scientists, scientific journals and papers in those journals he estimated that an exponential growth had occurred with a doubling period of about 15 years. The doubling period for expenditure on physics was probably considerably shorter than this because of the increasing cost of experimental apparatus as it became more complex.

By 1930 the USA, then as now the biggest research and development (R and D) spender and in both relative and absolute terms, was spending about 0.2% of its GNP on all scientific R and D. It was World War II, however, which was to provide an enormous boost, sustained well into the post-war period, to physics research funds. Physicists had been spectacularly successful not only in the Manhattan Project but also in other fields such as radar research, and took advantage of the high public esteem in which they were held to ask for and obtain funds for ambitious programmes, particularly in high-energy physics, nuclear physics and later radioastronomy.

The period since 1940 is often referred to as the period of big science, where 'big' refers not only to expenditure but also to total manpower, to the projects undertaken and to the experimental apparatus used. Only the largest and richest nations can afford to go it alone in some of the most expensive fields. For medium-sized or small industrial countries international cooperation has for some time been essential, in organisations and projects increasingly reduced to sets of initials, such as CERN, JET, ESPRIT, ESA and RACE.†

If one of the noteworthy features of modern physics is its internationalisation through its involvement in huge cooperative projects (though of course physics has always been international in a broader sense), then another is its militarisation. Military R and D accounts for more than half of all government R and D spending in the USA and the UK, and employs very large numbers of physicists. And it is worth remembering that the US Strategic Defense Initiative ('Star Wars') is the largest single technological project ever planned in the history of the human race.

The last decade has also seen in many countries and especially in the UK a shift in attitudes towards the provision of public resources for science. Cases arguing for a high level of support for physics research because of its pursuit of ultimate truth or its cultural value have fallen on ears less and less willing to listen, unless economic benefits can be demonstrated to be highly likely to flow from the research. Pressures have grown to fund only research that is industrially relevant—perceptibly 'strategic' or 'applied' rather than 'pure'. Thus in the UK over the last 20 years the percentage share of the Science and Engineering Research Council budget assigned to nuclear physics has declined while that given to engineering has increased.

Historians have considered many aspects of the issue of financial and other material support for research, but we may identify three main types of question for further study. The first type relates to the size of the science 'cake' and the size of the 'slice' given to physics. To what extent can we attribute leadership in physics by particular institutions and particular nations at different times to abundance of resources? Does quantity imply quality? Under what circumstances has physics enjoyed good support? How has it fared in relation to the other sciences? The second type of question, which is closely linked to the first, concerns the effects on the development of physics of resources being scarce and hence restraining growth, or relatively abundant, for some types of work rather than others? To what extent, in other words, have the individuals or authorities which have provided funds, rather

†CERN, Conseil Européen pour la Recherche Nucleaire; JET, Joint European Torus; ESPRIT, European Strategic Programme for Research and Development in Information Technology; ESA, European Space Agency; RACE, Research and Development in Advanced Communications Technologies in Europe.

than physicists themselves, laid down the agenda for research? Have military and industrial paymasters, for example, in some way distorted physics, diverting effort away from the goals of pure research? Lastly there is the whole question of physics education, both for the community at large and for the future physicist. Much work has already been done on the history of science education; but perhaps there is a need to draw together the experiences of different countries particularly in the field of the public understanding of science, and to ask how, if at all, in the past physics has interacted with other subjects in the curriculum, and how public consciousness of physics has changed.

Physics and technology

We regard the key technologies of earlier phases of human history as sufficiently important to use them as labels for those periods—the Stone, Bronze and Iron Ages. However, few historians now argue the case for pure technological determinism—the theory that the life and structure of a society are ultimately fully determined by the technology used by that society. Nevertheless no adequate account of a society can be given without paying attention to the technology it develops and uses. Technology interacts with economic, political and social factors to shape society; and, since the days of its origins, but particularly in the last hundred years, physics has had a leading role in shaping that technology.

To judge by the statements of many people prominent in the science policy field it is still a widely held belief that a straightforward linear progression exists from pure research in physics to applied research, then to development and finally to the manufacture of new products and the introduction of new processes. At first sight it seems a straight line leads from the cathode rays of J J Thomson (1856–1940) to the modern television set. But in fact the situation is much more complicated.

The story of the emergence of thermodynamics in the nineteenth century as a major field in physics illustrates a sort of 'reverse process'. Here the technology of the steam engine preceded the pure physics. Primitive steam engines in the early eighteenth century were primarily the work of self-educated engineers with craft-based rather than science-based knowledge, and they were improved largely by crude trial-and-error methods though some scientific input in the context of the caloric theory of heat came from Watt and others. The steam engine had reached a high level of development before the work of S Carnot (1796–1832) in the 1820s, and that of J P Joule (1818–89) and

Lord Kelvin (William Thomson 1824–1907) later, laid the theoretical foundations of modern thermodynamics.

The steam engine itself was invented, some historians have suggested, primarily because of the need of society for a source of power more concentrated than human and animal muscle, and less tied down to particular locations than the water wheel. This school of historians considers science as arising ultimately from economic forces and argues that even the 'purest' work of Newton must be seen in the context of the needs of seventeenth-century mercantile capitalism.

Simple linear models, however, linking pure physics, technology and society in one direction of development or the other, are quite inadequate to account for the complexities that can be discerned when anything more than a superficial analysis of particular case studies of technological innovation is undertaken. It is certainly true that without pure physics—the special theory of relativity or quantum mechanics for example—many of today's technological wonders could not have come into being. In this sense the broad basic theories of physics underpin most modern technological developments. Equally, however, without economic forces such as market demand, without potential sales on a large scale, the applied physicists' most brilliant ideas would not have materialised. Scientific or technological push and economic pull are both necessary for successful innovation. The broad line joining Thomson's cathode rays to the modern television set has, on closer inspection, a very complex fine structure. It is composed of branches, twigs and networks that divide up and rejoin; it is really only one of a series of similar lines, more or less parallel, which eventually converge. These other lines have starting points at Maxwell's theory of electromagnetic waves, or Hertz's experiments or the concepts of Faraday. These pass through clusters of ideas marked Bohr's theory of the atom, Schrödinger's wave equation and so on. But many lines can be traced back not to theories of physics but rather to technological and engineering challenges (such as, for example, how to make very pure semiconductors), the responses to which open up new theoretical fields or provide experimentalists with new tools and materials which extend the range of their investigations. Our notional diagram contains the names of Nobel prize-winning physicists but the finer print will reveal teams of minor figures and technicians who were equally indispensible, not to mention others with no scientific knowledge at all. Some of the historical linkages along the networks of lines seem well planned and predictable; others seem arbitrary, depending on chance meetings, fortuitous discoveries and accidents of fate.

As we have seen in the last section the provision of funds for physics research by government and industry is linked increasingly closely to the possibility of generating economically or militarily successful products and process innovations, and pure researchers are forced to find arguments in these terms to justify their work. The assessment of the relative importance of pure research on technological innovation is a controversial and complex operation. Different research studies have come up with different conclusions which often reflect the economic interests or disciplinary bias of the researchers.

In the 1960s in the USA, for example, the Department of Defense set up Project Hindsight to examine the extent to which pure and applied scientific research contributed to the production of new weapons systems. This study concluded that scientific research, and pure basic research in particular, were relatively unimportant as sources of ideas for weapons development, and that money earmarked for pure research would be better spent on applied or mission-oriented development work. The National Science Foundation responded to this playing-down of pure science by carrying out a study called 'Technology in retrospect and critical events in science' (TRACES), which found that a group of major twentieth-century innovations could be reduced to a series of key events of which 70 per cent were classifiable as 'pure research', 20 per cent were 'mission-oriented' and only 7 per cent were 'development and application' (Ronayne 1984).

Other studies have used different methods of approach and different selections of technology for analysis. All such studies are beset by semantic problems (how do you ultimately distinguish 'pure' from 'applied' research?) and by other difficulties in identifying, classifying and quantifying 'key events'. What emerges is a picture of great complexity in which ammunition for argument can be found by all sides.

Some historians have identified a gulf between pure and applied physics in English universities in the past, with leading professors sharing a gentlemanly belief in the innate superiority of the quest for pure knowledge and some distaste for the vulgar exploitation of physics in the market place. This is one aspect perhaps of an anti-industrial bias in English culture from the mid-nineteenth century onwards which was not present (or at least not present to the same extent) in Germany, France and the USA. At the moment in the UK the emphasis is swinging the other way, though it should be noted that Japan, which has built up a very strong technological position in many fields without a strong pure research base in the past, is now rapidly expanding its pure research programmes. How much of Britain's effort should go into pure physics? This is an intriguing and important question. Though much work has already been done further case studies of the relation between physics and technology in different countries at different periods can provide valuable food for thought about present policy making.

The origins of physics in society

Of all the areas involved in the consideration of the interactions of physics and society the study of the roots of the ideas of physics in society is perhaps the most difficult and least developed one. It is usually much easier to assess the impact on society of views and concepts that come from physics than it is to trace the origins in society of the forerunners and, as it were, the ancestors of these concepts. And so it is this study of the emergence of physics from complexes of pre-scientific, religious and philosophical thought and of the development of the physics of any particular period as part of the cultural and intellectual life of the period, that presents the biggest challenge of all to historians and epistemologists. What follows below raises questions rather than provides answers.

One set of questions surrounds the emergence of the fundamental concepts of physics—concepts of space, time, matter, electricity, force, transformation, conservation, scientific law and so on. Many of these concepts have of course undergone considerable modification even during the period of the mature development of physics of the last hundred years or so, and will no doubt be modified further in the future. These shifts of meaning and emphasis have been well studied by physics historians. What is less clear, however, is how the concepts in their primitive form came into being in the first place. To read even a few pages of Francis Bacon or Robert Boyle is to realise the enormous differences in background assumptions, modes of thought and vocabulary that separate the seventeenth century from the twentieth and to marvel at the gulf that has been bridged.

Physics textbooks tend to give a grossly over-simplified picture of the early history of physics when they mention it at all. Physics is presented as a body of knowledge which has started from a few clear and simple experiments and ideas, and then grown cumulatively and unproblematically as more and more 'facts' are discovered and added to it. The complexity of the thought in which the early ideas were embedded is ignored; in this respect it is significant that the large body of writing by Newton on religious topics and on alchemy has been almost totally neglected even by historians of science until relatively recently.

T S Kuhn, whose work has been very influential in the Anglo-Saxon world, identifies the birth of a 'real' science with the formulation and articulation of a 'paradigm', a set of theories, examples and rules generally accepted by the leading thinkers of the day in that field, which lays down guidelines for future research. But he has not devoted a great deal of attention to the period of confused pre-paradigm research when there are almost as many competing and incompatible theories as there are researchers. The French philosopher of science, Gaston Bachelard (1884–1962), has written (Bachelard 1972) of the struggle of modern physical thought to overcome what he has called 'epistemological obstacles' (such as, for example, animist ideas, purely verbal explanations, excessive emphasis on generality and other mistaken preconceptions). Bachelard has suggested a sort of psychoanalysis of objective knowledge to throw light on how 'errors' have been circumvented. Neither of these writers, however, has discussed in detail the linkages between specific developments in general culture and in the growth of the embryo that was later to become physics.

More work has been done at the macroscopic level—that is in trying to understand possible broad mutual influences between physics and two other major systems for organising thought (and life) that emerged slowly but more or less contemporaneously in Western Europe in the sixteenth and seventeenth centuries—namely Protestantism and capitalism. Suffice it to say here that there is no general agreement among scholars about either the extent or the nature of these influences, though there seems to have been a quite disproportionate number of leading English natural philosophers from Puritan backgrounds, right into the eighteenth century.

Reflections on the relations between religion and the rise of capitalism and modern science prompt other intriguing questions: why did science, and specifically physics, emerge in its modern form only in the seventeenth century and not earlier? And why in Europe rather than in China, say, which was in 1600 in many ways materially, organisationally and technologically superior to Europe? After all, gunpowder, the printing press and the magnetic compass—the three inventions Bacon cited in arguing (against strong opposition) that the civilisation of his time was higher than that of the ancient Roman Empire—all had their origins in discoveries made a considerable time previously in China.

For a further example of the sort of questions historians need to examine if we are fully to comprehend the social relations of physics, let us look at an historical era much closer to the present day—the first 25 or 30 years of this century. More than any other of comparable length of time, this period was one of revolutionary change in physics. The complacent 'certainties' of classical physics towards the end of the nineteenth century were overturned. Newtonian ideas of absolute time and space previously unquestioned were undermined by Einstein's theories of relativity. The Heisenberg uncertainty principle introduced notions of indeterminism that would have appeared unthinkable to physicists of 50 years earlier. The investigation of radioactivity led ultimately to a whole new and unsuspected field of study—nuclear physics.

Is it purely accidental that this was also a period of unprecedented revolution in the arts? Freud's psychoanalytic theories created reverberations in all corners of society. In the world of painting the post-impressionists, the cubists and the surrealists trampled the hallowed tenets of classical painting underfoot. In the literary world writers as varied as Proust, Eliot, Joyce, Kafka and Pound introduced radically new techniques and forms. Composers such as Schoenberg and Stravinsky broke with tradition in the musical world.

Much has been written on the history of physics between 1900 and 1930 and many volumes more about the arts in this period. And yet I know of no sustained study which tries to link 'the two cultures' and to examine physics in the general intellectual context of the day, apart from one praiseworthy though controversial attempt to locate the development of the uncertainty principle in the political and cultural context of Germany in the 1920s (Forman 1971).

One final question may be asked that relates to various themes considered above. Some writers, concerned with the way physics is bound up with and used for so many essentially destructive and repressive purposes in both capitalist and communist states in the modern world, and concerned too with the degradation of the environment by an increasing number and volume of pollutants, have suggested that physics emerged from an aggressive, domineering and masculine as opposed to a yielding, cooperative and feminine approach to nature. Most physicists are, and have been, men. Seventeenth-century writing on natural philosophy, it is argued, reveals a characterisation of nature as female and the process of research as the exploration and penetration of her innermost secrets.† The sexual imagery is sometimes explicit. Other authors have presented nature as a source of wealth provided by God for man to do with as he pleases. Could a different type of society in the seventeenth century and different types of natural philosophers have produced a different type of physics, less exploitative and violent, more in tune with the natural environment? Is masculine aggressiveness in some way built into our physics? Is some other type of physics still possible?

The scientific world view

The successes of seventeenth-century physics ultimately had the effect of establishing a profoundly new world view in Europe that largely if not totally displaced the religious outlook that had permeated all mediaeval life. This scientific world view was accepted only very

† See, for example, Easlea (1981).

slowly, and at first by only a small minority, but it has since gradually established a hegemony over more and more fields of thought and activity in a wide variety of cultures, though it is still challenged from time to time by religious fundamentalist movements and by humanist critiques of science.

The natural philosophy of the seventeenth century as opposed to that of earlier centuries placed a central emphasis on the bringing together of analysis, experiment and mathematics as a means of approaching natural phenomena. If it failed in many areas to break through to modern science, at least in the fields of mechanics and astronomy it achieved spectacular successes which later in the eighteenth century deeply influenced many non-scientific thinkers. It sought answers to questions relating to the natural world not in the Bible nor in biblical commentaries nor in the writings of established authorities who were seen as theologically acceptable, but rather in precise analyses of 'simple' concepts amenable to mathematics and experimental measurements. The work of Newton in uniting explanations of terrestrial and celestial motions under one and the same set of laws had a deep impression on far greater numbers of people than the select few who knew enough to appreciate fully its mathematical elegance and exactness.

Most of the seventeenth-century natural philosophers were deeply religious, but the ultimate effect of their work and that of their successors was to undermine the authority of the Church and of the Bible. From the discussions of the Copernican helio-centric system to those of the geological theories that underpinned Darwin's theory of evolution there was a long series of more or less direct confrontations with the Church. These usually resulted in the Church retreating from positions that defended the whole of the Bible as literal truth to acceptance of increasingly large parts of the scriptures as embodying spiritual verities only metaphorically, and providing little enlightenment about the physical universe.

The early physicists were instrumental in establishing the idea of human progress, though historians can no doubt trace other additional sources of this concept. From the idea of progress it is a short step to political demands for social and material improvements. Many religious teachings are presented as timeless and unchanging truths; even in turbulent times changes are regarded as superficial and limited—ripples on the surface of a largely static world, with a fixed social order given by God. By challenging religious teaching, however circumspectly, but above all by generating a completely new type of knowledge, a knowledge moreover that was open-ended and provisional, the natural philosophers opened up the possibility of futures for the human race that were radically different from its past. The

French *philosophes* in the eighteenth century and a broad range of other optimistic thinkers later spelled out the social and political implications of these ideas.

We have suggested that early physics generated sweeping changes of outlook in society; it is more accurate perhaps to say that it formed one important current of thought that interacted with others eventually to produce the swirling stream of modern culture, a stream which itself, like a river flowing over a bed of shifting sands, has both shaped and been shaped by economic and political events. With these limitations in mind we may identify three main features of the scientific world view promoted by the early physicists: it was mathematical, analytical and experimental.

The idea that numbers are the key to the understanding of nature dates back at least to the time of Pythagoras. Regularities in nature, it was asserted, could be linked to *mathematics* and studied by means of careful measurement. Emphasis on measurable quantities led to the increased importance for descriptive purposes of the primary qualities of bodies (size, shape, state of motion) and a relative neglect of secondary qualities (colour, smell, sound emitted) and other attributes which could not be quantified. The use of mathematics encouraged precision in thought, the refinement of vague concepts, and a move away from animistic and 'supernatural' notions. (The processes of the life cycle—generation, birth, growth, death and decay—were regarded as providing models for many natural phenomena we now see as having nothing to do with life—for example, the production of ores under the earth's surface.)

The new science was *analytical* in that it sought to study isolated, idealised and simple systems which could be described using a few variables only. One might think of Newton's particles, point-masses at a certain distance apart, mutually attracting one another but unaffected by the rest of the world. From these simple systems more complex ones could be built up that approximated better to certain parts of the real world. This sort of approach reflected the views of Descartes, but was at odds with methods which stressed the collection and classification of the varied and voluminous writings of scholars down the ages on any given subject, as a preliminary to the understanding of that subject. It is not easy of course to arrive at the simplifying assumptions which allow one to extract meaningful material from huge, heterogeneous aggregations of knowledge that lump together the physical properties of objects, literary descriptions of them, their attributes in various mythical systems, their role and significance in religion, animist beliefs relating to them and so on.

Inevitably, the use of mathematics and analysis encouraged an *experimental* approach, which rejected much traditional and book learning in favour of direct appeals to experience. Aristocratic disdain for craftsmen and others who worked with their hands was set aside, new experiments were devised to test hypotheses and better instruments designed to translate observations into numbers. Extraordinary observations listed in records from the past were treated with increasing scepticism if they could not be reproduced. The results of experiments were discussed, published and disseminated more openly and widely than previously. The work of the alchemists was continued but in a rather different spirit, a spirit of independent enquiry unafraid of insulting nature by tinkering with it.

Ultimately, it can be argued, the early natural philosophers were motivated by the desire to gain much greater control over nature than had been possible before, and their methods were adapted to this end. For our purposes here the most significant aspect of what we have called the new scientific world view was that it could be applied to many other fields of human activity as well as astronomy and mechanics. We may think first of its influence in accelerating the development of other branches of the natural sciences—chemistry, electromagnetism and the earth sciences in the nineteenth century, the life sciences in more recent years—but its effect went far beyond science. In some ways the history of the last 300 years can be seen as the history of the 'colonisation' of all sorts of different activities by scientific methods and attitudes. Thus the Industrial Revolution can be considered as the result of gradually applying scientific thinking to methods of production. The organisation of work, the management of people, the planning of technology, the bureaucracy of the state, the waging of war, agriculture, health care, education, sport (the list can be continued almost indefinitely) have all been, and are still being, transformed by analytical mathematical and experimental methods that have their origins in the physics of the seventeenth century. This is true whether we look at broad generalities or the specifics of, say, the marketing of a new soap powder or the effort of an athlete to improve his or her performance. Essentially the process is one of breaking down the problem into various elements, attempting to quantify and measure the significant variables, and testing a series of possible solutions that have not been thought of before. The computer is the ultimate instrument that incorporates this scientific world view.

This world view has of course its weaknesses, and its application has entailed costs as well as benefits. It ushered in a picture of a secularised and fragmented world of colourless, odourless matter in motion. Its reductionist philosophy squeezed out concepts of wholeness. It sought to banish dreams, emotions, mystery, all the elements of the poetic imagination, and in exorcising devils and ogres it did away with guardian angels too. It could not cope with concepts such as joy, love and beauty which are central to human life but not amenable to mathematical analysis. In both its strengths and its failures its impact has been profound.

Physics and society and the future

This brief account, though it has tried to examine some general questions, has inevitably looked at only a few of the complex processes occurring at the physics–society interface. Many relevant and important ideas have scarcely been mentioned—for example the concept of law as it applies in physics and in society, and the concepts of equilibrium or balance and of change whether gradual, explicable in terms of small perturbations and slight departures from equilibrium or swift and radical leading to a completely new state. Cause and effect, determinism, reductionism and the nature of explanation are other topics for the discussion of which space should perhaps have been found.

The main aim of this account of physics and society has been, however, to stress the importance of this subject. Specialists are no doubt needed at the frontiers of research, but even they should have some appreciation of the multitude of often unpredictable social consequences of science. It is now no longer morally acceptable, if it ever was, for scientists naively to absolve themselves from all responsibility for the uses to which their results are put.

The applications of physics, like those of many other fields of knowledge in the modern world, are rapidly changing and increasingly complex. These applications form parts of systems that are more and more interdependent. It is imperative that physicists be fully aware of the social, economic and political forces at work at national and international levels shaping the type of work they are required to do and the uses to which it is put.

The result of the increasing complexity and interdependence of the various technological interests at play in modern society can be perceived as greater control, wealth and efficiency if certain criteria are applied, but if other criteria are applied as just the opposite. A good image to illustrate this point (and one that is often used) is that of the astronaut in space exercising the power miraculously to live and move where none could go before, but at the same time totally vulnerable to any one of a large number of possible failures in his support systems. All individuals in the modern world are becoming more like the astronaut—with immense external power at their fingertips but no autonomy, and utterly lost if this power fails.

These sorts of considerations though obvious enough are often forgotten by the individual physicist in everyday life, faced at work with pressing small problems that must be solved. These problems loom large and the undesired side-effects of the solutions found for them seem remote and small. But when these side-effects are combined with those from the work of thousands of other physicists and scientists all over the world the synergistic impact may be very great. Physical dangers arise from the proliferation of more and more ingenious and destructive weaponry, from some large industrial complexes and from the degradation of the environment. Political dangers arise from the concentration of social and economic control made possible by new technologies. To combat these threats scientists need to be broadly educated and to be socially and politically aware. The history of physics, properly taught, can help students appreciate how obstacles to the understanding of nature have been overcome in the past, and this can in turn help them to overcome the obstacles they too experience and to gain insight into research in difficult areas at present. This is important in improving their understanding of physics. But even more important—if an adequate *social* dimension is provided in this history teaching, then a great many lessons relevant to society today can be learned from the interactions of physics, technology and society in the past, and this, I suggest, is the proper preparation for minds which will face the dangers and the challenges of the future.

Summary

What should one say to summarise, addressing the physicist who has little interest in history? Let me reiterate in different words some of the main points I have tried to make.

A small and dedicated group of people is interested in the study of the past for its own sake; these, I suppose, are the *real* historians. A much larger group is primarily interested in the past for the light it throws on the present and on possible developments in the future. Belonging to this latter group are those philosophers of science who wish to learn more about the nature of the scientific enterprise by studying how physics has evolved and developed. This group contains in addition students of technological innovation who are anxious to use physics effectively in the present to tackle a broad range of social and economic problems and who hope to gain insights from a study of the social relations of physics in the past. For both these groups the problem is to separate out those factors which are seemingly constant over the years and those which are specific to the present situation.

The main questions then to be asked are: in growing so enormously has physics changed in its essence? Is it now in danger of succumbing

Part 3

Physicists Look Back

to political and financial pressures that will corrupt its ideals of objective truth? This still leaves the biggest question of all: though it has enabled many people to live very comfortable lives, why has the immense power over natural forces that physics has placed in our hands not liberated human beings all over the world from starvation, poverty and oppression, but has instead raised the spectre of nuclear destruction and violent mass death? Only by looking at the history of physics *in society* can we begin to explain these things.

References

Bachelard G 1972 *La Formation de l'Esprit Scientifique* (Paris: Librairie Philosphique J Vrin)
Easlea B 1981 *Science and Sexual Oppression* (London: Weidenfeld and Nicholson) pp 65–89
Forman P 1971 Weimar culture, causality and quantum theory *Hist. Stud. Phys. Sci.* **3** 1–115
Ronayne J 1984 *Science in Government* (London: Arnold) pp 54–6

Short bibliography

Basalla G (ed.) 1968 *The Rise of Modern Science: Internal or External Factors?* (Lexington, MA: Heath)
Ben-David J 1971 *The Scientist's Role in Society: A Comparative Study* (New Jersey: Prentice-Hall)
Bernal J D 1969 *Science in History* vols 1–4 (Harmondsworth: Penguin)
Boyle C, Wheale P and Sturgess B 1984 *People Science and Technology: A Guide to Advanced Industrial Society* (Brighton: Wheatsheaf/Harvester)
Braun E and Macdonald S 1982 *Revolution in Miniature: The History and Impact of Semiconductor Electronics* (Cambridge: Cambridge University Press)
Easlea B 1973 *Liberation and the Aims of Science* (London: Chatto and Windus)
Freeman C 1974 *The Economics of Industrial Innovation* (Harmondsworth: Penguin)
Kuhn T S 1970 *The Structure of Scientific Revolutions* 2nd edn (Chicago: University of Chicago Press)
Marcuse H 1964 *One-Dimensional Man* (London: Routledge and Kegan Paul)
Mason S F 1962 *A History of the Sciences* (New York: Collier Macmillan) (originally published as *Main Currents of Scientific Thought*)
Ravetz J R 1971 *Scientific Knowledge and its Social Problems* (Oxford: Clarendon)
Ziman J 1976 *The Force of Knowledge* (Cambridge: Cambridge University Press)

12

On Attending to the Instrument Maker in Physics History†

Brian Gee

Introduction

On the occasion of the 21st Annual Exhibition of Apparatus of the Physical and Optical Societies, Sir Arthur S Eddington, in his presidential address, while referring to the reciprocal debts of the scientific instrument maker and scientific worker, remarked

> I do not know which to place first. I do not think one can be placed before the other: for the instrument maker provides the resources of the scientific worker, and the scientific worker provides the resources of the instrument maker. Nor do I think less of the instrument-designer because he sees by trained instinct what we can only worry about by laborious formulae. I look upon him as the master chess-player who grasps the strong and the weak positions of the game intuitively, whilst we inexperienced players have to analyse the situation by moving the pieces one by one.
>
> (Eddington 1931)

† Based on a presentation given at the Conference on Projects in the History of Physics 10 July 1985, Oxford.

Such expressions showing the respect of the user for the maker raise more questions than might be fruitfully answered in any definitive way in a short chapter. Nevertheless, the prospect that instrument makers should be recognised and remembered for their contribution to the scientific enterprise is one which must not be passed over lightly. Clearly Eddington appreciated the symbiotic relationship between the science-based researcher and the technology-based instrument maker but how balanced was his view? Did he really suppose that their contributions were of equal merit, or did he insist that one species is superior in driving the relationship onwards?

Questions of this nature touch upon the old chestnut of the science–technology relationship and the problem of which leads which. Solving that enigma, however, is not the intention here although its presence in this context is hardly avoidable. If the instrument maker, and particularly the philosophical instrument maker in the case of experimental philosophy (physics), had some quintessential role to play then, presumably, it is a worthwhile task to seek out these instrument making individuals for special investigation. Unfortunately this is easier said than done. Scrutiny of the usual sorts of published literature scarcely reveals what needs to be known in any supportive argument for the instrument maker. This chapter therefore, is more one of contemplation and intended provocation than example. As background it begins by seeking out their emergence and continues by considering their ability, status, and role. Finally it proposes a model of relationships and ends with a postscript on attempts currently being made to map out a knowledge of makers through trade sources.

Emergence of the philosophical instrument maker

Instrument making in England arose out of the needs of mathematical practitioners whose prime concerns lay in the problems surrounding calendar making, gunnery, fortification, navigation and surveying: that is, problems requiring immediate solution rather than philosophical disquisition. A trade in the manufacture of relevant instruments of measurement had already begun in the reign of Henry VIII and, in some significant way, this was supported by immigrant craftsmen, particularly clockmakers, who had arrived in England as religious refugees. Certainly by the mid-seventeenth century, when invention abounded, makers were already adept in the manufacture of slide-rules, universal mechanical joints, watch and balance springs, mural quadrants, wheel barometers and even devices for optical telegraphy.

It was inevitable that this trade should flourish with the arrival of new instrumental principles such as the isochronic behaviour of the pendulum and optical arrangements for extending sense perception into the macroscopic and microscopic worlds. Indeed, advances of this kind in the seventeenth century were at the very heart of the 'new learning' and instrumental demands presented by experimental natural philosophers caused some upturn in the trade. One of the earliest pieces of evidence indicating how the trades had profited from the new trends was given by Robert Boyle in his *The Usefulness of Experimental Philosophy* (1671) (Boas Hall 1965). He told

> inventions of ingenious heads do, when once grown into request, set many mechanical hands a-work, and supply tradesmen with new means of getting a livelihood, or even enriching themselves.

He also observed that (Boas Hall 1965)

> since [Galileo's] death several others have had profitable work laid out for them, by the newer directions of some English gentlemen, deeply skilled in dioptricks, and happy at mechanical contrivances; insomuch that now we have several shops that furnish not only our own virtuosi, but those of foreign countries.... .

Thus well before the close of the seventeenth century there existed an interdependence between maker and user over matters of experimental philosophy. Indeed, less well remembered makers, such as John Marshall, James Mann, John Yarwell, Christopher Cocks and Richard Reeves, were sought out variously by better remembered men like R Hooke, R Boyle, S Collins, S Pepys, I Newton, J Flamsteed and J Gregory. Some indication of the customers of those mentioned will be found in Taylor (1954) and in Simpson (1985).

The rise of Newtonian experimental philosophy in the first half of the eighteenth century and the move to diffuse a knowledge of its methodology, its findings, and its applications is moderately well known (Turner 1973, Bennet 1985). The case of the elder Hauksbee as instrument maker, with his drive to investigate the 'active principles' manifest in electricity, chemistry, capillarity and other obscure domains as well as his determination to diffuse a knowledge of the subject through courses of experiments offered from his premises, is usually considered as a starting point for the collaboration of the trade with regard to Newtonian philosophy although others since have offered their own peculiar mix of contribution and diffusion. By the fifth decade, as hinted in Campbell's book *The London Tradesman* (1747) the mathematical instrument makers had already diversified their customary trade in globes, orreries, scales, quadrants, sectors, sundials and air pumps, to include 'the whole Apparatus belonging to Experimental Philosophy' whilst others had found specialism in philosophical machines such as air pumps or electric machines (Campbell 1747) (see figure 12.1). Thus instrument makers like James Ayscough at the Great Golden Spectacles in Ludgate Circus could be found advertising 'all Sorts of Optical, as well as Mathematical and Philosophical Instruments', Walter Gough at Middle Row in High Holborn, claiming to be a 'Real Electrical-Machine and Air-Pump Maker', Stephen Davenport 'Against the Distillary' in High Holborn, supplying 'All Machines and Instruments respecting...the conclusions in Experimental Philosophy made according to the latest improvements' and the elder Francis Watkins at Charing Cross, making special provision for 'such persons as are provided of an electrical machine and a requisit apparatus' by offering directions in 'how to put them in order, and [how to] perform with them the experiments usually exhibited in publick...' (see figure 12.2).†

The recognition for a separate department for the philosophical instrument maker, implicit in Campbell, became explicit in the latter half of the eighteenth century when the trade took off in a most spectacular way. It was, of course, a period of growing commercialisation for science when its applications became a dominant theme by those itinerant lecturers who diffused experimental philosophy further afield. The 'polite learning' of earlier years, when the renowned Benjamin Martin travelled the length and breadth of England dispensing what he called 'knowledge *a la mode*' in favour of Newtonian philosophy, was fast replaced by 'useful knowledge' of a more practical kind which was concordant with the economic values expressed by various ascendent industrial and business leaders in country towns where populations were fast nucleating.

It is not the purpose here to dwell on the importance of individuals or their trade in the eighteenth century although it is important to recognise their contributions as a relevant factor in the economic and social change normally associated with that peculiar phenomenon called the English Revolution (Turner 1976, Bennet *et al* 1985 *passim*). Overall evidence for delineating the philosophical instrument maker as an indicator of growth may be derived from the vast two-volume survey of mathematical practitioners of Tudor, Stuart and Hanoverian England undertaken by the late Professor E G R Taylor (1954, 1966). Entrants to the instrument trade are recorded from a miscellany of sources for the two main periods 1485–1714 and 1714–1840. Taylor's characterisation, rather than definition, of what constitutes the categories of a 'mathematical', or 'optical' or 'philosophical' instrument maker before the mid-eighteenth century is, in most cases,

† From trade cards in the Heal Collection (British Library Reference Division) 105.4/105.5, 105.45, 105.33 respectively; see also Watkins (1747).

CHAP. LV.

Of the Mathematical and Optical Inftrument, and Spectacle-Maker.

SECT. 1. Of the Mathematical Inftrument-Maker.

THE Mathematical-Inftrument-Maker makes all kind of Inftruments conftructed upon Mathematical Principles, and ufed in Philofophical Experiments: He makes Globes, Orrerys, Scales, Quadrants, Sectors, Sun-Dials of all Sorts and Dimenfions, Air-Pumps, and the whole Apparatus belonging to Experimental Philofophy. He ought to have a Mathematically turned Head, and be acquainted with the Theory and Principles upon which his feveral Inftruments are conftructed, as well as with the practical Ufe of them. He employs feveral different Hands, who are mere Mechanics, and know no more of the Ufe or Defign of the Work they make, than the Engines with which the greateft Part of them are executed; therefore the Mafter muft be a thorough Judge of Work in general.

His Bufinefs and Genius.

SECT. 2. Of the Optical-Inftrument and Spectacle Maker.

The Optical-Inftrument-Maker is employed in making the various forts of Telefcopes, Microfcopes of different Structures, Spectacles, and all other Inftruments invented for the Help or Prefervation of the Sight, and in which Glaffes are ufed. He himfelf executes very little of the Work, except the grinding the Glaffes: He grinds his Convex-Glaffes in a Brafs Concave Sphere, of a Diameter large in proportion to the Glafs intended, and his Concave-Glaffes upon a Convex Sphere of the fame Metal: His Plane-Glaffes he grinds upon a juft Plane, in the fame Manner as the common Glafs-Grinder, mentioned Chap. XXXII. Sect. 4. He grinds them all with Sand and polifhes them with Emery and Putty. The Cafes and Machinery of his Inftruments are made by different Workmen, according to their Nature, and he adjufts the Glaffes to them.

It is a very ingenious and profitable Bufinefs, and employs but a few Hands as Mafters. The Journeymen earn a Guinea a Week, and fome more, according as they are accurate in their Trade. Such a Tradefman defigned for a Mafter ought to have a pretty good Education, and a penetrating Judgment, to apprehend the Theory of the feveral Inftruments he is obliged to make, and muft be a thorough Judge of fuch Work as he employs others to execute. A Youth may be bound to either of thefe Trades any time between thirteen and fifteen Years of Age, and does not require much Strength.

Wages.

Figure 12.1 Campbell's definitions of the instrument maker (1747).

dependent upon the intended use of the artefacts known to have been made by the maker. For example, a person who made a microscope is usually termed an optical instrument maker, and so on. From about 1760 the picture is clearer and the categorisation appears to be concordant with the styles offered by the makers themselves on their trade cards.

Figure 12.2 Trade card of Benjamin Martin (post 1760). (Courtesy of the Science Museum, London.)

Numerical totals for *metropolitan* mathematical, optical and philosophical instrument makers have been counted for each decade starting with that in which the Royal Society was founded. Despite certain difficulties surrounding Taylor's floruit dates her entries are, nevertheless, invaluable evidence for the *preliminary* insight into the emergence of the philosophical instrument maker. These are displayed in figure 12.3. Taylor takes a craftsman's working life to begin at the age of 21, that is, on completion of apprenticeship. It is not likely that young makers at this age would be trading on their own account.

Therefore, it must be borne in mind that the subsequent analysis will, inevitably be an overestimate. It may thus be seen that by the beginning of the eighteenth century the number of makers had doubled and that this remained roughly constant at approximately 50 named mathematical instrument makers and 35 named optical instrument makers until *c.* 1770–9.

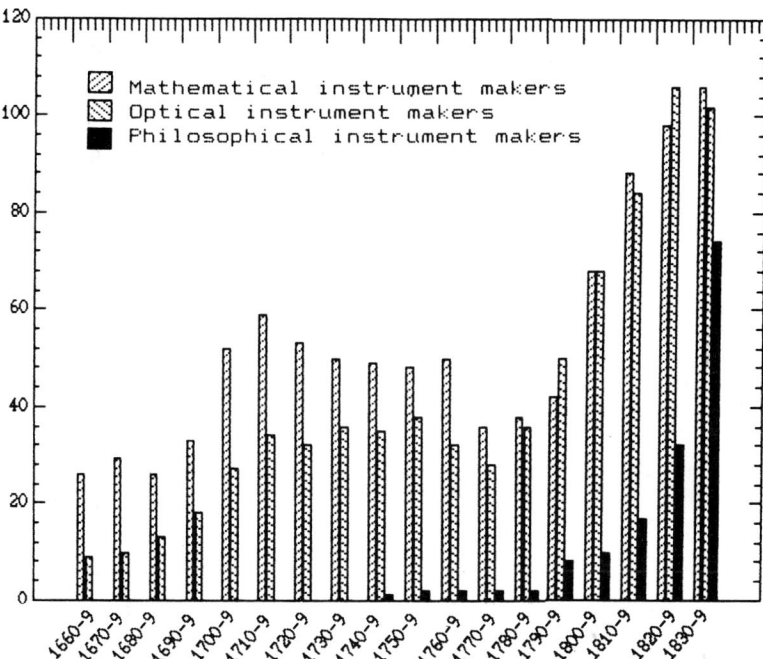

Figure 12.3 Estimates of the trade size based on known makers (1660–1840). *Note*: The data arranged here do not represent the numbers of new entrants to the trade decenially arranged as originally presented by Taylor but the actual totals for each decade. Thus, a mathematical and optical instrument maker (combined) with floruit dates 1740–85 is counted twice (as optical and mathematical maker separately) and included in the totals for each of the five decades in which the business operated. This obviously provides an overestimate but allows clarity and does not alter the trend.

The apparent drop in numbers between 1760–9 and 1770–9 may reflect some restructuring of the trade. It was a time for rationalising production with new tools including lathes and dividing engines, and new production methods with instances of the division of labour. Amongst those taking on a greater workforce at this time were the well known establishments of Nairne, Ramsden and Adams (Daumas 1972). These larger employers may well have caused smaller firms to go out of business. But the trend for growth appears to have re-asserted itself and the incidence of those makers adding the term 'philosophical' to their trade signature (figure 12.4) begins to take off in a somewhat 'more than' exponential fashion.

Figure 12.4 Broadside of Chadburn Brothers, scientific instrument makers, prepared for the Great Exhibition (1851). (Courtesy of the Science Museum, London.)

By the first decade of the nineteenth century, then, the philosophical instrument making aspect of this nascent industry was represented by at least some 70 named persons (not businesses). An indication of the continuing trend throughout the Victorian era for businesses (not individuals) is shown in figure 12.5 and is based on counts from the trade directories. This graph shows some peaking by 1860 and some decline during the Great Depression in the 1880s/90s. It is interesting to compare these simple statistics tentatively with known economic trends, using British overseas trade as a barometer of advancement. Over the first half of the eighteenth century this grew at one per cent per annum which—in view of the stagnation of population increase—represents a fairly substantial rate of progress for a pre-industrial economy. A take-off appears to have begun in the mid-century rising to about 2.5 per cent per annum by the beginning of the nineteenth century, checked only by the American War of Independence. Export trade then peaked in the third quarter of the nineteenth century at about 4 per cent per annum and, thereafter, fell to about 2.5 per cent per annum for the three decades prior to World War I (Deane and Cole 1967). The parallel of these statistics with the trends indicated (above) would suggest some positive correlation although the task of validating the figures for the instrument trade yet requires a greater reservoir of information.

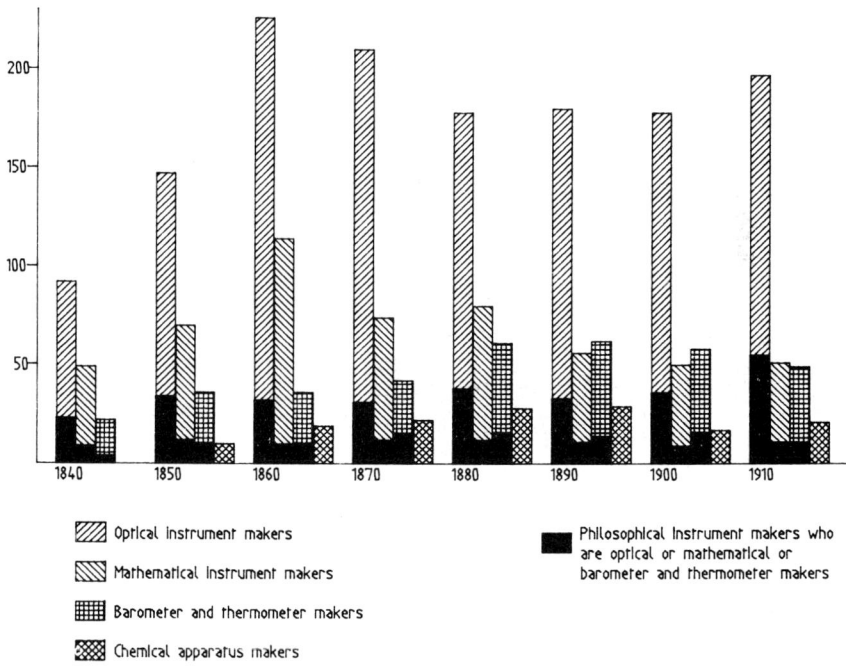

Figure 12.5 Numbers of instrument making businesses (1840–1910). Based on Kelly's *Post Office Directories*.

There appears a further correlation between these figures and the instrument making workforce as depicted in the occupational statistics for the decennial census surveys between 1841 and 1901. This is currently under investigation by the author. Some representation of occupational statistics for the years 1841 and 1881, originally analysed by Charles Booth in 1886, has been undertaken by Willem Hackmann (1985a).

From this brief introduction to the trade it is noted that a fraction of each branch of the trade claimed to deal in 'philosophical' supplies in one form or another. Thus in seeking those commercial makers who are of potential interest in the history of physics this becomes a key expression in the identification of the relevant makers from trade cards, directories or catalogues. But before giving notice of current research in this direction it is worthwhile returning to the issue raised at the start of this paper relating to the nature of the makers.

Ability, status and role

At first sight it would appear that Eddington (in the opening paragraph) is suggesting that the communities of the scientific research worker and technological instrument maker are to be held as mirror-image twins. However, he continues: 'But there is another less direct debt which the instrument maker owes to the scientific worker, [namely] the stimulus to satisfy new requirements.' Nowhere does he indicate correspondingly that intellectual profit might result from instrumental innovation or design improvements advanced by the maker. Thus it is inferred that his model of symbiotic communities, the research worker and the instrument maker, contains a superior species which drives the relationship onwards.

Other remarks telling of the relational aspects of maker and user may be found in earlier years. For example, amongst the trade definitions employed in the 1851 census it was stated that:

> They [the instrument makers] are engaged in the *higher class* of mechanical and chemical arts; are intimately connected with artists and men of science; *from whom frequently, either directly or indirectly, they derive materials, direction, or inspiration*. They multiply copies of original works. The matter in which they deal comes from the animal and vegetable as well as the mineral kingdom; but it is no longer living. They do not breed animals nor grow plants. They make things, and use tools or machines; they either in their works employ matter of various kinds in combination, or the works involve skill that fixes the mind and diverts attention from the material of which it is composed. The subdivision of labour is extensive. The masters often possess *great intelligence*; and have *large capitals* embarked in their materials, machines, buildings and products.
>
> (Office of Population Censuses 1977)

It is of course expected that the 'men of science' must take initiative in the knowledge-seeking enterprise and, therefore, they must have some knowledge of the instrument makers' work, however rudimentary, in order to lead. Correspondingly their status is often considered to be 'above' that of the instrument maker because of their ability in abstract thought and investigation of problems of the scientific kind. That hierarchy, in which 'brain work' is given a higher status and reward than 'hand work', has a history stemming from the long tradition of Platonist philosophy. Yet it is a mistake to imagine that the technological process, involving instrument design, is of a lesser intellectual standing for, as the above definition tells, the masters also had 'great intelligence'.

The importance of having due regard for the instrument makers' intelligence, besides their dexterous or psychomotor skills and technical know-how, was well understood in the mid-eighteenth century when experimental philosophy became thoroughly established. To become a mathematical instrument maker, especially one who was diversified into the philosophical domain, it was first necessary to comprehend intellectually constructed classes of scientific knowledge in order to carry out related work. This meant that the training process was far more demanding than that of the ordinary craftsman or mechanic who usually learned to perform simple manipulations by precept or imitation of a master. Therefore, although practicalities were the base line of people of this trade rather than theoretical speculation, it was, nevertheless, essential for them to keep abreast of the developments of science: in this respect they verged on the domain of the scientific community itself.

At least this much was appreciated by the already mentioned Campbell who, in 1747, informed parents of the various skills and abilities requisite of their offspring when entering the various trades. 'All too often', he warned the misguided 'fond mothers' and 'doating fathers', 'a child is bound, that is, chained to a Trade, to which Nature never designed him [which hence] furnishes the Pulpit with Cobblers instead of Divines. (Campbell 1747, p. 2). Indeed the schooling of youth was taking great strides in the early decades of the eighteenth century and parents were strongly encouraged to consider the 'Capacity', 'Disposition', and 'Constitution' of their sons before binding them to 'Seven Years Slavery'. For example, Campbell advised that a glass grinder could be 'any common labourer', a pump maker required 'more Strength in his hands than judgment in his Head', and a clockmaker required 'no great Strength nor much Education'. (Campbell 1747 pp 179, 247, 252 and *passim*).

To become a mathematical instrument maker, however, Campbell (1747 and figure 12.1) stipulates:

> He ought to have a Mathematically turned Head, and be acquainted with the Theory and Principles upon which his several instruments are constructed, as well as the Practical Use of them.

Intelligence and accomplishment, then, could raise an aspiring apprentice above the mere 'Mechanics...[who] know no more of the Use or Design of the Work they make, than the Engines with which the greatest Part of them are executed'. In essence a high degree of criticality and judgment was required if mastership was the pinnacle of ambition. Again the compiler of trades informs

> Such a Tradesman designed for a Master ought to have a pretty good Education, and a penetrating Judgment, to apprehend the Theory of the several Instruments he is obliged to make, and must be a thorough Judge of such Work as he employs others to execute.

In fact bright juveniles suited for this trade could be found in mathematical schools such as that at Christ's Hospital in the City of London where it is likely that those in *loco parentis* counselled for future careers. There are several connections of instrument makers with this school, from the time of John Rowley in 1724/5 to the time the school moved to Horsham.†

In considering the problem of relation between maker and user it is relevant to inquire how instrument makers themselves perceived their role. Excavating evidence of this sort is particularly difficult. Nevertheless, when such evidence has appeared instrument makers have, not unexpectedly, projected an image telling of their *service* to science. For example, some remarks concerning the new 'instrument technologist', contained in the opening issue of the journal *Instrument Practice* just after World War II, indicate a dual aspect to the trade. On the one hand there was the *making and improving* aspect in which the maker was portrayed as a 'supermechanic' at transforming 'fragile, awkward newborn' articles of the scientist into commercial 'foolproof' form of rugged construction, practical utility, accuracy, sensitivity, etc. On the other hand there were *technical support and management* aspects for which makers were required to lend their expertise to the 'selection, installation, supervision, maintenance and general management' of instruments (*Instrument Practice* 1947). Of course, neither role represents anything new for the instrument technologist of the twentieth century which was not apparent in former times.

A useful summary statement of the same dual role may be discerned in the following list of 'duties' incumbent upon the instrument maker

---

†I am indebted to V K Chew (Research Fellow, Science Museum, London) for this information. Account ledgers of Christ's Hospital school (fd 1673) (Guildhall MS 12/874, City of London). For an introduction to this unique school see Plumley (1976).

issued by the old microscope making firm of Watsons on the occasion of their centenary in 1937. They specified that a maker should be able:

> To associate with those whose profession, calling for the continual use of scientific instruments, enables them to advise and suggest new designs, or modifications to old.
> To keep a constant watch for any new process or method which will serve the purpose of improvement in construction and use of an instrument.
> To be ready with a helping hand for those with an optical difficulty or doubt.
>
> (Watson 1937)

Clearly, then, the internal perception of the trade was that it was a service industry, that is, one working in tandem with the scientific enterprise. So much might be expected by definition but is it correct to believe that the initiative for development should always come from science? Given the intelligence and ability of many of its representatives and their close association with the scientific enterprise it is reasonable to suppose that the philosophical makers initiated developments *of their own choosing*, certainly as business people but possibly also as technologists rather than scientists.

Complexions of the maker: towards a model of relationships

Errors of perception

It seems that there might be an error of perception in treating instrument makers as 'lesser' or 'marginal' men of science.† Such diminutive expressions are not so much false as misleading. In the knowledge-seeking *scientific* process it might, indeed, be rare to find an instrument maker adding to or successfully formulating theoretical knowledge. The error of perception lies in the hidden assumption that instrument makers belong to the same community as the scientists. According to the contemporary definitions just introduced, and from common knowledge, this is manifestly not so. Historically instrument makers descend from and are tied to their trade in the practical arts by the genealogy of master and apprentice. In other words, they belong to a different community entirely.

†For example, D H Sadler in the Foreword to Taylor (1966) refers to mathematical practitioners as 'lesser men' and Shapin and Thackray (1974) characterise them as 'marginal' men of science.

For example, Hackmann (1985b) suggests that men like J Short, J Ramsden, E Troughton and other well known makers were 'scientists in their own right' and that there was 'little to distinguish' them from men like J Priestly or J Cuthbertson. The argument (above) does not deny that there are exceptional men of the trade but does rather hint of a not insignificant distinction in their nature from the scientist.

Within the trade itself at one extreme there were large numbers of artisans whilst at another extreme there were masters with sufficient ability and comprehension to capitalise upon and even take part in the various scientific movements. Individuals in the latter extreme, which are of prime interest here rather than the anonymous workforce, are particularly worthy of study, not least because they are likely to have formed a bridge between the two communities alluded to above. This also was true for the rising business leaders whose education was not necessarily obtained internally by apprenticeship. Therefore it may be far too narrow a view to state that their 'direction' and 'inspiration' was derived *entirely* from the scientific community although that they *frequently* did so, makes the earlier census definition fitting. Adopting such a stance would entail a second error, also of perception, namely that technological acts and achievements only follow in the wake of science. A pertinent summary of the alternative perceptions appears in Barnes (1982). Any attempt to solve the enigma of whether science precedes technology or vice versa is perhaps futile although it is important to note that the motivations of the maker are at times otherwise than in the *service* of science.

Characteristics of the maker–user communities

Before inquiring further into the position of the instrument makers it may be helpful to point out some of the distinctions and characteristics which exist between the sub-cultures across which the instrument makers and scientific researchers straddle. The most distinguishing feature lies in the spectrum of knowledge itself. Briefly science has its goals in *knowing* and *extending knowledge* while the practical arts lean towards *know-how* and *extending the state of the art*. In other words it should be possible to distinguish between practitioners of two broad kinds: namely the 'experimenter' in science and the 'improver' in the practical arts. Such a view was proposed by Multhauf (1959). Distinctions of this kind are not unknown and, for convenience, G Sarton's long standing 'systematised positive knowledge' represents well the hierarchical structure of abstract concepts associated with scientific knowledge (*epistēme*) and C Singer's 'systematic discourse on the useful arts' represents correspondingly the knowing associated with procedures involving dexterous skills (*technē*), that is, the making and doing (Sarton 1954, Singer *et al* 1954). Philosophical instrument makers, who have their base in the trades, are orientated to a good measure in the practical arts *qua* technology although they transcend these confines to thrive in symbiosis with the experimental philosopher *qua* scientist. It might also be noted that experimenters mirror their counterparts when they transcend their base in science by becoming involved in the design and improvement of new instruments. In this case the experimenters' activities are of a technological nature rather than scientific.

It is not the intention here to deny the practical arts/technology of any thought component. Indeed, thinking is an essential part of any creative activity. A difficulty arises when the thinking and discovery associated with technology is claimed as a different mode of knowledge. This was the line taken by E Layton Jr (1974), which brought a rebuttal and reply from A R Hall (1978). The view taken above is that practising individuals, possessing human potential, both cognitive and psychomotor, may display both attributes. Their motivations for doing so will, naturally, show preferences, often according to their occupations. Hence the 'scholar'–'craftsman' division of old is not so unusual nor unexpected. The subject is well explored in educational psychology.

It is worth elaborating on the major distinction between knowledge and know-how just a little further in order to build up a working model of the knowledge spectrum. The trade definition from the 1851 census offers a convenient starting point because it describes the function of the maker to be that of 'multiplying and copying original works'. This process would stretch from the bench worker, involved in numerous operations (planing, milling, drilling, soldering, tapping, assembling and so on) to the master who might design and orchestrate the whole production process. The end product of this process is, of course, the instrument itself or multiple copies of the instrument. The bench workers, even if ignorant of how to use the end product, would, nevertheless, 'know' the instrument by its parts, its shape, and its assembly. That is, they would know the instrument from *practical knowledge*. The same instrument, removed from the makers' workshop and placed in the scientific laboratory enters a different process entirely. In this case it is the scientific practitioners, technicians in their own craft at teasing out observations, who 'know' the instrument from the point of view of *experimental knowledge*; they, too, might be excused if they are ignorant of its precise mode of construction. For the theoretician the observational fruits of the instrument are valued more than the article itself for its utility lies in anchoring *theoretical knowledge* to nature itself. It often happens of course, that the user is both experimentalist and theoretician just as the maker might conceivably be the designer and fabricator.

A second distinguishing feature relating to the knowledge spectrum, and useful in building up an appropriate model, lies in the system of *rewards and recognition* accepted by these separate yet complementary communities. There is much justification in the generalised statement that the majority of makers are little remembered for their writings. Apart from obvious exceptions like Benjamin Martin, George Adams Jr or John J Griffin who were prolific in diffusing science and therefore not archetypal instrument makers, most makers only raised their pens when preparing a new catalogue or issuing bills of sale, or, at best, when preparing a descriptive pamphlet for some instrument. This reluctance or unwillingness to publish adds further reinforcement to the concept of two communities. Such a distinguishing feature between scientists and technologists was analysed by the late Derek de Solla Price who labelled the respective communities *papyrocentric* and *papyrophobic* (Price 1965a, 1965b, 1982). In brief, scientists positively seek to state claims to private intellectual property through publication as each stage of research develops and therefore belong to the former category; instrument makers more easily fit the latter category because of their aversion during development to divulging their methods and processes. Yet even without publication makers' achievements do eventually become known although the stance taken is opposite to that of their scientific counterparts. That is, they guard any inventive property in a jealous way by not giving public release to each developmental step and only providing *demonstration* or *exhibition* once an idea or improvement or new process is ready for public release.

Common-ground: balancing the model of two communities

So far the impression given has swayed towards an asymmetrical model in which the maker is a follower in an enterprise led by the scientist. But, as a consequence of this view it is all too easy to relegate the position of the maker to marginality. The problem, of course, is that of comparing dissimilar occupations by the same system of values. Some balance or symmetry might be restored by taking a position in which the dual-serving role is disregarded and, instead, instrument makers are examined through their personal motivations. Questions such as 'Do they derive fascination by the feeling that they can manipulate and control physical phenomena with instruments and mechanisms?' or 'Do they perceive and enjoy an aesthetic in instrumental design?' or 'In what sense do they reflect on the changes brought about by their instrumental designs?' and 'Are there goals to which they ought to proceed?' each allow for instrument makers having idealistic goals. Indeed, it is only through such questions that an understanding of the instrument makers' true position can be reached.

Now the viewpoint sketched out above, even if simplified and untenable in certain instances, has the merit of suggesting that it is the instrument itself which provides the common ground between maker and user. Not only is it implicit that the processes of the two communities should be set in tandem, with the instrument maker working towards the business of science, but also it is possible for there to be some flow in the other direction. The common ground upon which these processes meet also provides a weak link for the exchange of information from which instrument makers adopt new scientific principles and incorporate them in their design repertoire. One example will suffice to illustrate this point: when the London-based American instrument maker Joseph Saxton took on Faraday's principle of electromagnetic induction to develop a spark-producing apparatus he had no further recourse to science in order to lure bigger sparks from his magnet. That is, despite the non-availability of Ohm's conductivity theory in England at that time he was, nevertheless, able to achieve the required spectacular spark by cut and try methods which led him to winding a multiparallel wound coil.

Finally it is worth noting, on behalf of the maker, that something of the bringing together of these two processes from the distinct communities was recognised and symbolised by the instrument maker E M Clarke who, early in the 1840s, adopted an heraldic engraving of the scholar and the craftsman to mark their respective contributions in instrumental achievement.† The scholar, dressed in cap and gown and with quill in hand, and the craftsman, dressed in apron with hammer in hand, stand either side of a pedestal on which appears a revered artefact, the tangible outcome of their joint conceptions. The motto *Mente et Manu* serves to emphasise that both mind and body were equally necessary for the advancement of humankind. Such imagery is of course reminiscent of the Baconian objective for uniting science and the trades towards some human beneficial purpose.

Another emblematic scholar–craftsman design of the recently formed Worshipful Company of Instrument Makers, signifies the

---

†A version of this symbolism was displayed in tiles on the building of the Royal Panopticon of Science and Arts in Leicester Square (London) and featured on Clarke's prospectus literature for that institution. Westminster Reference Collection: [Royal Panopticon of Science].

scholar–scientist as the source of instrumental ideas. Here Faraday and Newton, suitably robed in academic regalia and symbolically clutching artefacts by which they are remembered, the induction ring and the reflecting telescope, stand beneath the head of Minerva, the Roman goddess whose attribute is wisdom and knowledge. But lest it should be forgotten how empirical knowledge comes into being, their heraldic motto reads *Sine Nobis Scientia Languet* meaning 'Science cannot flourish without us' (Younson 1978).

Clearly this perception has validity in recognising those makers whose names are associated with some historic machines such as the elder Hauksbee's air pumps, Cuthbertson's electric machines, H Pixii's magneto-electric machine, or the younger Watkins's electromotive engines and, perhaps, in some measure it encapsulates the essence of what Eddington meant in his conception of a joint enterprise.

Postscript on resources for 'knowing' the instrument makers

There is no conclusion to a chapter which sets out to open up, for those interested in the history of physics, a consideration of the position of the instrument maker in scientific and technological advance. There is, however, a need to offer some guidance in resources or starting points: some have been used in preparing this chapter although the field is vast indeed. In general terms instrument makers are 'known' through their products or sometimes through their association with scientists. The former necessitates a fund of encyclopaedic knowledge about the history of instruments and the latter requires the good fortune of recorded evidence from diaries, letters, laboratory notebooks, etc. Such approaches to knowing the makers are useful although somewhat chancy. Nevertheless a perusal of the *Bulletin* of the recently formed Scientific Instrument Society will quickly indicate the wealth of information which is currently in progress by interested persons from a variety of backgrounds including university departments, museums and the antiquarian world.†

In recent years a number of researchers have begun to examine, quite systematically, the various primary sources of trade material relevant to the instrument trade. Records of the City Guilds, trade directories and trade cards all yield important information about trade

† The Scientific Instrument Society was formed in 1983 to bring together those with a particular interest in the history of instruments. It may be contacted through its Executive Secretary, Howard Dawes, PO Box 15, Pershore, Worcestershire WR10 2RD, UK.

structure, individuals and their businesses. (Brown 1979, Calvert 1971, Crawforth 1987).† Also Project SIMON, a major undertaking originally based at the Museum for the History of Science (Oxford), has begun to tie together the threads of what is known in the form of an Index of Instrument Makers (Crawforth 1987). Another project under the auspices of the National Museums of Scotland and Royal Dublin Society, explores Irish instrument makers (Morrison-Low and Burnett 1989). Both promise, in their respective ways, to offer a valuable reference point for future researchers.

While the aforementioned sources will provide names, dates, addresses and other miscellaneous details and checks, they may not provoke curiosity or inspire speculation on physics history in the way offered by yet another largely neglected source, namely the trade catalogue. In terms of searching for an instrument maker's involvement or contribution this source has particular relevance to the physical domain. Until recently such extant catalogues have remained uncoordinated, unlisted and generally inaccessible for study in the history of physics. In bibliographic terms trade catalogues were always classed as ephemeral, therefore disposable, and hence they are now rare indeed. Few copies remain in present-day companies, especially those which have suffered wartime blitzes, removals, take-overs and office clear-outs. Nevertheless, as a result of a wide-scale search, at home and abroad, over 1500 catalogues have now been located between the mid-eighteenth century and 1914 (Anderson *et al* forthcoming).

It is believed that, if employed wisely, such ephemeral sources might be extended beyond the customary use by museum curators for deriving the occasional facts of business history or in dating some artefact. Catalogues allow, as might be expected, an insight into what was available at a given time, that is, they offer a window on what once was state of the art in laboratory instruments and apparatus. The first appearance of an item in a trade catalogue often signals that research and development had reached the point of commercial viability. More importantly it is sometimes possible to trace the kind of information required to point to the involvement of an instrument maker in the scientific process. For these, and other reasons, trade catalogues are worthy of the attention of those who wish to provide that extra dimension to physics history in recognition of the philosophical instrument maker.

† Dr A Chaldecott (formerly Science Museum, London) and David Wright (Science Museum, Department of Medicine) have compiled lists in this respect. See Wright (1987).

References

Anderson R G, Burnett J and Gee B forthcoming *Handlist of Instrument Makers Catalogues to 1914* (Edinburgh: National Museums of Scotland)

Barnes B 1982 The science–technology relationship: a model and a query *Soc. Stud. Sci.* **12** 166–9

Bennet J 1985 *Science and Profit in 18th-Century London* ed. J Bennet *et al* (Cambridge: Whipple Museum for the History of Science) 5ff

Boas Hall M 1965 *Robert Boyle on Natural Philosophy* (Bloomington, IL: Indiana University Press) p. 161

Brown J 1979 Guild organisation and the instrument-making trade, 1550–1830: the Grocers' and Clockmakers' Companies *Ann. Sci.* **36** 1–34 and *Mathematical Instrument Makers in the Grocers' Company (1688–1800)* (London: Science Museum)

Calvert H R 1971 *Scientific Trade Cards* (London: HMSO)

Campbell R 1747 *The London Tradesman* (London: Garner). Reprinted in facsimile 1969 (Newton Abbot: David and Charles) pp 253–4

Crawforth M A 1985 Evidence from trade cards for the scientific instrument industry *Ann. Sci.* **42** 453–554

—— 1987 Makers and dates *Bull. Sci. Instrum. Soc.* no 13 2–8

Daumas M 1972 *Scientific Instruments of the Seventeenth and Eighteenth Century* transl. M Holbrook (London: Batsford) 311–20

Deane P and Cole W A 1967 *British Economic Growth 1688–1959* (Cambridge: Cambridge University Press) 28ff

Eddington A S 1931 Physical and Optical societies twenty-first annual exhibition of apparatus: opening address by the President *Proc. Phys. Soc.* **43** 120–3

Hackmann W 1985a in Nineteenth century trade in natural philosophy instruments in Britain *Nineteenth Century Scientific Instruments and their Makers* ed. P R De Clercq (Leiden: Rodopi) pp 53–91 esp. 77 ff

—— 1985b Instrumentation in the theory and practice of science: scientific instruments as evidence and as an aid to discovery *Ann. Ist. Mus. Storia Sci.* **10** 87–115, in p. 98

Hall A R 1978 On knowing and knowing how to... *Hist. Tech.* **3** 91–103

Instrument Practice 1947 **1** 9 (editorial)

Layton E Jr 1974 Technology as knowledge *Technol. Cult.* **15** 31–41

Morrison-Low A and Burnett J 1989 Vulgar and Mechanick: The Scientific Instrument Trade in Ireland 1650–1921 (Dublin: Royal Dublin Society and National Museums of Scotland)

Multhauf R 1959 The scientist and the 'Improver' in technology *Technol. Cult.* **1** 38–47

Office of Population Censuses and Surveys 1977 *Guide to Census Reports, Great Britain 1801–1966* (London: HMSO) xcii–xciv

Plumley N 1976 The 'Royal Mathematical School with Christ's Hospital' *Vistas Astron.* **20** 51–9

Price D de Solla 1965a Networks of Scientific Papers *Science* **149** 510–15;

—— 1965b Is technology historically independent of science? *Technol. Cult.* **6** 553–68

—— 1982 The parallel structures of science and technology in *Science in Context* ed. B Barnes and D Edge (Milton Keynes: Open University Press) 164–76, at 169

Sarton G 1954 *An introduction to the History of Science* vol. 1 (Baltimore: Carnegie Institution of Washington) p. 3

Shapin S and Thackray T 1974 Prosopography as a research tool in the history of science: The British scientific community 1700–1900 *Hist. Sci.* **12** 1–28 in p. 12 ff

Simpson A D 1985 Richard Reeve—the 'English Campani'—and the origins of the London telescope making traditions *Vistas Astron.* **28** 357–65

Singer C, Holmyard E J, Hall A R and Williams T I 1954 *A History of Technology* vol. 1 (Oxford: Clarendon), p. vii

Taylor E G R 1954 *The Mathematical Practitioners of Tudor and Stuart England: 1485–1714* (Cambridge: Cambridge University Press)

—— 1966 *The Mathematical Practitioners of Hanoverian England: 1714–1840* (Cambridge: Cambridge University Press). Note: This survey includes over 2200 brief notes (including occasional addresses, business styles, floruit dates, etc) on surveyors, navigators and other practitioners including instrument makers of various sorts.

G L'E Turner 1973 A very scientific century in *Martinus Van Marum: Life and Work vol. IV Van Marum's Scientific Instruments in Teyler's Museum 1713–1776* ed. E Lefebvre and J G De Bruijn (Leyden: Noordhoff) pp 3–38, in pp 12–32

—— 1976 The London trade in scientific instrument making in the 18th century *Vistas Astron.* **20** 173–82

Watkins F 1747 *A Popular Account of the Electrical Experiments, Hitherto made Publick* (London) p. 3

Watson W and Sons 1937 *Centenary* (London: Watson and Sons) p. 21

Wright D 1987 letter to *Bull. Sci. Instrum. Soc.* no 15 pp 14–15

Younson E 1978 *The Worshipful Company of Scientific Instrument Makers* (London: Worshipful Company of Scientific Instrument Makers)

13

Reflexions on Crystals: Instrumentation for Crystal-Structure Determination

H Lipson

Introduction

In 1930 I graduated in physics at Liverpool University, a department not noted for its research. But I wanted to do research and Dr R W Roberts—the only member of staff with some influence—suggested that I join C A Beevers, who had graduated a year earlier, in his work on crystal-structure determination.

This was a new subject; it had started only in 1912 and had made no progress during the war years, 1914–18. So there was not much to learn, but neither was there much guidance. The literature was sparse—books by the Braggs (Bragg and Bragg 1915) and by Wyckoff (1924) and an erudite paper by Astbury and Yardley (1924) (later Lonsdale) on the 230 space groups.

We had some primitive apparatus, including a home-made gas x-ray tube, and we soon learned how to take single-crystal photographs and how to analyse them, basically from a paper by Bernal (1927). But what could we do with the results?

Fortunately, Liverpool was not too far from Manchester, where Professor W L Bragg's department was the world centre for the subject. We approached him diffidently, but we found him very helpful; he introduced us to his experts, particularly R W James and W H Taylor. We soon had some success and went on to solve some problems that had even beaten Manchester.

This chapter gives an indication of what the subject was like when we started and how it developed over the years.

The beginning

In 1912 M von Laue (1879–1960), in Munich, had the idea that a crystal could be used as a diffraction grating to establish the nature of x-rays. Being a theoretician, he had to find experimentalists to carry out the work, and two, W Friedrich and P Knipping, set up the apparatus and were successful. At one stroke they proved that x-rays *were* waves, and opened up means of examining matter on an atomic scale—an activity that has affected all branches of science and is still producing results of immense significance.

Laue fully worked out the theory of diffraction by a three-dimensional grating (Friedrich *et al* 1912). He showed that each order of diffraction was quantified by three digits, in place of one for a one-dimensional grating. But he was not able to fit his theory to the results obtained by his colleagues, and was not therefore able to apply his theory in practice; no one knew the nature of the x-ray spectrum.

The gap was filled by W L Bragg (1890–1971) who was a research student at Cambridge. He did not have the theoretical ability of Laue, but he proposed a much simpler theory, based upon the idea that each order of diffraction could be regarded as a reflexion from a set of planes within the crystal (Bragg 1913); this theory could be easily understood and therefore applied.

Of course, the orders of diffraction are *not* reflexions, but the idea is so beautiful that the word has persisted, as shown by the title of this article.

Bragg's theory allowed each problem to be reduced to a set of one-dimensional problems, and he immediately applied his ideas to the first crystal structure, sodium chloride (NaCl). This was a great breakthrough; it showed that NaCl was composed of ions of Na^+ and Cl^-, not of molecules as most chemists thought.

This approach set the subject on its feet. But NaCl was a simple structure; for more complicated structures a more general method was needed, allowing us to specify the three integers for each order of diffraction for a crystal. Over the years this problem has been entirely solved. The first, very simple, apparatus giving results difficult to interpret, was gradually replaced by more complicated apparatus that gave simpler results.

The Laue method

The first photograph (figure 13.1) was taken with a fixed x-ray beam falling upon a fixed crystal, giving a diffraction pattern on a fixed plate. The crystal was $CuSO_4.5H_2O$, chosen, I suppose, because it is a common substance in laboratories, although Laue made some sort of theoretical justification for it. It had very little symmetry, and so more symmetrical crystals were tried; properly oriented they showed beautiful symmetry (figure 13.2) related to that of the crystal. But how could one assign indices to the spots?

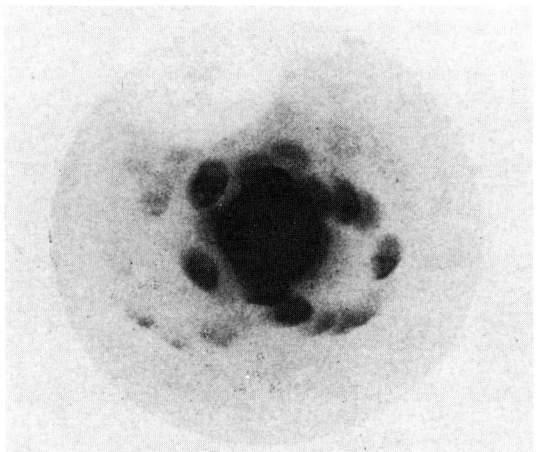

Figure 13.1 The first x-ray diffraction photograph (from Friedrich *et al* (1912) p. 390), courtesy of the Bodleian Library.

The chief problem was that nothing was known about the x-ray spectrum. Laue—most amazingly—envisaged a spectrum of fixed wavelengths; he tried one, two, three, four and five, accounting for more and more of the spots, and then gave up. Only when W H Bragg (1862–1942), with the assistance of his son, W L Bragg, using an ionisation spectrometer (Bragg and Bragg 1913) found a mixture of continuous and characteristic wavelengths, was the problem solved.

The relationship between the Braggs was interesting, and Beevers and I were fortunate in having an opportunity to meet both of them. W L gave us an introduction to his father on one of our visits to

London and he greeted us very warmly in spite of our being two raw research students and he was President of the Royal Society. He asked us if we would like to see round the Royal Institution (of course we would!) and to our surprise he did not summon a member of his staff, but took us round himself. I still remember his enthusiasm about Faraday's laboratories and his apparatus.

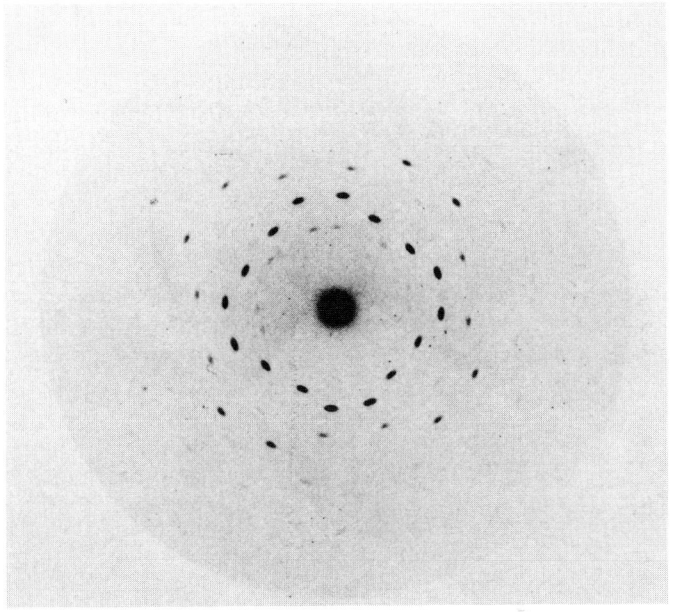

Figure 13.2 A photograph showing symmetry (from Friedrich et al (1912) tafel II, fig. 5, opposite p. 390), courtesy of the Bodleian Library.

W H and W L Bragg were quite different personalities. We can see how important is the cooperation between people with different sorts of abilities; W H was the good sound eminent physicist, whereas W L was the man with intuition. The idea of x-ray reflexion came to him in the grounds of Trinity College, Cambridge, where he was a student of J J Thomson's and should not have been thinking of such things.

Their solution raised other problems. A spot with indices h,k,l, produced with a wavelength λ would overlap with one with indices $2h,2k,2l$ produced with $\lambda/2$ and so on. The indexing of the spots was therefore not unequivocal.

R W G Wyckoff (1920) was the prime expositor of the Laue method. He took photographs with different voltages on the x-ray tube, so controlling the range of wavelengths in the spectrum. But it needed a genius like Wyckoff to make full use of this idea, and not many other people could follow him.

Rotating-crystal method

The Braggs' work suggested a better approach. If the crystal were rotated during the exposure, the continuous radiation would be drawn out into streaks and the characteristic radiation would appear as a spot on the streak (figure 13.3). Such spots could thus be unambiguously identified.

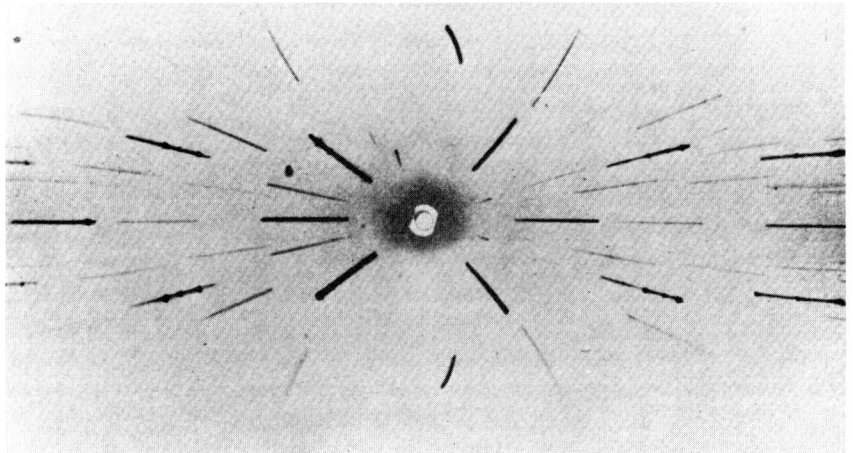

Figure 13.3 White radiation streaks with characteristic spots. (From Henry *et al* (1951) p. 57, courtesy of the authors.)

Moreover, if the crystal were rotated about a crystallographic axis—an edge of the unit cell—and if the film were in the form of a cylinder (figure 13.4) the characteristic spots would be in straight lines (figure 13.5), which were called layer lines. The spacings of these lines would give directly the length of the edge of the unit cell.

This was a great step forward, in spite of the extra instrumental problems. But it was still not known at which orientation of the crystal each spot was produced. This problem was partially solved by the next step.

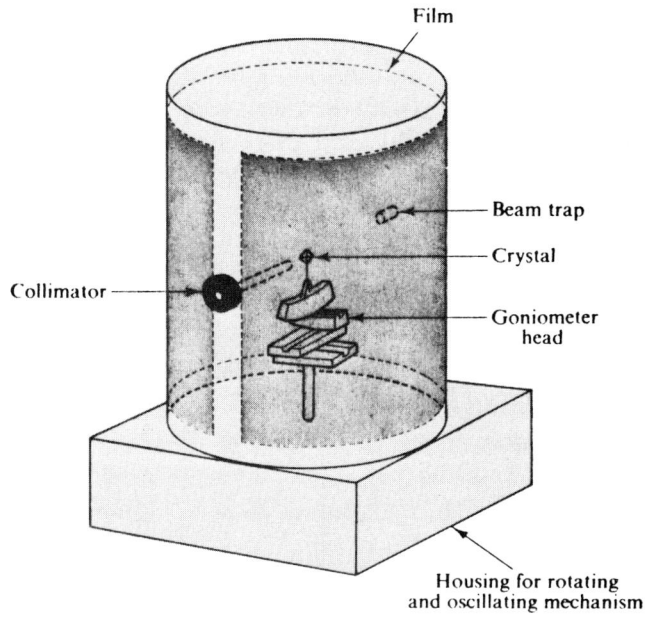

Figure 13.4 An oscillation camera. (From Wolfson (1970) p. 135, courtesy of the author.)

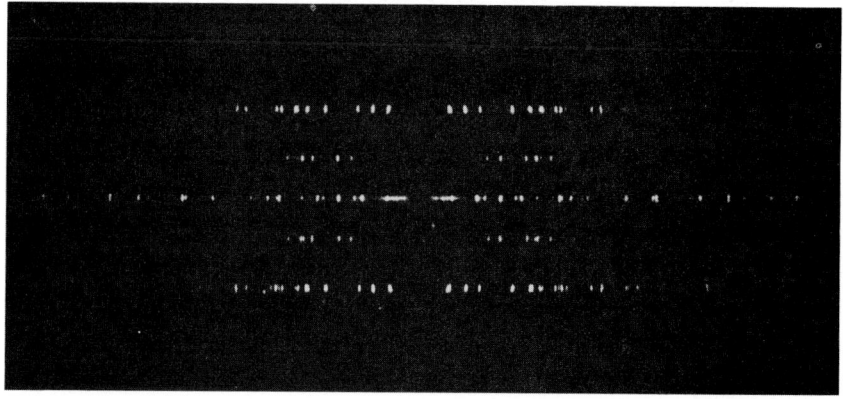

Figure 13.5 Layer lines. (From Nuffield (1966) p. 285.)

The oscillating-crystal method

Instead of rotating the crystal completely, let us oscillate it through a finite angle; the smaller the range of oscillation, the greater the certainty of indexing the spots, but the larger the number of photographs required—a not unimportant factor in those impoverished days! The most favoured angle turned out to be 30°, but for special reasons 5° was sometimes used. For 30°, seven photographs were needed to cover the entire range of 180°; some overlap was desirable.

This was the main method used when I started my work in 1930. The interpretation of oscillation photographs was not easy and required the use of a special discipline laid down by J D Bernal (1901–71) in 1927 based on a fundamental concept introduced by P P Ewald (1921) known as the reciprocal lattice. A special chart, known as a Bernal chart (figure 13.6) was required.

The basic difficulty was that we were having to produce three-dimensional results from a two-dimensional film. The solution to this problem was produced by K Weissenberg (1924).

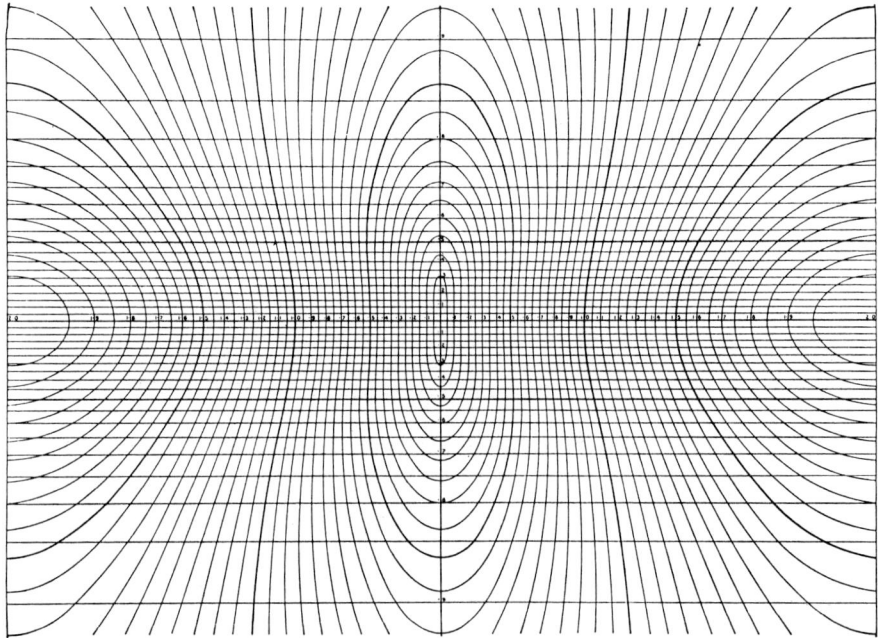

Figure 13.6 A Bernal chart. (From Henry *et al* (1951) p. 50, courtesy of the authors.)

The Weissenberg method

Weissenberg's method is so obvious that it is surprising that it was not thought of before, particularly by Bernal, who was noted for his originality.

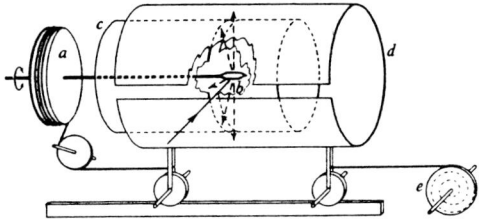

Figure 13.7 Component parts of a Weissenberg goniometer. (From Henry *et al* (1951) p. 81, courtesy of the authors.)

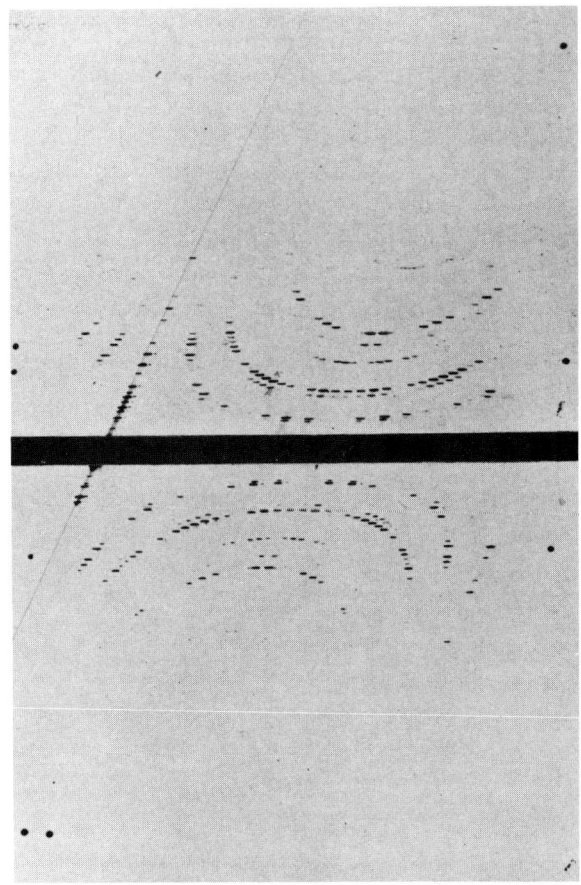

Figure 13.8 A Weissenberg photograph. (From Henry *et al* (1951) p. 90, courtesy of the authors.)

The idea was to treat each layer line separately: each was isolated by a screen and the film moved parallel to the axis (figure 13.7). At the zero layer line, for example, one index is zero, and therefore we need to find only two indices; these could easily be read off from another chart.

Weissenberg cameras were normally horizontal because it was easier to move a carriage horizontally in a controlled way. Separate photographs were required for each layer line (figure 13.8), but this was a small price to pay. In any case, in the 1930s most work was two dimensional (projections of structures on to planes) and so non-zero layer lines were not often required. Charts could be made for interpretation, but the experts did not even need these!

The precession method

The only drawback was that the Weissenberg photograph bore little relation to the reciprocal lattice, and introduced some difficulties for large unit cells, which gave spots very close together. M J Buerger (1944) set himself the problem of producing a photograph like the reciprocal lattice; his method was called the precession method (not the same as a physicist's precession.) The photographic plate (figure 13.9) had to be planar and had to follow the movement of the crystal. A typical photograph is shown in figure 13.10.

But there was a cost. The method could only be used with small angles of diffraction, and so required either small wavelengths or

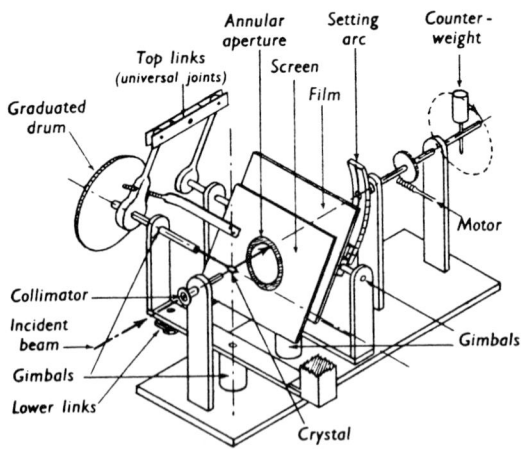

Figure 13.9 The precession camera. (From Henry *et al* (1951) p. 135, courtesy of the authors.)

crystals with large unit cell. This suited the Americans who, for historical reasons, used mainly Mo Kα radiation ($\lambda = 0.71$ Å), rather than the European Cu Kα radiation ($\lambda = 1.54$ Å), but nevertheless the method became very popular.

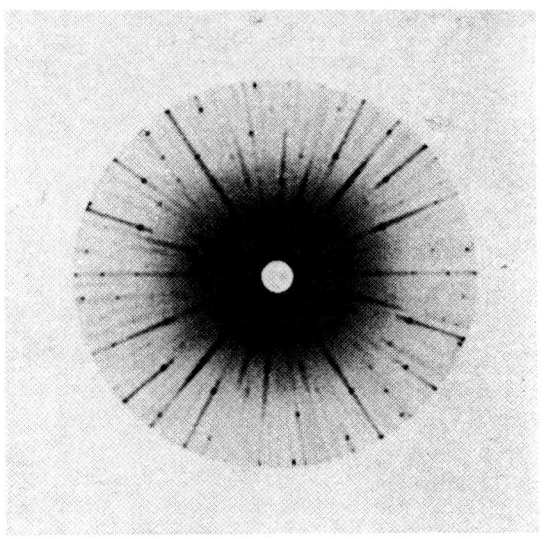

Figure 13.10 A precession photograph (from Azàroff (1957) p. 415), courtesy of the Bodleian Library.

The future

Has finality then been reached? Not necessarily. With the production of very intense sources of continuous radiation, Laue photographs can now be taken with exposures of the order of a millisecond, but it would take too long to describe the work here.

Nor have I attempted to describe the extension of the Braggs' ionisation spectrometer which can now be used automatically to give quantitative measurements of all the possible x-ray reflexions from a crystal. Bernal once prophesied that ultimately we might be able to drop a crystal in a hopper; it would be orientated automatically; all the reflexions would be measured; the structure would be derived by theoretical methods; and a paper ready for publication would emerge. Fortunately (?) we have not yet arrived at this state of affairs!

I hope, nevertheless, that my concentration on photographic methods has been worthwhile.

References

Astbury W T and Yardley K 1924 VI. Tabulated data for the examination of the 230 space-groups by homogeneous X-rays *Phil. Trans. R. Soc.* A **224** 223–57 and plates 5–24

Azároff L 1957 A new method for measuring integrated intensities photographically *Acta Crystallogr.* **10** 413–17

Bernal J D 1927 On the interpretation of x-rays, single crystal rotation photographs *Proc. R. Soc.* A **113** 117–60

Bragg W H and Bragg W L 1913 The reflection of X-rays by crystals *Proc. R. Soc.* A **88** 428–38

—— 1915 *X-rays and Crystal Structure* 1st edn (London: G Bell and Sons)

Bragg W L 1913 The diffraction of short electromagnetic waves by a crystal *Proc. Camb. Phil. Soc.* **17** 43–57

Buerger M J 1964 *The Precession Method in X-ray Crystallography* (New York: Wiley)

Buerger M J *et al* 1944 The photography of the reciprocal lattice *American Society for X-ray and Electron Diffraction Monograph* no I

Ewald P P 1921 Das 'reziproke Gitter' in der Strukturtheorie *Z. Kristallogr.* **56** 129–56

—— (ed.) 1962 *Fifty Years of X-ray Diffraction* (Utrecht: International Union of Crystallography)

Friedrich W, Knipping P and von Laue M 1912 Interferenz-Erscheinungen bei Röntgenstrahlen *Sitzungsber. Bayerischen Akad. Wiss. München. Math. Phys. Klasse, Jahrg. 1912* 303–73

Henry N F M, Lipson H and Wooster W A 1951 *The Interpretation of X-ray Diffraction Photographs* (London: Macmillan) (2nd edn 1960)

Nuffield E W 1966 *X-ray Diffraction Methods* (New York: Wiley)

Weissenberg K 1924 Ein neues Röntgengoniometer *Z. Phys.* **23** 229–38

Woolfson M 1970 *An Introduction to X-ray Crystallography* (Cambridge: Cambridge University Press)

Wyckoff R W G 1920 The crystal structure of sodium nitrate *Phys. Rev.* **16** 149–57

—— 1924 *The Structure of Crystals* (New York: The Chemical Catalog Company)

14

From Research to Development: The Early Development of Mechanical Parts of Electron Probe Instruments†

D J Unwin

In the nineteenth century science began to be an accepted subject and was being taught in the universities. Some scientific research was carried out in the newly opened Cavendish Laboratory at Cambridge and researchers were having difficulty in obtaining their apparatus.

In 1881 Horace Darwin (1851–1928), son of Charles Darwin the great natural historian, formed a partnership with Albert George Dew-Smith (d. 1903), a wealthy amateur, to form the Cambridge Scientific Instrument Company (CI Co.). This company specifically set out to design and manufacture scientific equipment for the university departments, but in a short time was supplying apparatus to scientists all over the world. It soon gained a wide reputation for craftsmanship and become the cradle of the scientific instrument industry through the world.

† Based on a presentation given at the conference on Experiments and Instruments, 20 February 1985, IOP Headquarters, 47 Belgrave Square, London.

In the years following World War II the company was under joint manager directorship, the last (W A Apthorpe) retiring in 1957 at the age of 71.

The new managing director Harold C Pritchard was the first to be appointed from outside the company. He was 49 years old and had a period as Chief Superintendent of the Woomera Rocket Range in Australia as well as other industrial experience. From 1956 the company had been under the chairmanship of Dr Percy Dunsheath who also had valuable industrial and research experience. Pritchard joined at a time of low profits and he was looking for products, particularly those already developed by research organisations, which could be made under licence.

One of these, a new advanced technology, came his way almost by chance through one of his old colleagues who was at that time Research Director of the Tube Investments (TI) research laboratory at Hinxton Hall near Cambridge. The outcome of the contact was a decision in 1959 for the company to undertake the commercial manufacture of the electron probe x-ray microanalyser which Drs P Duncombe and D A Melford had developed and made at TI as a research tool for their own use.

Production drawings were prepared from Dr Melford's design sketches and a prototype instrument was completed in the CI Co. research workshops in six months, just in time to be shown at the CI Co. London office at the same time as the Physical Society Exhibition.

The instrument called the 'Microscan' was an almost exact copy of the TI instrument and involved no development by the Company.

Because of its origin as a research tool it was not designed for large batch production and manufacturing costs were rather high. Nevertheless it sold well, 11 in the first year and over 80 in the first five years. Many new techniques had to be learned by the manufacturing teams as some of the technology, vacuum engineering for example, was quite different from anything previously used in the company.

In 1959 the company opened new research and development laboratories at Cambridge. The building had four floors, a drawing office at the top, chemistry and gas analysis next, physics with electronics on the first floor and the mechanical laboratory which also included the research workshop and stores on the ground floor.

The scanning electron probe microscope

Shortly after the opening of the new laboratories Pritchard had become interested in increasing the involvement with electron probe technology and had come to an arrangement with Professor C W Oatley of

Cambridge University Engineering Laboratory to develop for commercial manufacture the scanning electron probe reflection microscope they were working on. This project was very different from the Microscan in that much more development was to take place at CI Co. and as a result the physics and mechanical sections of the research department in particular became very involved with probe work.

In a scanning electron probe instrument a beam of electrons provided by a heated tungsten filament is accelerated by extra high tension (EHT) voltage and focused on the specimen by electromagnetic condenser and final lenses to a very small spot. Scanning coils in the final lens deflect the beam so that it scans a small raster.† In the microanalyser x-rays are emitted from the specimen surface which pass to an x-ray spectrometer where they are dispersed by a crystal set at the appropriate Bragg angle for the element under examination to a proportional counter. The signal from this is used to modulate the brightness of a cathode ray tube (CRT) display beam in synchronism with the electron beam so producing an x-ray image (figure 14.1).

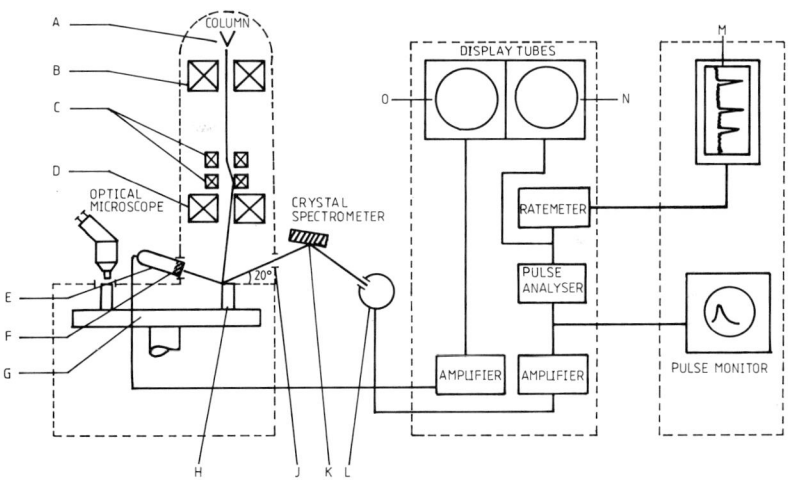

Figure 14.1 Schematic diagram of Microscan: A, electron gun; B, condenser lens; C, scanning coils; D, final lens; E, photomultiplier; F, scintillator; G, stage; H, specimen. (From CI Co. *Catalogue* 1961, D J Unwin files.)

† A raster is the pattern of closely spaced lines formed on the face of the cathode ray tube when frame and line scanning voltages are applied simultaneously. The image is formed by modulating the beam brightness as it moves over the screen.

Another CRT display beam is modulated by the signal from a scintillation counter, the phosphor of which is energised by reflected electrons. This produces the electron image which also forms the basis of the scanning electron microscope. The CRT display can be scanned slowly and so build up a complete image on a photographic plate to produce a permanent record.

An electron beam has a much shorter wavelength than visible light so has a better limit of resolution. The beam must pass through a vacuum free of contamination which requires the whole system to be evacuated.

The first electron microscopes operated by transmission of the beam through a very thin specimen which imposed limitations on the type of sample that could be examined. The scanning electron probe reflection microscope overcomes this problem with other advantages such as greater depth of focus enabling irregular surfaces to be examined without refocusing (figure 14.2).

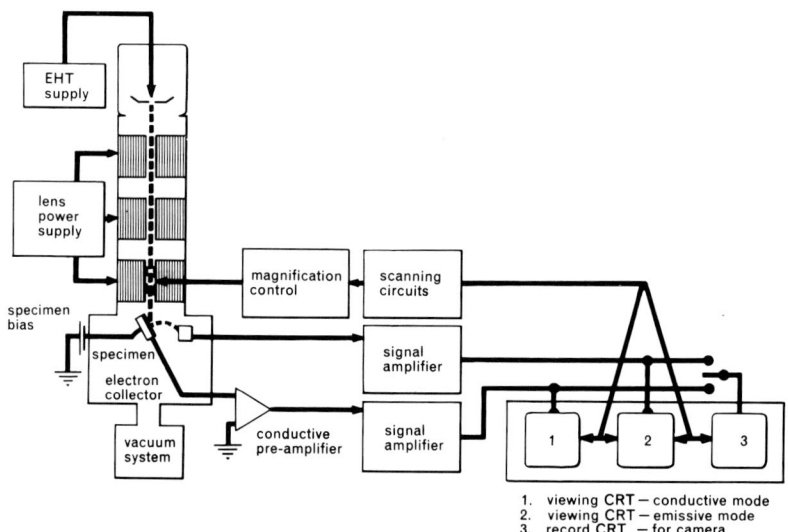

Figure 14.2 Schematic diagram of scanning electron microscope. (From CI Co. *Catalogue* 1964, D J Unwin files.)

The organisation of research and development

By the time the electron probe development work was to take place the CI Co. research department was organised into a number of groups operating as a matrix with extremely flexible cross contact between members of the different groups (figure 14.3).

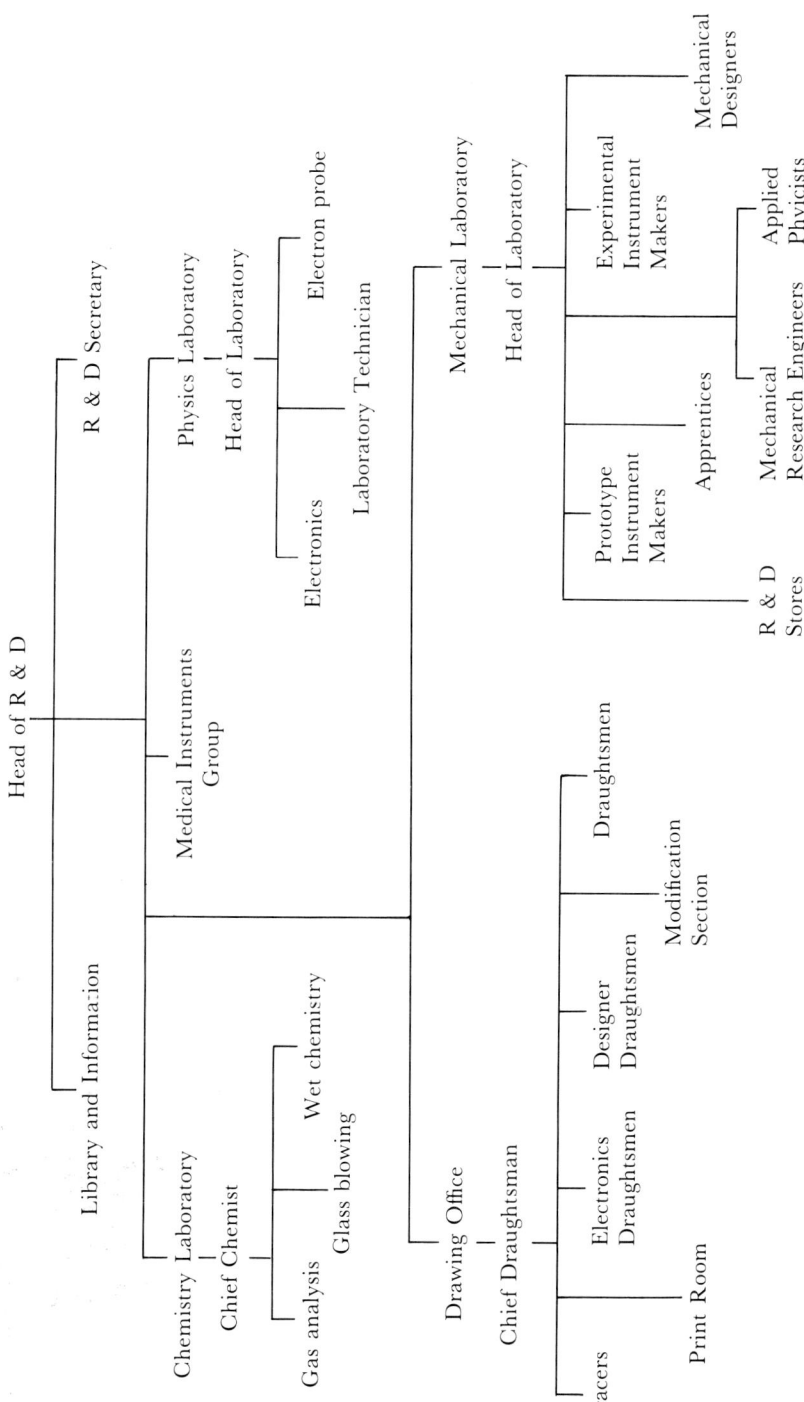

Figure 14.3 Cambridge Instrument Co. Ltd, research and development department organisation chart about 1960. (From DJ Unwin files.)

Any development project would be led by the person most suited, in the case of electron probe projects one physicist in the physics laboratory dealt with the electron microscope and another with microanalysis. Their roles involved coordination of the various subsections of the projects being undertaken by other members of the group, for example, electronic development, the drawing office and the mechanical laboratory. The major part of the electron probe mechanical development was undertaken in the mechanical laboratory, known colloquially as the Mech Lab, using its own staff or in conjunction with the drawing office.

Mech Lab staff represented a wide range of qualifications and skills, as figure 14.3 shows. This situation was ideally suited to the probe projects with their many diverse mechanical and applied physical problems.

Graduate mechanical research engineers and an applied physicist dealt with vibration, structure stiffness, specimen preparation and similar problems.

Mechanical designers were generally recruited from experimental instrument makers. They were chosen for their aptitude for design and putting ideas down on paper together with the ability to guide several experimental instrument makers. Specimen stages, gearboxes and spectrometers were typical examples of the designers' work. Some of them became extremely knowledgeable in the area in which they worked.

Experimental technicians were 'green-fingered' experimental engineers with a flair for dealing with difficult and unusual problems such as special coils, crystal manipulation, testing or checking methods for example. Like designers they became specialists in a particular field and were therefore working on several different projects at one time.

Experimental instrument makers were craftsmen able to work from sketches or verbal instructions to make experimental models. They usually worked with a research engineer or a designer and were often left to scheme out detail for themselves. Some of these people were selected for their meticulous attention to accuracy and detail so that the performance of their work was dependent entirely upon the design and not influenced by any manufacturing discrepancy. Spectrometers were an example of the work these people undertook. At the other extreme there were experimental craftsmen who were able to build up very rapidly a crude but adequate model to illustrate if some idea was feasible but with no pretence at accuracy. It was interesting that the meticulous worker was quite unable to do this type of work whilst the rough and ready worker was no use on accurate jobs. Nevertheless they both fulfilled an essential need in the laboratory.

Prototype instrument makers were selected for their integrity, skill and craftsmanship. Their job was to work accurately and intelligently to drawings, but if they noticed an error, machining difficulty or possible improvement in a component they were making they were expected to discuss it with the designer before proceeding. These people were often working with first production drawings from the drawing office, some of which had been made from the experimental or first prototype models.

The drawing office (DO) had separate sections dealing with electronic and mechanical drawings. Electronic chassis, racks, plinths, displays, control panels and vacuum systems, for example, were designed on the drawing board, the draughtsman working in conjunction with the physicist or the Mech Lab. They also prepared production drawings from experimental models or first prototypes. Tolerance fixing, always a tricky path to tread, was also a DO duty. Generally a production drawing would be checked by a prototype instrument maker producing the component.

To enable it to deal with the diverse problems presented the Mech Lab facilities were comprehensive. Apart from the usual desks, laboratory benches and drawing boards there was a constant-temperature room, optical room, vibration generation and analysing equipment, Talyrond† roundness/flatness and other measuring and testing apparatus.

In addition the workshop area was equipped with good quality precision machine tools and supporting equipment able to deal with every type of work encountered. Throughout the 1960s the Mech Lab expanded due to the increase in mechanical development work both in the probe and other fields. Towards the end of the decade it became part of a larger R&D technical services group ultimately incorporating the drawing office as well (figure 14.4).

Detailed development work

The obvious point to start development of the scanning electron microscope (SEM) was to take the Microscan column as this provided a ready-made electron optical system. It needed some modifications to suit the different requirements of the SEM. The physics laboratory was responsible for the overall planning with the DO and Mech Lab closely involved. There was considerable liaison between the CI Co. staff and Professor C W Oatley's team at the University Engineering Labora-

† Talyrond was the proprietary name for the precision metrological instrument designed and made by Rank Taylor Hobson Ltd, later Rank Precision Industries Ltd of Leicester, for accurately measuring circularity and flatness.

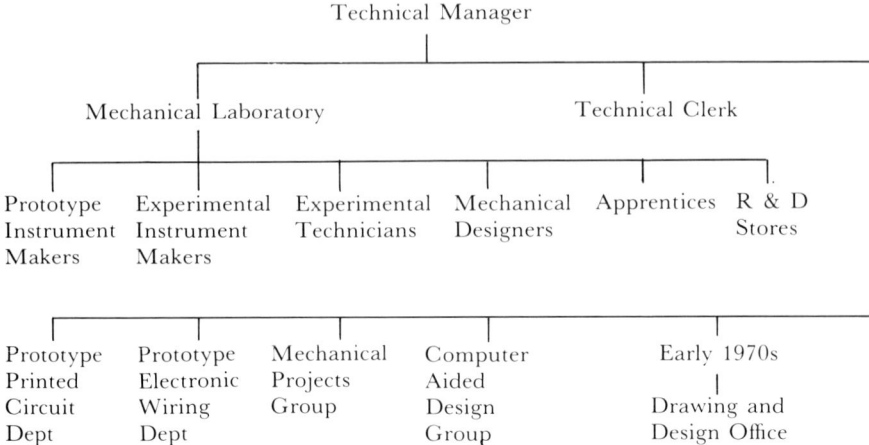

Figure 14.4 Cambridge Instrument Company Ltd, R&D technical services group organisation chart, late 1960s. (From D J Unwin files.)

tory. After the appropriate parts had been extracted from the Microscan production a start was made by the research department in 1961.

The modifications necessary to the final lens contributed to the satisfactory resolution of 50 nm achieved by the experimental instrument which was completed in 1962. There was a great thrill amongst all involved when the first picture of a diatom appeared. Needless to say there were a number of serious problems to be solved, of which vibration was one. Any relative movement between the column and the specimen degrades the resolution. The rigidity this demands, together with support for a specimen 12 mm diameter having X, Y and Z linear movements and rotation posed considerable design problems on the specimen stage. An experimental stage was designed and constructed in the Mech Lab but whilst it was adequate for experimental work it was quite unsuitable for a commercial instrument, partly due to the limitations imposed by the Microscan specimen chamber.

The vacuum system consisted of a backing pump and oil diffusion pump which, although coupled to the column by a flexible bellows, was mounted within the plinth on which the instrument was mounted.

Both pump and ground vibrations were transmitted to the column so degrading the performance.

One of the objections to the Microscan column was the need to realign the components each time it had been taken down for cleaning. This was a fairly frequent occurrence as dirt such as oil from the pump contaminated the bores and caused deterioration in performance.

The realigning operation was irritating, time consuming and required some knack on the part of the operator. Experience with the experimental instrument showed us that the more stringent requirements of the SEM made it essential that we find a solution to this problem.

The Microscan specimen chamber was rectangular and being of brass provided inadequate screening. Other designs which provided more convenient side access ports and were made of steel were considered, one of the original sketches made at a design discussion is reproduced in figure 14.5.

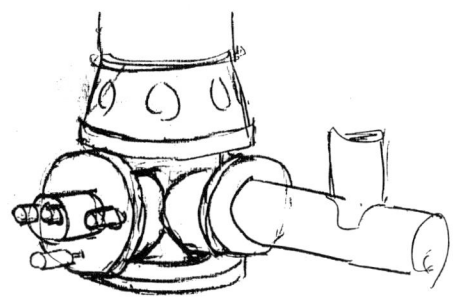

Figure 14.5 Reproduction of original design sketch of a SEM specimen chamber proposal, 1963. (From D J Unwin files.)

The final form was a square with a large circular port in each face. Each of the four corners was angled off at 45° and had two smaller round ports one above the other. The specimen chamber was made from a slab of mild steel flame cut to shape and machined all over. It was immediately given the nickname 'threepenny bit', a name which, like the design, remained unchanged throughout the production life of the instrument.

Designing the lenses involved a considerable amount of calculation, all executed on a desk calculator as computers did not become available until some years later. After the physicist had defined his requirements these had to be interpreted into a practical form which could be manufactured. As the pole pieces of the lenses formed part of an electron optical system the quality of the spot was dependent upon the accuracy of the components in the same way as in the light optical system. This imposed geometric tolerances on machined components comparable with those used on glass optical components and the soft iron parts had to be designed in a form which would enable the desired accuracy to be achieved. An example is the final lens. The sketch reproduced in figure 14.6 was drawn on the desk by the mechanical

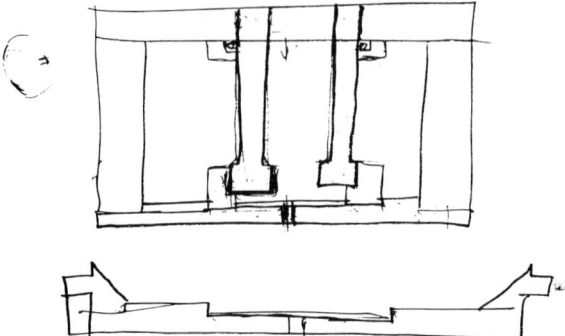

Figure 14.6 Reproduction of original design sketch of Stereoscan final lens, 1963. This design was used until the mid-1970s. (From D J Unwin files.)

designer during a discussion with the physicist and was to become the basis of the lens design used throughout the production run of the 'Stereoscan' as the SEM was named. The central zone of the bottom plate needed to be flat to within a few Newton rings whilst the bore to be less than $0.3\,\mu$m out of round. Similarly, the upper pole piece needed to be of the same order and also mounted precisely with respect to the lower plate. When the Mech Lab approached various manufacturers for advice about lapping surfaces to these degrees of accuracy they were inundated with representatives all anxious to show us how to do it. Unfortunately they were all very experienced in producing surfaces of this type on hard materials but electron optical lenses have to be made of soft iron and their techniques did not produce the desired results. They all left despondent and unable to help us. Left on their own the Mech Lab were able to devise means of producing the required surfaces on the soft iron by lapping using carefully chosen laps and certain types of lapping compounds. Producing the finish was only half the problem as this had to be combined with circularity, straightness and flatness. Measuring the flatness of the small central zone of the plate was relatively easy using glass proof planes although the larger surfaces were a bit of a problem. Determining circularity of the bores was more difficult. At that time the Rank Taylor Hobson (RTH) Talyrond was the best tool available for checking circularity but it was expensive. The nearest machine to Cambridge was at Leicester so we had to take our work 80 miles each time we needed to check the result of a test lapping. This made progress slow and tedious. When Mr Pritchard heard of the problem he told me to go to RTH and see if I could get a Talyrond quickly and if so to order it on the spot. Armed with the order, a visit to Leicester enabled me to extract a machine to be delivered in a week or two. After it was installed and commissioned in the Mech Lab constant-temperature room we were able to solve the lapping problem within a week because we were able to check the trend after every few minutes of working. Later the representatives of the firms supplying the lapping materials came back to try to find out how we were getting the desired results. However, we were not forthcoming and the methods were known only to the few people doing it.

To isolate the column from ground and pump vibrations the whole column assembly was mounted on the top plate of the plinth. This plate, which was very stiff, was coupled to the plinth framework by large low-natural-frequency coil springs. These were compressed to the correct working position by the downward force exerted by the flexible vacuum connection located in the centre of the bottom of the specimen chamber (figure 14.7). To design this system knowledge of the amplitude and frequency of the disturbing vibrations was needed. For many years CI Co. had been making a successful recording universal vibrograph which was well suited to measuring frequencies up to about 100 Hz, the range in which we were interested. The instrument was used to provide data necessary to design the isolation system. Later the

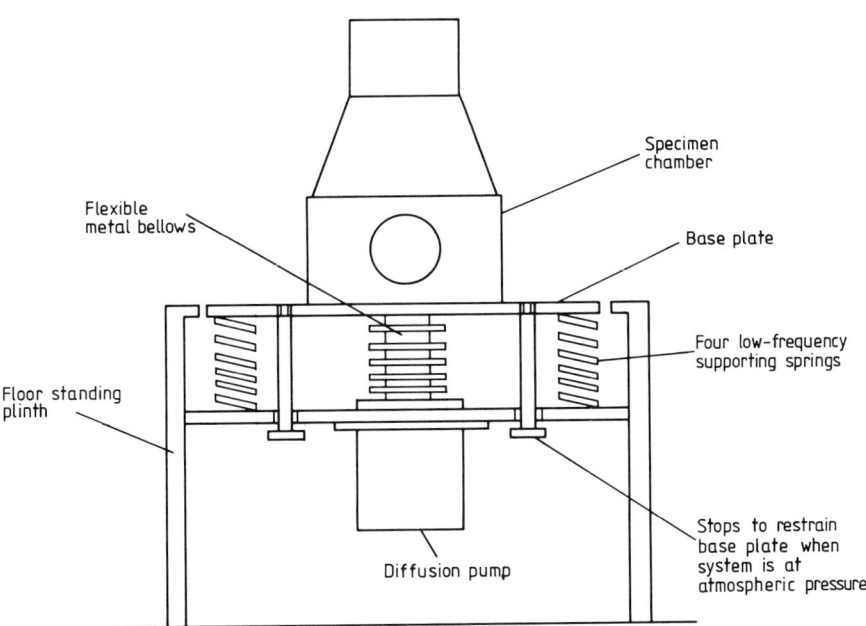

Figure 14.7 Diagram of Stereoscan column isolation mounting, 1963. (From D J Unwin files.)

service department had a vibrograph allocated to them which they used to advise customers of the suitability of their site when installing Microscans and Stereoscans.

By choosing very tight geometric tolerances for column components and the mating faces it was possible to design a prealigned column which would need no adjustment after reassembly. The clamping rings of the Microscans were unnecessary as the column was held rigid when under vacuum by the atmospheric pressure acting outside. Simple clamps were fitted to the outside to hold the joints sufficiently close to seal the O-rings during pump-down. Production instruments were always despatched in a partially pumped down condition to ensure that the column was rigid. To ensure the geometric concentricity, parallelism, circularity and flatness of the components needed the development of special manufacturing techniques in the laboratory workshop to make the first prototype. The lathes used for this work were first checked to make sure that they turned circular and flat, that is the mandril was free of cam action or end movement, by testing specially made test pieces on the Talyrond. Before any new machine was purchased the manufacturer was required to allow us to turn and check a test piece on our Talyrond. It was interesting to discover that it was not the most expensive lathes which were best in this respect.

While the engineering work was going on the applied physicist had to fit up the evaporating plant so that non-conducting specimens could be aluminised to dissipate the charge that would build up on the surface.

Production and subsequent developments

The first prototype was satisfactorily completed in June 1964 and sold to Du Pont USA (figure 14.8). When their representative was trying the instrument out at Cambridge he was so excited that he cut a piece out of his shirt and asked for it to be prepared so that it could be examined in the microscope. Elaborate arrangements were made to transport the instrument to their site in the USA by air and road. All this was to no avail, however; when the crate was opened by our physicist he found the plinth/column assembly badly damaged. When the remains were returned to Cambridge we had the job of deciding what may have happened for the insurers. It was almost certain that the crate had been slid down a ramp from the back of a lorry when in transit from the airport to Du Pont site.

To the Mech Lab also fell the task of making a replacement instrument. This was achieved in three months, largely by subcontracting a lot of the work. The exercise was used to prove to the production departments that the tight tolerances achieved on the prototype column were possible on a production basis. Microscan components with which they were familiar were far less demanding than those of the Stereoscan and they felt that they would not be able to make them economically.

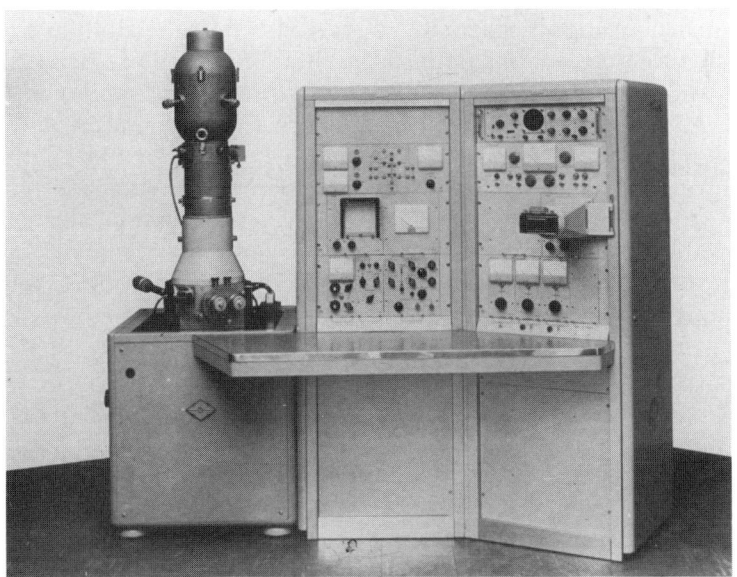

Figure 14.8 The first prototype Stereoscan, June 1964. (CI Co. photograph, D J Unwin files.)

By subcontracting the work to a firm of precision engineers who had no difficulty whatever making the parts for an acceptable price, we were able to show the works that manufacturing was economically possible. When manufacture of the Stereoscan started the Mech Lab provided support by transferring the technology to the production teams.

Learning from the mistakes of the experimental specimen stage, a new one to fit the steel chamber was designed in the Mech Lab. To enable the specimen to be changed the stage had to be easily removed from the chamber. Orthogonal traverses were required to the specimen of 12 mm each direction and 10 mm vertically. In addition the specimen had to be tiltable from horizontal to vertical and to have 360° rotation. Large-diameter micrometer heads were fitted to enable the specimen to be positioned to 0.01 mm. Backlash-free movements were essential. Simple stubs 12.5 mm diameter were designed to take the specimens. These stubs slipped into a socket with kinematic locations on the tilting cradle of the stage.

The new specimen stage was arranged so that the specimen-supporting assembly was clamped to the underside of the bottom lens plate when in use, thus reducing relative movement and loss of definition. After one batch of instruments had been made this was found to be an unnecessary complication and the much simpler system of bottom clamping illustrated in figure 14.9 became standard for all types of stage used throughout the life of the Stereoscan. Since Darwin's time the principles of kinematic design† and controlled constraints had been used by CI Co. for precision apparatus, and these principles were used in the specimen stage. Some of the Mech Lab mechanical designers became very experienced in stage design and continued to develop the basic stage to incorporate many extra degrees of freedom for special requirements.

One of the problems which consumed a lot of effort was the assessment of bearing materials suitable for use in a vacuum. Useful advice was forthcoming from NASA but there were significant differences between the space application and the electron probe system. Space applications have a vacuum 'pump' of infinite capacity and contamination is not a problem. The probe system has limited pumping capacity, whilst outgassing can increase pump-down times unacceptably and contamination degrades the beam involving unnecessary cleaning of the column. Some of the materials were new to us and the techniques of machining, moulding and manipulation had to be mastered, then the methods transferred to the production departments.

All brass components in the vicinity of the electron beam had to be non-magnetic. This needed the introduction of special brasses with no ferrous inclusions. Processes were developed to clean the finished components of ferrous particles picked up from the tools during machining and tests for the initial acceptance of the raw material and acceptance of the finished parts.

An interesting development was the aperture holder. Apertures consisted of three small platinum discs 3 mm in diameter with central

---

† Kinematic design, sometimes called 'geometric' design, is a concept of design first proposed by J Clerk Maxwell in 1876. He stated that any rigid body has six spatial degrees of freedom. If the rigid body is to have N degrees of freedom it must have $6N$ points of constraint with reference to a second body. Hence if a rigid body is constrained in more than six ways it will become strained or distorted and subjected to internal stress. The three-leg stool is a well known example whereas the four-leg chair is over-constrained. This concept was used extensively in Cambridge designs.

holes of a few micrometres in diameter. They were located in the small space between the upper and lower pole pieces of the final lens. In addition to the three positions the slide had a fourth blank position which provided means to seal off the column so that it was only necessary to let the specimen chamber down to atmosphere when changing specimens.

The slide was precisely located in any of the four positions and moved by an external lever. Two external knobs provided a small amount of orthogonal movement to the aperture discs which were held in kinematic recesses in the holder slide. The whole assembly had to be easily withdrawn from the column to enable apertures to be replaced.

Platinum apertures were originally purchased from outside suppliers but as the numbers required increased supply became a problem. We devised a method of coining these from sheet platinum thus easing the supply. Fortunately the scrap price of the platinum waste was almost as high as the new sheet so that offcuts and work spoilt during the experimental activities could be sold at little loss.

The phosphor tip of the scintillator required some development to produce the shape and polish, whilst work was also required to select the adhesive to fix the tip to the Perspex light pipe. Adhesives used had to be suitable for working in a vacuum and the refractive index to match that of the light pipe material.

Due to the extreme sensitivity to light of the photomultiplier and the high voltages used, the housing involved elements of design not familiar to us. This, as in many other areas of the work, meant that the team had to learn many new 'tricks of the trade'. It became clear that our concepts of light traps had to be abandoned and the photomultiplier housing designed as a hermetically sealed chamber.

Laboratory involvement did not finish when full production of the instrument had started. Electroplating or painting could only be used as protection on non-fitting surfaces due to thickness variations. The need to keep the unprotected mild steel and soft iron surfaces scrupulously clean permitted corrosion to take place. We were able to find and introduce a phosphor nickel chemical plating system which enabled us to overcome the problem on many of the surfaces. Unlike electroplate, the thickness deposited was uniform over all surfaces to within 10 per cent and the plater could guarantee a deposit thickness to within 10 per cent of that specified. This enabled close-tolerance surfaces to be machined a specified amount over or under size then plated up the correct amount to bring the surface to the required dimension and still maintain the tolerance. All mild steel and soft iron components except a few pole pieces were subsequently plated in this manner. Another associated problem involved corrosion arising from the finger marks of assemblers. We found out that the perspiration of young men up to about age 24 and women at certain times was very corrosive. The trouble was cured by providing people in these groups with cotton gloves to wear when assembling critical column components.

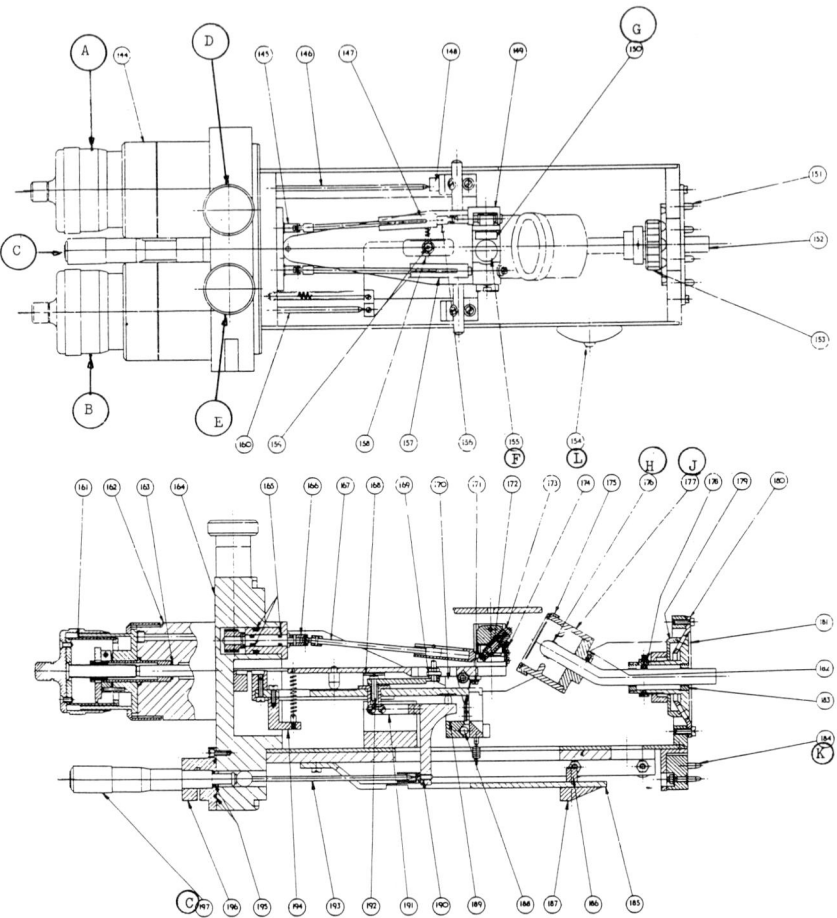

Figure 14.9 Standard specimen stage: A, *Y* traverse micrometer head; B, *X* traverse micrometer head; C, *Z* movement micrometer head; D, specimen tilt control; E, specimen rotation control; F, specimen stub; G, focusing stub; H, scintillator tip and light pipe; J, electron collector body; K, connector plug; L, EHT connector. (From CI Co. instruction manual, D J Unwin files.)

As the electron optical performance of the instrument was improved the slag inclusions in the mild steel column components began to become significant. The lens bodies were machined from billets of good commercial quality mild steel in which quite large randomly distributed slag inclusions occurred. A component could not be verified as satisfactory until it was completely machined, by which time it was expensive to scrap. Ultrasonic or x-ray examination did not prove satisfactory in revealing the smaller inclusions. Discussions with a steel producer enabled us to negotiate the supply of a vacuum remelted steel in which the inclusion size did not exceed 0.025 mm. It was considerably more expensive than ordinary mild steel and less easy to machine but stopped the costly scrapping of finished lens bodies.

An interesting example of the combined efforts of the physicist and the Mech Lab experimental engineer was the development of the three coil assemblies. These consisted of beam-bending coils, stigmator coils† and scanning coils. All the completed assemblies had to be encapsulated for use in the vacuum. It was important that the encapsulating material exerted no distorting forces during curing or subsequently and had no plasticiser in it to leach out when under vacuum. The experimental engineer worked out and made jigs, tools, fixtures and methods to machine from solid the special shaped ferrite formers, wind them with the wire coils and encapsulate the whole assembly. Scanning coils were wound and preformed on a specially made tool. The core was a very thin non-magnetic brass tube with the coil-locating lugs of epoxy resin cast on the circumference in a specially developed mould. Again after mounting and fitting connectors the whole was encapsulated to prevent any movement of the coils.

The Geoscan

While the Stereoscan work was going on microanalyser development was also proceeding. Pritchard had come to an arrangement with Dr J V P Long of the Department of Mineralogy and Petrology, University of Cambridge, for the company to work with him on the development of a microanalyser designed primarily for geological specimens. This instrument became known as the 'Geoscan'. It was a very different instrument from the relatively simple Microscan, having two programmable precision fully focusing x-ray spectrometers, pre-aligned column, variable-speed specimen movements, automatic standard selection and other advanced facilities. The specimen chamber volume was small, which permitted rapid pump-down after

---

† The stigmator coil is an electromagnetic coil which when energised changes the shape of the electron beam. Several are spaced equally round the axis and can be energised independently. Their purpose is to correct beam deficiencies such as astigmatism.

changing the large interchangeable combined specimen and standard holder. The first prototype was completed in 1964.

Some of the mechanical development work was of a similar nature to that of the Stereoscan and went on in parallel, often being undertaken by the same experimental engineer.

Possibly the most demanding project was the fully focusing spectrometer. The required Bragg angles had to be preselectable, repeatable to within six seconds of arc and within two minutes of arc absolute accuracy (figure 14.10). Three crystals had to be quickly interchangeable without letting the chamber vacuum down to atmosphere. The original idea for providing the movements of the counter carriage was suggested to us by the way in which the level luffing crane† maintains a

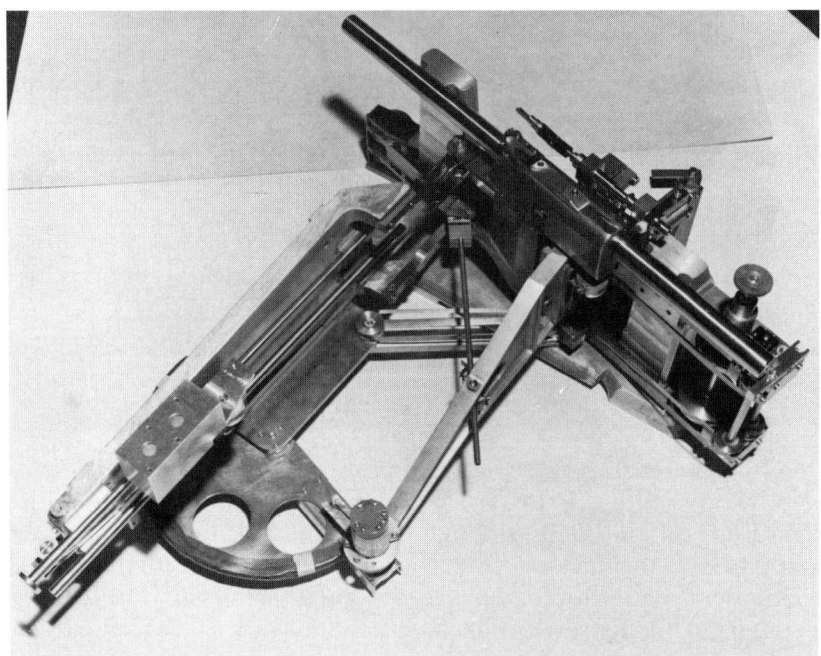

Figure 14.10 Geoscan fully focusing spectrometer. The specimen is located in line with the main track at centre top. The counter carriage without counter or slit is at bottom left. The main drive to the gear wheel is at right centre. (CI Co. laboratory photograph, D J Unwin files.)

† The level luffing crane is a particular type of crane, often seen in harbours, such that when the load is moved towards or away from the crane, called 'luffing', the height of the load remains unchanged, unlike many other types of crane.

level load regardless of the luffing angle (figure 14.11). Various movements of the two carriages and the crystal which maintain the $L1 = L2$ and $\theta 1 = \theta 2$ relationship were controlled by a number of interconnected flexible beryllium–copper tapes. Kinematic or semi-kinematic design principles were used throughout to ensure continued accuracy during long use. Whilst the first experimental model spectrometer components were fabricated of brass, light alloy castings were used for the first prototype and subsequent instruments. To maintain the necessary accuracy the castings had to be stabilised after rough machining and finish-machined at a constant known temperature to ensure precise dimensions. As accurate positioning of the movements was needed a 400 Hz AC servo-system was used, a positional encoder driven from a precision-geared quadrant provided angular position. Checking component parts, straightness of tracking, repeatability and absolute angles involved the adaptation and development of optical measuring devices. Lasers were not available at that time.

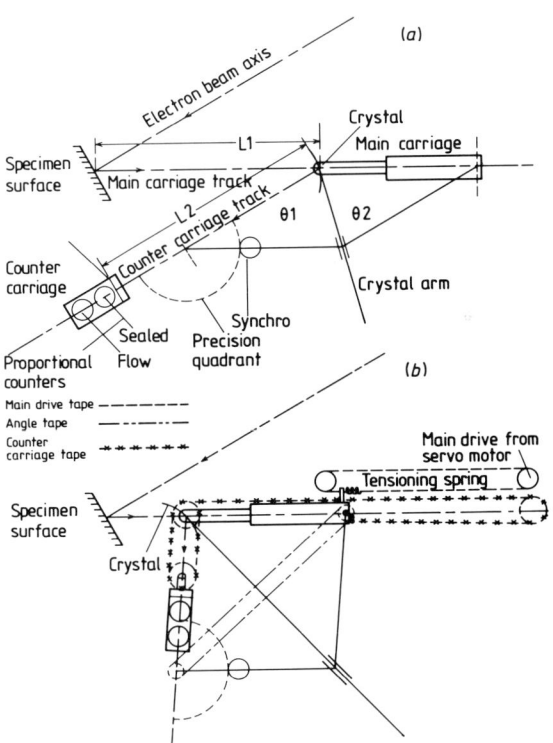

Figure 14.11 (*a*) Schematic layout of Geoscan spectrometer. (*b*) Schematic layout of connecting flexible tape drives. (From D J Unwin files.)

A spectrometer command unit of comparable precision was also needed. This has means to preset Bragg angles, sweep or step scanning over a range of angles and speeds. Designed by us, these were ultimately manufactured by a firm specialising in high-precision gearing who also provided the precision quadrant for the spectrometer.

An interesting problem came to light some years later when the Geoscan was being upgraded to the Microscan Mk5. To permit computer control we were replacing the spectrometer analogue positioning system by a digital system using a circular grating and stepper motors. When modified the spectrometer would no longer repeat to within six seconds of arc which had been easily achieved with the AC servo-system. We found that the 400 Hz servo-motors induced just sufficient vibration in the mechanical system to overcome static friction and so reduce hysteresis. It was solved by artificially introducing high-frequency vibrations.

One of the experimental engineers was involved in devising methods of preparation of the large crystals for the fully focusing spectrometer. These had to be cut, bent and cemented to the curved holder, then ground to a precise curvature. Some of the crystal materials were very difficult to manipulate requiring considerable expertise. After we had been working on the project for a while we asked the advice of an expert from Warren Spring Laboratory. When he left us he complimented our engineer and said that he felt that there was little he could teach us.

Special crystal cutting equipment was obtained and a custom-built machine for grinding the curved surfaces of the holder and the mounted crystal was built in the Laboratory.

Making the stearate crystals was a challenge to both the chemist and the experimental engineer who had to devise a special bath for producing the film and depositing it on the mount.

The orthogonal traverses of the specimen stage were to be operated in both the manual and automatic modes covering the relatively large area of 45 mm × 80 mm (figure 14.12). As the automatic system had to have constant and high-velocity modes and positional mode, a DC servo-system with tacho feedback and precision screws was used. Hysteresis had to be kept to an absolute minimum so considerable ingenuity went into the anti-backlash mechanical design.

The proportional counters were fabricated of stainless steel with Kovar glass-to-metal seals at each end. To avoid problems with flux residues a vacuum brazing furnace was installed; this was later to be transferred to the production department in the same manner as much of the other specialised equipment that had been obtained for development purposes. One counter was gas filled and sealed whilst the other had gas passing through it. Both had thin Mylar windows which had to be vacuum sealed, more of a problem than had been anticipated. The gas supply tubes to the flow counter had to be of very flexible plastic to avoid biasing the spectrometer position. No plasticiser could be permitted in the tube material as this might leach out when under vacuum and cause stiffening. This happened on one occasion when an incorrect batch of tube was supplied.

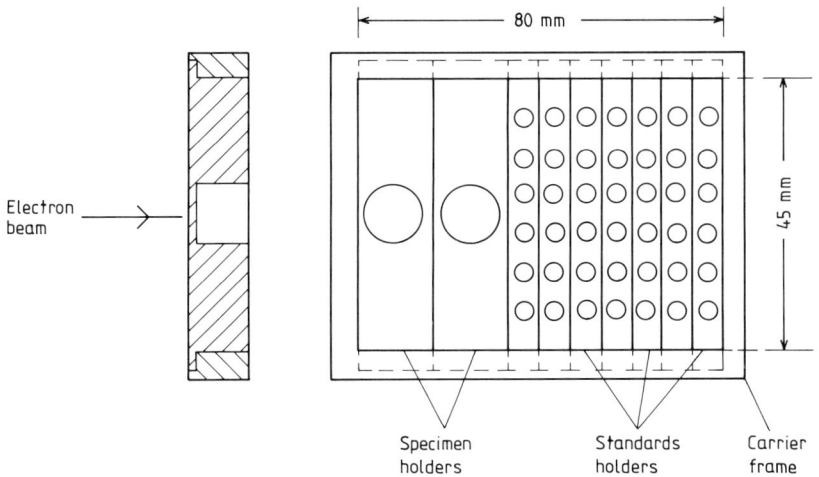

Figure 14.12 Diagram of Geoscan specimen and standards holder with interchangeable holders (From D J Unwin files.)

Unlike the Microscan and Stereoscan, which had vacuum enclosures of wrought materials, the greater size of the Geoscan main frame and spectrometer chamber necessitated the use of light alloy castings. Finding an economical method of making them vacuum tight and with a smooth surface on the vacuum side proved to be difficult. Vacuum impregnation was not wholly reliable, certain solvent-free epoxy paints were more successful.

The oil-filled electron gun was of different design to that used on the Microscan and Stereoscan which were air insulated. It used a large ceramic insulator designed in cooperation with a ceramics manufacturer. Providing the accurate surfaces for the O-ring seals required some development.

Problems of oil leakage were also experienced with the filament holders, which were a casting of filled epoxy resin. These were cast in moulds then a groove machined to receive the O-ring seal. The leak occurred where the filler particles were exposed by the machining operation. The problem was not cured until means were found to accurately mould the O-ring groove and eliminate all machining.

All the guns used a standard tungsten wire filament or 'hairpin'. For many years these were a bought-in item but supply difficulties prompted us to investigate the possibility of manufacturing our own. A Kovar seal base was designed and a source located whilst bending jigs were made to bend the tungsten wire to form the filament. Spot welding of wires of special materials had been a process used at Cambridge for many years and the equipment was adapted to weld the 'hairpins' to the base wires.

Some years later the Mech Lab became involved in developing means to machine lanthanum hexaboride pins which were to replace tungsten filaments in a new generation of guns. Considerable assistance was provided by the Royal Radar Establishment (RRE) at Malvern on this work, which also supplied us with suitable equipment.

Converting a laboratory technique into a production process was also undertaken by the experimental engineer doing the work.

In addition to examining the specimen under the influence of the electron beam a sophisticated light optical viewing system was developed working in cooperation with R & J Beck Ltd. It provided means of viewing both sides of transparent specimens, also viewing opaque specimens from the beam side (figure 14.13), all with high- or low-power objectives.

In addition simultaneous viewing of the specimen surface for probe-induced luminescence was provided. The high- and low-power objective lenses were carried on a slide mounted between the final electron lens and the specimen. This slide, which could be operated from outside the chamber had a third position which carried the scanning coils for use in the probe mode. Limitations of space, long light paths and difficulties of vacuum sealing contributed to a very demanding project ultimately carried to a very satisfactory solution.

An amusing incident occurred in the Mech Lab when the first instrument was being assembled and tested. Unlike the Stereoscan the Geoscan beam is horizontal about 0.75 m from the ground. The partially assembled but working instrument was being demonstrated to a party of international company representatives who were gathered around. As an aside the technician demonstrating said that exposure to the beam in the region of the groin was believed to cause males to become sterile. It was interesting to observe which of the party moved away and which moved in closer! Actually there was no risk as the column was adequately screened.

Dr Long used the first Geoscan to examine some of the rock brought back from the first Moon landing.

The many new mechanical processes introduced by the research team during the electron probe development work, some already mentioned, included precision machining of lens components, lapping accurate surfaces, ferrite machining, ultrasonic machining, spark erosion machining, friction welding, vacuum brazing, epoxy resin casting, crystal cutting, cementing and precision radius grinding, aperture making as well as many specially devised measuring, checking and inspection systems.

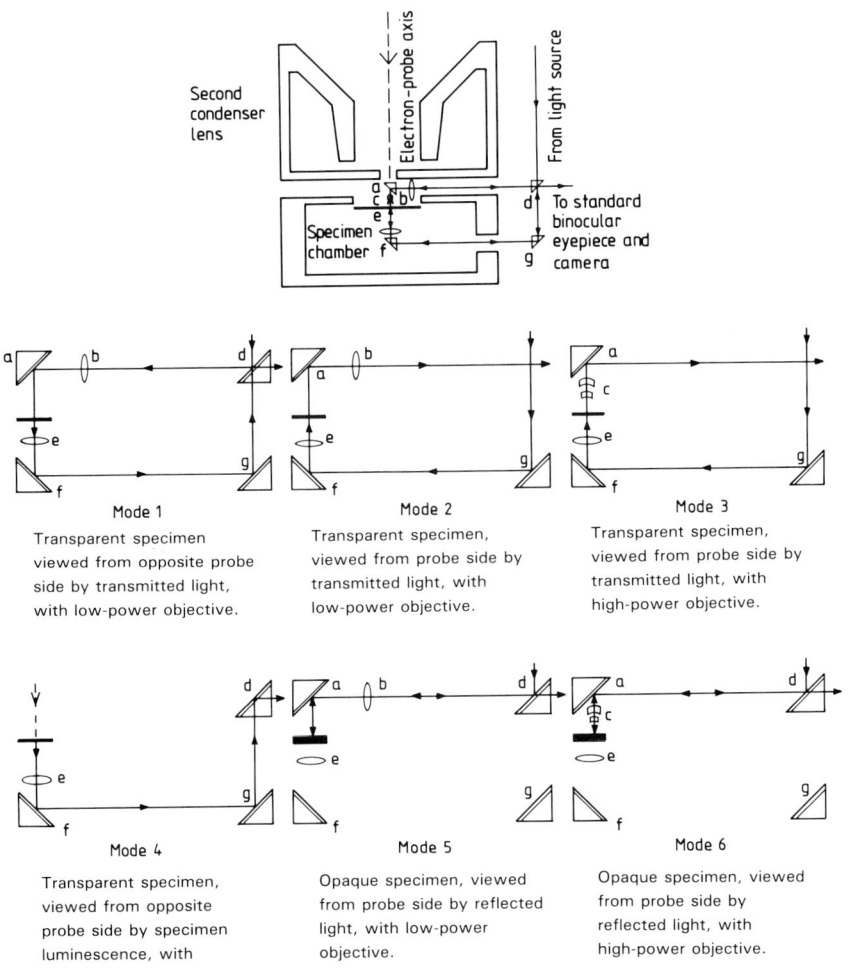

Figure 14.13 Geoscan light microscope viewing system showing six modes of operation. Binocular viewing head, low-power magnification ×160, high-power magnification ×600. b, e, low-power objective lenses. c, high-power lens. a, d, f, g prisms. a, b, c and d are interchangeable and removable under vacuum. (From CI Co. *Catalogue* 1967, D J Unwin files.)

Conclusion

After the successful launching of Stereoscan Mk1 and Geoscan Mk1 electron probe development work still continued. However, the work was one of improvement and devising new versions which were variants of the original designs so lacking the challenges, problems of breaking new ground and the satisfaction of creating the first commercial scanning electron microscope in the world.

Further information

Readers interested in the history of the Cambridge Instrument Company Ltd are referred to the excellent book *Horace Darwin's Shop* written by M J G Cattermole and A F Wolfe and published by Adam Hilger, Bristol, 1987.

The original Tube Investments prototype instrument and an early example of the Stereoscan Mk1 can be seen in the Optical Gallery of the Science Museum, South Kensington, London.

A Microscan Mk1 of the first CI Co. batch, originally sold to Pilkington Glass Co Ltd, can be seen at the Cambridge Museum of Technology, Riverside, Cambridge.

The D J Unwin files concerning the mechanical development of the electron probe instruments together with a great deal of other archival material relating to the Cambridge Instrument Company are preserved in the Manuscripts Room, Cambridge University Library, West Road, Cambridge, located under the class heading 'Cambridge Scientific Instrument Company archive'.

15

Meteors and Meteor Showers: An Historical Perspective 1686–1950†

David W Hughes

Early views on meteors

Meteors are starlike objects that appear irregularly to shoot across the sky. Their paths are linear and they are usually visible for, at most, a second or two. Nowadays we know that they are caused by minute, sand-grain-sized, interplanetary dust particles hitting and burning out in the Earth's upper atmosphere. The word 'meteor' comes from the Greek μετεωρον (plural μετεωρα) which is now used as the root of the word describing lower-atmospheric phenomena—meteorology. Aristotle (384–322 BC) had placed fire as the highest sphere of our earthly world and all transitory luminous phenomena not taking place in the celestial realms of eternal aether were banished there—and were called meteors, 'the higher things'.

Aristotle (1952) described two types of earthly exhalations in his *Meteorologica*. The first, a vapour, rose up from water; the other was dry, windy and smoky and issued from crevices. This 'fire' rose to the top of the sublunar region 'immediately beneath the circular celestial motion' and was inflammable. According to Aristotle 'whenever then conditions are most favourable this composition bursts into flames

† Based on a presentation given at the conference on the History of Atmospheric Physics, 2 April 1986, at the City University, London.

when the celestial revolution sets it into motion'. The celestial phenomena so produced depended on the quantity and location of the inflammable material and the speed of its ignition†. Slow burning resulted in comets, fast burning in meteors. Other conditions produced aurorae and lightning and haloes. Aristotle emphasised the fact that meteors were not far away, 'a proof of which is the fact that the speed of their movement is comparable with that of objects thrown by us, which seem to move faster than the stars and Sun and Moon because they are close to us' (Aristotle 1952 p. 35). To quote Burke (1986), 'Aristotle's writing style is pedantic and didactic. Nevertheless, it exudes self-assurance and confidence. His descriptions had just enough verisimilitude to make them palatable and convincing to naturalists for many centuries.'

Seneca (c. 4 BC–AD65) describing fireballs, wrote that 'our sight does not discern their passing but believes the entire path is on fire whenever they fly' (Seneca 1971).

The persistence of Aristotle's ideas is exemplified by Isaac Newton (1642–1727) who wrote in his *Optiks* (1730) 'sulphureous steams, at all times when the Earth is dry, ascending into the Air ferment there with nitrous acids, and sometimes take fire cause lightning and thunder and fiery meteors'.

The first serious challenge to Aristotle's meteor hypothesis was proposed by Edmond Halley (c. 1656–1743), Astronomer Royal. Unlike Aristotle, Halley continually tried to *measure* quantities. The height of the Earth's atmosphere is a typical example. Halley published, in 1686, a paper entitled 'A discourse of the rule of the decrease of the height of the mercury in the barometer, according as places are elevated above the surface of the Earth; with an attempt to discover the true reason of the rising and falling of mercury, upon the change of weather'. In this he calculated that the atmosphere only extended to a height of some 40 to 45 miles. So if meteors were ignited exhalations, they *had* to occur below this height.

Observations of fireballs, such as that seen by G Montanari (1633–87) on 21 March 1676, led Halley (1714a) to write a paper entitled 'An account of several extraordinary meteors or lights in the sky'. Addressing the Royal Society, and clearly thinking of Aristotle's words he said (Halley 1714a)

> It may deserve the Honourable Society's Thoughts, how so great a Quantity of Vapour should be raised to the very Top of the Atmosphere, and there collected, so as upon its Accension or otherwise Illumination, to give a Light to a Circle of about 100 Miles Diameter, not much

† For modern definitions of meteors, meteoroids, fireballs, detonating meteors and meteorites see p. 291 below (ed).

inferior to the Light of the Moon; so as one might see to take a Pin from the Ground in the otherwise dark Night. 'Tis hard to conceive what sort of Exhalations should rise from the Earth, either by the Action of the Sun or subterraneous Heat, so as to surmount the extream Cold and Rareness of the Air in those upper Regions: But the Fact is indisputable and therefore requires a Solution.

What sort of vapour was it? The Montanari fireball passed over Italy with a hissing and rattling noise. It looked bigger than the Moon. Halley wondered what substance could be so impelled and ignited,

there being no Vulcano or other Spiraculum of subterraneous Fire in the N.E. parts of the World, that we ever yet heard of, from whence it might be projected?

Reports of meteors and fireballs were not uncommon and Halley stressed

... that this sort of luminous Vapour is not exceedingly seldom thus collected.

He also noted that 'it could not be less swift than 160 miles in a minute of time'. And this was of the same order as the Earth's orbital velocity. He wondered 'how such an impetus could be impressed on the body thereof, which by many degrees exceeds that of any cannon ball; and how this impetus should be determined in a direction so nearly parallel to the horizon'.

All this led him to propose the extraterrestrial origin of meteors (Halley 1714a):

I have much considered this Appearance, and think it one of the hardest things to account for, that I have yet met with in the phenomena of meteors and am induced to think that it must be some Collection of Matter form'd in the Aether, as it were by some fortuitous Concourse of Atoms, and that the Earth met with it as it passed along in its Orb, then but newly formed, and before it had conceived any great Impetus of Descent towards the Sun.

That interplanetary space contained objects with random orbits was only just coming into vogue. Fifty years earlier, in 1665, Robert Hooke (1635–1700) had concluded that space was empty. He had been attempting a laboratory analogue of the formation of lunar craters. First he dropped round bullets into a viscous mixture of clay and water and noticed that the resulting craters were similar to those seen on the Moon. But the formation of lunar craters by impacting objects was ruled out because 'it would be difficult to imagine whence these bodies should come'. Hooke returned to another experiment in which he heated dry gypsum powder in a pot. This released water vapour that rose in bubbles leaving behind replicas of lunar craters. His conclusion was that the craters were volcanic. (The first satisfactory scientific lunar impact theory was propounded by G K Gilbert in 1893.)

Most comets were thought to have random orbits. Twenty one out of the twenty four orbits calculated by Halley (1705) were random. This indicated that cometary impacts were possible and Halley suggested at a Royal Society meeting on 12 December 1694 that the 'Choc of some comet or other great body, striking against the Earth after any sort ... would produce a new axis and poles of diurnal rotation, and a new length of the day and year, but, above all, so great an agitation of the waters as may account for all these strange marine things found on ye tops of hills and deep underground'. He even went on to suggest that the Caspian Sea might have been the depression occasioned by such a collision.

Isaac Newton had cometary orbits being perturbed so that some comets eventually fell into the Sun; 'so fixed stars that have been gradually waisted by the light and vapors emitted from them for a long time, may be recruited by comets that fall upon them' (Newton 1934).

It is clearly a large step from fuelling stars and forming seas, to fireballs and meteors, but at least interplanetary space was becoming less empty.

Science, however, moves ahead only slowly. On reading Halley's 1714 paper natural philosophers did not immediately recognise impacts with extraterrestrial 'fortuitous concourse of atoms' as being the cause of meteors. Even Halley wandered back into the arms of Aristotle. He had watched the Reverend John Whiteside perform experiments in which vapours of gun-powder rose in vacua and shone in the dark. He also hypothesised about the aurora borealis of 6 March 1715 (Halley 1714b). Maybe 'vapours are suddenly produced by the fall of water upon the nitrosulphurous fires under ground', and these then rise, having 'a tendency contrary to that of gravity'. Maybe the cause was a 'magnetic effluvia' of a 'much more subtle nature'. So aurorae, an Aristotelian *meteoric* phenomenon, might have a quasi-Aristotelian cause.

Halley returned to a serious consideration of meteors in 1719 and again tried to base his hypothesising on facts. A 'wonderful luminous meteor' was seen from all over England on the night of 19 March 1719 (Halley 1719). According to Hans Sloane (1660–1753) the meteor '... seemed in brightness very nearly to resemble, if not to surpass that of the body of the sun in a clear day' (Halley 1719, p. 979). It was 'whitish with an eye of blue' and seemed to pass across the sky from the Pleiades to Orion's Belt, leaving a reddish-yellow trail which persisted for more than a minute. Sloane saw it from Bloomsbury Square. It was also observed from Oxford and Worcester. The elevation of the meteor, as seen from these different vantage points, enabled Halley to calculate its height. This turned out to be 74 miles (119 km). The speed of the causative meteorite was also estimated to be 300 miles per minute (8.0 km s^{-1}),

> which is a Swiftness wholly incredible, and such that if a heavy Body were projected horizontally with the same, it would not descend by its Gravity to the Earth, but would rather fly off, and move round its Centre in a perpetual Orb, resembling that of the Moon.
> (Halley 1719, p. 988).

The speed is probably an underestimate, being well below the Earth's escape velocity. But Halley was clearly undecided as to the cause of the fireball, and seems quite happy to return to Aristotelian rising vapours,

> the unusual and continu'd heats of the last summer in these parts of the world, may well be suppos'd to have excited an extraordinary quantity of vapour of all sorts; of which the aqueous and most others, soon condens'd by cold, and wanting a certain degree of specific gravity in the air to buoy them up, ascend to a small height, and are quickly returned in rain, dews etc. Whereas the inflammable sulphureous vapours, by an innate levity, have a sort of vis centrifuga, and not only have no need of the air to support them, but being agitated by heat, will ascend in vacuo boileano, and sublime to the top of the receiver, when most other fumes fall instantly down, and lie like water at the bottom; the experiment whereof was first shown me by the Reverend Mr Whiteside at Oxford and was lately made before the Royal Society. By this we may comprehend how the matter of the meteor might have been raised from a large tract of the earth's surface, and ascend far above the reputed limits of the atmosphere; where being disengaged from all other particles, by the principle of nature that congregates homogenia visible in so many instances, its atoms might in length of time coalesce and run fortuitously together, as we see salts shoot in water; and gradually contracting themselves into a narrow compass, might lie like a train of gunpowder in the ether, till catching fire by some internal ferment, as we find the damps in mines frequently do, the flame would be communicated to its continued parts, and so run on like a train fir'd.
> This may explain how it came to move with so unconceivable a velocity; ... it was not a globe of fire that ran along, but a successive kindling of new matter... .
> (Halley 1719, pp 989–90)

If the meteor was 'above the reputed limits of the atmosphere' there was another problem—observers of the March 1719 meteor also heard sounds 'like a broadside followed by a rattle like small arms fire'. Halley knew that the transmission of sound was greatly diminished in the artificial vacuum of the air pump and found it very difficult to envisage how sound could be transmitted through what he thought was almost vacuous space.

The paradigm shift from vapours to extraterrestrial bodies occurred slowly. Halley clearly started it, but as Beech (1990) points out, his original mechanism was vague and poorly described—and he drifted away from his first idea. Meteors had to wait for meteorites. As soon as it was realised that meteorites were not thunderstones or volcanic ejecta and were actually extraterrestrial rocks, the relationship between meteorite entry and fireballs led to a relationship between a less energetic process, the ablation of cosmic dust, and the production of meteors.

1794–1819: Meteorites—fireballs—meteors

This is not the place to go into the detailed history of meteoritics, a task which has been superbly accomplished by John G Burke (1986). Suffice it to say, by the 1790s the number of reliable, well documented reports of stones and irons falling from the sky had increased substantially. By 1803 scientific acceptance of meteorites as extraterrestrial bodies was almost complete. Much of the credit for this must go to Chladni and Howard.

Ernst F F Chladni (1756–1827) took heed of what Georg C Lichtenberg (1742–99) had told him, 'that if all circumstances about fireballs were considered they could be thought of not as atmospheric but as cosmic phenomena, or as bodies which came from outside'. A year later, in 1794, Chladni published a book whose German title may be translated as *Concerning the Origin of the Mass of Iron Discovered by Pallas and Others Similar to it*. He collected together about two dozen fireball reports and asserted unequivocally that the fireballs consisted of compact and heavy matter, which came from cosmic space. The end product was usually a meteorite fall, and the Pallas iron, from Siberia, was a typical example. He carefully rejected ideas that associated fireballs with the aurora borealis, lightning and the ignition of hydrogen or oily vapours. Their random direction of approach suggested to him small masses of matter distributed and travelling through space, and then coming under the influence of the Earth's gravitational force and subsequently entering the atmosphere. 'There severe friction created excessive heat and electricity, which caused them to become incandescent and molten, to produce gases in their interiors, and to expand to enormous sizes until they were ruptured by explosions' (see Burke 1986, p. 43). Chladni noted that it was just as reasonable to postulate the existence of small material bodies in space as to deny their presence, since neither condition could be proved on *a priori* grounds. He suggested perceptively that the material might be rock that failed to accrete into the major planets during their formation, or be ejector from some intraplanetary collision processes, impact fragments or debris from internal explosions.

Edward C Howard (1774–1816) was a proficient chemist who turned his attention to the analysis of meteorites. He collaborated with Count Jacques-Louis Bournon (1751–1825), a mineralogist. His most important conclusions were that the meteoritic iron in stony meteorites contained nickel (ranging from 9 per cent in the meteorite Tabor up to 28 per cent in Benares); meteoritic stones had black iron-oxide crusts and contained pyrites, and all had ground masses of similar chemical composition. The fact that iron meteorites and stony iron meteorites contained nickel, pointed to them also being the 'bodies of meteors'. Howard (1802) concluded that all meteorites had fallen from space, and his comprehensive research paper won over many other scientists to his views.

The last few sceptics were converted by Jean-Baptist Biot (1774–1862) who in 1803 presented a full account of the fireball and associated multiplicity of falling meteorite stones that fell at L'Aigle in Normandy on 26 April 1803. Biot noted that the fall zone was elliptical (axes 10 and 4 km) and that the zenith angle of the trajectory (the angle made by the trajectory with the local vertical) was $22°$.

So by the first decade of the nineteenth century stones falling from interplanetary space and fireballs in the atmosphere were believed to be closely linked.

1798–1820: Meteor heights

Practical meteor astronomy (as opposed to simply collecting data about fireballs) started in 1798 with Heinrich W Brandes (1777–1854) and Johann F Benzenberg (1777–1846) who, at that time, were students at the University of Göttingen. The plan was to observe simultaneously a specific meteor from two different locations and thus determine its height in the atmosphere and its trajectory. The idea had originated from Chladni, and Lichtenberg (the professor at Göttingen) had decided that two of his students could carry it out. The students positioned themselves 15.25 km apart. Between 11 September and 4 November 1798 they observed 402 meteors in total, of which they judged that 22 were identical (i.e. they had appeared at the same time and had similar magnitudes and other characteristics). The numerical results for 21 of these meteors were given by H A Newton (1830–96) (Newton 1864a). In all but four of these cases only the altitude of disappearance is given and these varied 152 to 7 miles (245 to 11 km). Both these extremes are clearly much in error. More importantly, however, was the fact that the mean value of the disappearance height came to 61 miles (98 km). It was also apparent that, with meteors being at such distances, the real velocities were tens of kilometers per second and were actually comparable with planetary velocities.

Unfortunately all was not unalloyed joy; two meteors had been observed that seemed to rise rather than descend. Conclusions were drawn that there were two types of meteoroids, atmospheric and cosmic. Benzenberg also guessed that the cosmic ones might be little stones erupted in former times by lunar volcanoes.

Brandes continued this work in the 1820s, while he was professor of Mathematics at Breslau. He organised a network of amateur meteor observers, but the unpredictability and inclemency of the weather led to a tailing off of interest.

J M Schaeberle (1853–1924) produced a convenient and accurate method for analysing dual observations (Schaeberle 1895) and he applied this to observations of the 1894 Perseids.

12 November 1799: Meteor storms and radiants

The first meteor storm to be carefully observed and described by a competent scientific writer was the Leonids in 1799. F W H A von Humboldt (1769–1859), the famous German naturalist, and his companion G J A J Bonpland (1773–1858) (a French natural historian) were at Cumana in South America. They had apparently risen early to enjoy the morning air (Humboldt and Bonpland 1822). For four hours thousands of falling stars were noticed. Bonpland stated that there was not a space in the firmament equal in extent to three diameters of the Moon that was not filled at every instant with bolides and falling stars. Meteors were seen up to sunrise and it was noted that the meteor storm had been observed at other places in South America, some up to 700 miles from Cumana.

Andrew Ellicott (1754–1820) witnessed the display at sea, off the coast of Florida, at latitude 25°N (Ellicott 1803). The woodcut shown in figure 15.1 gives an impression of what he saw. His journal (Ellicott 1803) recorded

> November 12, 1799 about 3 a.m. I was called up to see the shooting of the stars, as it is commonly called. The phenomenon was grand and awful; the whole heaven appeared as if illuminated by sky-rockets, which disappeared only by the light of the Sun after daybreak. The

meteors, which at any one instant of time appeared as numerous as the stars, flew in all possible directions, except from the Earth, towards which they all inclined more or less.

Figure 15.1 A woodcut showing the Leonid meteor storm as seen off the coast of Florida on the night of 12 November 1799 (see Dunkin 1891).

Leonids had been seen before (see, for example, Newton (1864a) and Imoto and Hasegawa (1958)), recorded November star showers taking place in AD 902, 931, 934, 1002, 1101, 1202, 1366, 1533, 1602, 1698 as well as 1799, 1832, 1833, 1866, 1867 and 1868. J A Condé (1765–1820) recorded (Condé 1854–5) that on the death of King

Ibrahim bin Ahmad, in the middle of October (old style) 902

> an infinite number of stars were seen during the night, scattering themselves like rain to the right and left, and that year was known as the year of the stars.

An Arab writer wrote, again of AD 902,

> in this year there happened in Egypt an earthquake lasting from the middle of the night until morning: and so-called flaming stars struck one against another violently while being borne eastward and westward, northward and southward; and none could bear to look towards the heavens.

We can also read that in AD 1202 (see Newton 1864a)

> on the night of Saturday, on the last day of Muharram, stars shot hither and thither in the heavens, eastward and westward, and flew against one another, like a scattering swarm of locusts, to the right and left.

What is especially important about the 1799 display is that von Humboldt, for the first time, drew attention to the fact that the meteors seemed to originate from only one point on the celestial sphere—a point which has subsequently become known as the radiant (Humboldt and Bonpland 1814).

12 November 1833: more Leonids

The great Leonid meteor swarm of 12 November 1833 attracted immense popular attention and dragged meteors from scientific obscurity. A spectacular display occurred, meteors being seen so frequently that it reminded one witness of a snowstorm. Professor Thomson of Nashville, wrote (Olmsted 1834, pp 138–9)

> About an hour before daylight I was called to see the falling meteors, it was the most sublime and brilliant sight I have ever witnessed. The largest of the falling bodies appeared to be the size of Jupiter or Venus when brightest. The sky presented the appearance of a shower of stars and omens of dreadful events. I noticed the appearance of a radiating point which I conceived to be the vanishing point of straight lines as seen in perspective. This point appeared to be stationary. The meteors fell to the Earth at an angle of about seventy-five degrees with the horizon, moving from the east towards the west.

> A South Carolina cotton planter described his experience as follows

(see Chambers 1893)

> I was suddenly awakened by the most distressing cries that ever fell on my ears. Shrieks of horror and cries for mercy I could hear from most of the negroes of the three plantations, amounting in all to about 600 or 800. While earnestly listening for the cause I heard a faint voice near the door calling my name. I rose, and taking my sword, stood at the door. At this moment I heard the voice still beseeching me to rise, and saying, 'O my God, the world is on fire!' I then opened the door, and it is difficult to say which excited me most,—the awfulness of the scene or the distressed cries of the negroes. Upwards of 100 lay prostrate on the ground, some speechless, and some with the bitterest cries, but with their hands raised, imploring God to save the world and them. The scene was truly awful, for never did rain fall much thicker than the meteors fell towards the earth; east, west, north, and south, it was the same.

S Strickland (1804–67) stated (Strickland 1853)

> These luminous bodies became visible in the zenith, taking the northeast in their descent. Few of them appeared to be of lesser size than a star of the first magnitude; very many among them seemed larger than Venus. Two of them in particular appeared half as large as the Moon. I should think, without exaggeration, that several hundreds of these beautiful stars were visible at the same time, all falling in the same direction, and leaving in their wake a long stream of fire.

Note that the word large is often used to express brightness. During the 1833 display D Olmsted (1791–1859) (Olmsted 1834) and A C Twining (1801–84) (Twining 1834) again stressed that the meteors appeared to be radiating from a point and this was recognised as being a perspective effect caused by the Earth intersecting a parallel stream of particles.

Olmsted stressed that the meteors were entering the atmosphere from cosmic space.

Only two eyewitness illustrations of the 1833 shower seem to exist. The 'halo and snake' (figure 15.2) comes from Mr Henry J Pickering, one of the editors of *The Old Countryman* (a New York weekly) 'who witnessed the scene and under whose direction and instruction the prefixed representation was made'. This woodcut appeared in the 20 November 1833 edition and was then reproduced in various newspapers and journals such as *Mechanics' Magazine* (1833) and *The New York Journal of Commerce* 27 November 1833) (see also *The Telescope* October 1934, p. 80).

Many observers noted that the meteors 'were like the ribs of a gigantic umbrella' (see figure 15.3). And the snake in figure 15.2 is not as fanciful as it seems. Professor Denison Olmsted wrote from Yale (on

13 November 1833)

One ball that shot off in the north-west direction, and exploded near the star Capella, left, just behind the place of explosion a phosphorescent train of peculiar beauty. This line was at first nearly straight but it shortly began to contract in length, and dilate in breadth, and to assume the figure of a serpent folding itself up, until it appeared like a small luminous cloud of vapour.

(Olmsted 1834, pp 365–6)

Figure 15.2 An eyewitness engraving of the 13 November Leonids executed under the direction of Henry J Pickering, one of the editors of a New York weekly magazine called *The Old Countryman* (see the November 1833 edition).

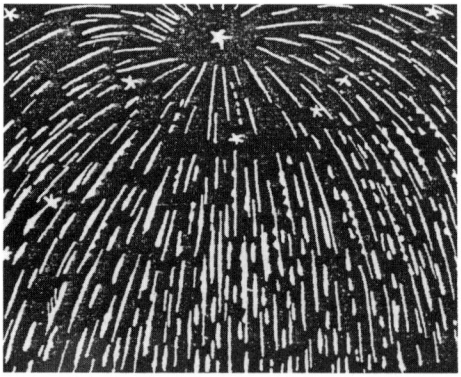

Figure 15.3 This Leonid engraving appeared in the November 1833 edition of the *Mechanics' Magazine* (New York) p. 288. The meteor shower was a subject of much public and scientific interest and the editor of *Mechanics' Magazine* wrote 'We shall close our account of this extraordinary phenomena by introducing another drawing and soliciting the favour of communications of any of our friends and correspondents, tending to elucidate the subject'.

The enormous number of meteors is underlined by the fact that François Arago (1786–1853) (Arago 1857) computed that no less than 240 000 meteors were seen during a period of seven hours above the horizon of Boston.

By far the most reproduced image of an ancient meteor shower is shown in figure 15.4 and it represents the Leonids of November 1833.

Figure 15.4 The Leonid meteor storm of 13 November 1833. This shower has been recorded at about 33 year intervals since AD 902. The observations of the 1833 shower confirmed von Humboldt's finding, that the meteors originated from only one point on the celestial sphere. The artist of the illustration is probably Karl Jauslin (M Barlow Pepin, private correspondence) and it was engraved by Fritz Voellmy especially for *Bible Readings for the Home Circle*, 1888. It appears on p. 66 and was still being reproduced in the 1908 edition.

It was engraved more than 50 years after the event by someone who, it appears, wasn't yet born when the shower occurred. Apparently its debut was in a religious text to illustrate a biblical message. And we still do not know clearly the name of the artist but it was probably K Jauslin (Hughes 1987).

It was first published in *Bible Readings for the Home Circle* (1899). This book, which addresses over 2800 questions on religious topics, uses the Leonid shower woodcut as an illustration of Old Testament prophecies fulfilled.

The Leonids on that cloudless night were thought by many to be a divine sign. R M Devens (1876) lists the 1833 Leonids shower among the 100 great and memorable events in American history. Devens writes:

> During the three hours of its continuance, the day of judgement was believed to be only waiting for sunrise, and, long after the shower had ceased, the morbid and superstitious still were impressed with the idea that the final day was at least only a week ahead. Impromptu meetings for prayers were held in many places, and many other scenes of religious devotion, or terror, or abandonment of worldly affairs, transpired, under the influence of fear occasioned by so sudden and awful a display.

Reference was made by some to *Matthew 24:29*: 'Immediately after the tribulation of those days shall the sun be darkened, and the moon shall not give her light, and the stars shall fall from heaven, and the powers of the heavens shall be shaken'; *Revelations 6:12–13*: 'And I beheld when he had opened the sixth seal, and, lo, there was a great earthquake; and the sun became black as sackcloth of hair, and the moon became as blood; and the stars of heaven fell unto the earth, even as a fig tree casteth her untimely figs, when she is shaken of a mighty wind'; and *Isaiah 34:4* '... and the heavens shall be rolled together as a scroll: and all their host shall fall down'.

1830–63: Periodic showers and the death-knell of atmospheric meteors

The remoteness of meteors and the difficulty inherent in interpreting their physical and chemical nature is underlined by the fact that even in the early 1860s scientists were still debating their atmospheric as opposed to interplanetary origin. One of the main factors responsible for the final transition was the realisation that meteor showers were periodic.

Arago wrote (see Walker 1843):

Chance, one can say, rather than scientific ingenuity, led to the establishment of a fact which guarantees henceforth to shooting stars a very important rank in our planetary system. More and more there is confirmed the existence of a zone composed of millions of small bodies, whose orbits meet the plane of the ecliptic at a point which the earth occupies every year from the 11 to the 13 of November. It is a new planetary world which is beginning to be revealed to us.

December (Geminids), November (Leonids), August (Perseids) and April (Lyrids) were being recognised as times of enhanced meteor activity (see, for example, Olbers (1837)). The constancy of the Perseids as opposed to the variability of the Leonids worried some. Astronomers assumed, however, that the 'asteroids' responsible for the August Perseids, were strewn all around an elliptical orbit whereas those responsible for the November Leonids were concentrated in a small segment. H W Olbers (1758–1840) (Olbers 1837) correctly surmised that the Leonids had a periodicity of about 34 years and that following the 1799 and 1833 storms one would have to wait until around 1867 for another. To quote Burke (1986, p. 78):

> In the late 1830's the observations of periodic meteor showers became almost a sacred duty for astronomers. Almost all pressed friends and relatives and, when possible, students into service during the nights when showers were expected to appear ... One can only speculate as to how many dozens or hundreds of young people learned the rudiments of astronomy or came to appreciate the majesty and vastness of the universe from such experiences. This also explains in part how astronomy became a popular science in the nineteenth century.

The fact that the radiant of the meteor shower moved only slightly with respect to the stellar background, as the Earth rotated, convinced the vast majority of astronomers that meteors had an interplanetary origin. Only L Quetelet (1796–1874) seemed to be out of step (Quetelet 1839). He was, however, the first (E Herrick (1841) following shortly after) to search through ancient records for meteor shower observations.

P H Boguslawski (1789–1851) (Boguslawski 1867), G Erman (1839) and S Walker (1843) all attempted to calculate the orbits of the meteoroids. Even though they assumed that the node was at one astronomical unit ($= 1.496 \times 10^{11}$ m) from the Sun and they knew the time of occurrence, and the position of the radiant, their erroneous attempts at estimating the geocentric velocity meant that their efforts did not meet with success. Olbers and Walker did realise, however, that Leonid meteoroids were in retrograde orbits.

The major step forward was due to H A Newton (1863) of Yale College, Princeton. He had been studying Quetelet's book *Physique du*

Globe (Quetelet 1861) and especially the chapter devoted to shooting stars.

> Doubts seem to have arisen in the mind of the distinguished author respecting the origin of these phenomena. Yet from this same chapter can be drawn a simple, and, as we think, a very strong argument, that the star-showers, at least, are caused by the entrance into our atmosphere of bodies revolving about the sun. And if this be admitted for the shooting stars of the annual periods, probably no one will deny that the sporadic meteors have a similar character and origin.
>
> The return of the August and other showers on fixed days of the year might *possibly* be due to meteorological changes. But if the magnetism, the heat, the electricity, or the other properties of the atmosphere produce these annual phenomena, the period should evidently be the *tropical* year. On the other hand, if rings of bodies revolving about the sun are met by the earth in April, August, November etc., thus causing these showers, the cycle should be the *sidereal* year.

H A Newton proved quite convincingly, from the dates of the earlier showers, that the true period between their annual maximum activity was 'not widely different from the sidereal year' and thus that meteors are of cosmic origin. Showers which don't suffer greatly from gravitational perturbation and thus nodal progression, should change their date of maximum activity by 0.01417 days per year, i.e. one day in about 70 years. Luckily the April Lyrids, August Perseids and November Leonids falls into this category. The effect on the Perseid shower has been investigated by Hughes and Emerson (1982).

Nowadays the Perseids maximise around 12 August. In AD 36 maximum occurred on 15 July.

Using the data assembled in his 1863 paper Newton (1864b) concluded that the Leonid meteor shower was caused by the Earth passing through a shower of meteoroid particles in November and that the most dense part of the Leonid meteor cloud was only met at intervals of approximately 33.25 years. Newton thus boldly predicted that there would be a spectacular display of Leonids in 1866. Unknown to Newton, Olbers had independently reached this conclusion some years earlier.

13–14 November 1866: the expected Leonids

The display duly happened as predicted. Robert S Ball (1840–1913) related (Ball 1886)

> The night was fine; the moon was absent. The meteors were distinguished not only by their enormous multitude, but by their intrinsic

magnificence. I shall never forget that night. On the memorable evening I was engaged in my usual duty at that time of observing nebulae with Lord Rosse's great reflecting telescope. I was of course aware that a shower of meteors had been predicted, but nothing that I had heard prepared me for the splendid spectacle so soon to be unfolded. It was about ten o'clock at night when an exclamation from an attendant by my side made me look up from the telescope, just in time to see a fine meteor dash across the sky. It was presently followed by another, and then again by more in twos and in threes, which showed that the prediction of a great shower was likely to be verified. At this time the late Earl of Rosse (then Lord Oxmantown) joined me at the telescope, and, after a brief interval, we decided to cease our observations of the nebulae and ascend to the top of the wall of the great telescope whence a clear view of the whole hemisphere of the heavens could be obtained. There, for the next two or three hours, we witnessed a spectacle which can never fade from my memory. The shooting stars gradually increased in number until sometimes several were seen at once. Sometimes they swept over our heads, sometimes from the east. As the night wore on, the constellation Leo ascended above the horizon, and then the remarkable character of the shower was disclosed. All the tracks of the meteors radiated from Leo. Sometimes a meteor appeared to come almost directly towards us, and then its path was so foreshortened that it had hardly any appreciable length, and looked like an ordinary fixed star swelling into brilliancy and then as rapidly vanishing. Occasionally luminous trains would linger on for many minutes after the meteor had flashed across, but the great majority of the trains in this shower were evanescent. It would be impossible to say how many thousands of meteors were seen, each one of which was bright enough to have elicited a note of admiration on any ordinary night.

Ball might have found it impossible to say how many meteors were seen, others didn't. Nearly 9000 were counted at the Royal Observatory Greenwich 'and it is supposed that at least another thousand may have escaped the attention of the eight observers employed'. The observed meteor rate is shown in figure 15.5.

Many reports of the 1866 Leonid activity were published in *Monthly Notices of the Royal Astronomical Society* **27** (pp 17–56, 65–79) and we shall quote a few. J R Hind (1823–95) (Hind 1866) assisted by three observers saw the meteor stream from Twickenham:

From 1 a.m. to $1^h 7^m 5^s$ no less than 514 were counted, and we were conscious of having missed very many, owing to the rapidity of their succession. At the latter moment there was a rather sudden increase, to an extent which rendered it impossible to count the number, but after $1^h 20^m$ a decline became perceptible. The *maximum* was judged to have taken place at about $1^h 10^m$, and at this time the appearance of the whole heavens was very beautiful, not to say magnificent. Beyond their immense number, however, the meteors were not particularly remarkable either as regards brilliancy or the persistence of the trains, few of which were visible more than three seconds.

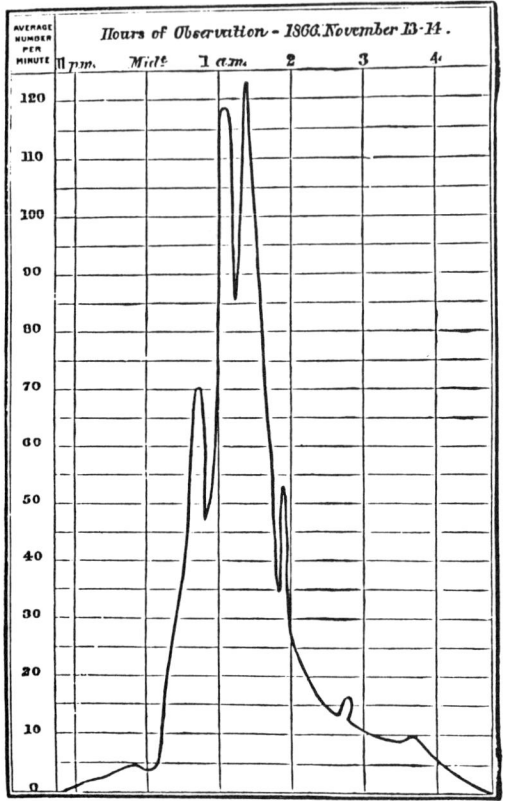

Figure 15.5 A diagram showing the average number of meteors per minute as observed from the Royal Observatory, Greenwich on the night of the great Leonid meteor storm, 13–14 November 1866 (see *Mon. Not. R. Astron. Soc.* **27**).

Sir J F W Herschel (1792–1871) (Herschel 1866) wrote:

I have mentioned the case of one of the meteors leaving a train visible during $2^m\,40^s$. A much more striking instance occurred in the case of one which exploded close to the three bright stars in Aries. The train left at first was very bright, and very nearly parallel to the line joining Alpha and Beta. It remained visible for no less than 6 minutes, during which time it drifted slowly to the southwards over a space of 8 or 9 degrees, and at the same time gradually changed its direction, so that just before its disappearance it was at right angles to its original position.

These observations obviously give a foretaste to the meteor wind method of measuring upper atmospheric circulation.

C Piazzi Smyth (1819–1900) (Smyth 1866) concentrated on meteor trains and noted that

> these trains usually lasted only from two to three seconds, were of an exaggerated long elliptical form, so as to be nearly invisible close to the bright head, and to be broadest near the middle of their length; and such broader central part of the train often remained abundantly visible, long after the head and chief part of the length of the train had entirely disappeared; so decidedly too was this a material fact, and not any optical impression caused by the overpowering brightness of the head during the time it was visible, that often, on turning to a new part of the sky, the last expiring traces of such short central part of a meteor track were seen, proving that a meteor had just passed that way, and had been missed by the observer.

The physicist's eye was also apparent:

> some years since, an abundant display of meteors was compared to flakes of snow in a snow storm; but that simile was by no means descriptive of the leading impression of the scene on the 13th ... their was a velocity, a certainty and an almost apparent purpose in the motion of every meteor ... they gave the impression of some ethereal description of rockets but endued with the speed of canon balls.

Meteor colour also aroused interest. Herschel reported that the majority of meteors seen were white but some were orange coloured and emerald green. Humboldt found two thirds to be white and the remainder yellow, yellowish red and green. At Greenwich the majority were registered as blue, the remainder being white, bluish white, yellow, red and green. The birth of meteor chemistry is touched on by E Dunkin (1891).

> Numerous analyses of meteors, portions of which have fallen to the ground, have been made by chemists, both English and foreign. It is supposed that the material forming the thousands of shooting stars visible on extraordinary occasions is of the same nature as that composing an ordinary aërolite. If so, their chemical elements consist principally of matter similar to that distributed throughout the Earth's crust, for analysis has shown that they contain indications of the presence of iron, copper, zinc, nickel, sulphur, carbon, and probably of many other substances found in the Earth.
>
> It is scarcely probable that the constituents of the great streams of meteors of August and November are composed of the same kind of metals found, by analysis, in some of the fallen masses of meteoric iron-stone, discovered after the explosion of a detonating fire-ball. Prof. Herschel has remarked that the relative abundance of carbon gases in

stony aërolites, and the ready evolution of these gases from them at a moderate heat, distinguishes these bodies from siderites, or iron meteorites; and the analogy which presents itself with the results of spectroscopic observations of comets, in which the lines of the carbon spectrum are the most conspicuous, suggest a similarity of origin as well as of the structure and composition of the materials of comets to the stony matter enclosing disseminated iron met with in the rock-like substance of ordinary meteorites. This apparent chemical coincidence of the structure of comets and ordinary meteors leads us to consider it a most important research, the results of which were announced in 1867 by M Schiaparelli, of Milan.

The Leonids have a geocentric velocity of 72 km s^{-1} so the excitation energy is high.

Sir John Herschel (1866) took great pains to determine the radiant point.

During the superb display of meteors on the night of November 13, 1866, my attention was particularly directed to the determination of the exact situation in the heavens of the radiant point of their courses, which

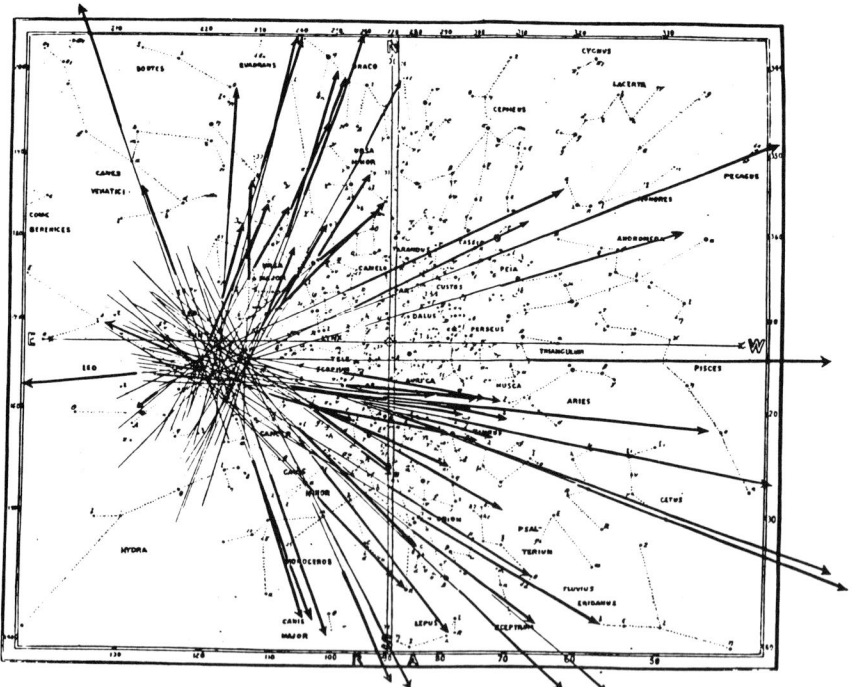

Figure 15.6 Tracks of the meteors seen from Greenwich during the night of 13–14 November 1866 showing the radiant of the Leonids.

has been commonly stated to be coincident with that of the bright star Gamma Leonis. This, however, was certainly not the case; and I am enabled to say with perfect confidence that their courses diverged, with a very remarkable degree of agreement, from a point considerably higher in declination and less advanced in right ascension ... Several circumstances enabled me to fix on the point with full assurance of its being the true radiant: First, the frequent out-shooting from a very near proximity to it, and once in several different directions, of a volley of meteors. Secondly, on two or three distinct occasions a meteor appeared in this very point, and in all these cases it was motionless, devoid of a train, and, on its extinction, left only a small nebulous light to mark the place of its appearance.

The mean of the observations made by several astronomers that year gave the radiant coordinates as RA $149°$ $12'$ and dec $+23°$ $1'$ (see figure 15.6). An accurate knowledge of the radiant position is essential when it comes to making a precise calculation of the orbits of the meteors.

1861–76: Meteor streams and comets

D Kirkwood (1814–1895) (Kirkwood 1861) broached the theory that 'meteors and meteoric rings are the debris of ancient but now disintegrated comets, whose matter has become distributed around their orbits'. Similar ideas had been expressed before Kirkwood's time but there were no available data to give support to these hypotheses so the ideas were usually disregarded as idle speculation. Chladni's work (1819) was a perfect case in point, but to quote G V Schiaparelli (1867b):

> Chladni wrote when little was known of the theory of the movement of meteor-systems. By his skillful manner of treating the subject, he was evidently working with a mind beyond his age.

Meteor orbits were calculated for the first time in the late 1860s. Remember that the first calculation of the orbital parameters of a comet was carried out in 1686 by Isaac Newton and his method was published in the *Principia* (1687). Edmond Halley (1705) extended this work and calculated the orbits of 24 comets.

1867 saw U J J Le Verrier (1811–77), Schiaparelli and J C Adams (1876–1956) independently publish orbits for the meteoroids in the Leonid stream. The input data required for the calculation were the position of the observer at the time of maximum shower activity, the coordinates of the radiant at that time and either the orbital period of the meteoroids or their geocentric velocity. The first two are easily found. Most of the calculators used the findings of H A Newton

(1864a,b) who concluded that the cluster of meteoroids responsible for the Leonids must describe either $2 \pm 1/33.25$, or $1 \pm 1/33.25$ or $1/33.25$ orbital revolutions per year. (Note that H A Newton, himself, favoured $1 + 1/33.25$ revolutions per year, equivalent to a period of 354.621 days, and an intersection occurring near the shower's orbital aphelion. The $1/33.25$ value required an intersection near perihelion.)

Each selected period resulted in a different orbit and Adams (1867) calculated the effects of planetary perturbation on each of the orbits. With a period of 33.25 years the increase in the nodal longitude during one orbit was a total of 28 minutes of arc, i.e. 20 minutes of arc due to Jupiter, 7 to Saturn and 1 to Uranus.

According to H A Newton's analysis of historical data (1864a) the observed change was 29 minutes of arc per 33.25 year period. This agreement firmly established the period.

Let us now turn to comets. The orbit of Comet 1866I, which had been discovered by Guillaume Tempel (1821–89) from Marseilles on 19 December 1865 and independently by Horace P Tuttle from Harvard, Massachusetts, on 5 January 1866, had been calculated and published by T von Oppolzer (1841–80) (von Oppolzer 1867). Almost to a man C F W Peters (1844–94) (Peters 1867), Schiaparelli (1867b) and von Oppolzer (1867) realised that the comet and the stream had similar orbits.

Giovanni Virginio Schiaparelli (1835–1910) was regarded as the leader of the field (Chambers 1910) and he was presented with the gold medal of the Royal Astronomical Society in 1872 for his research.

The similarities between the two orbits can be judged from the following table which comes from the work of Adams.

	Ring. of Nov. meteors	Comet 1866I
Perihelion passage	1866, Nov. 10.09	1866, Jan. 11.16
Period	33.25 years	33.18 years
Mean dis. from Sun (Earth's = 1)	10.340	10.325
Eccentricity	0.904	0.905
Perihelion distance	0.989	0.977
Inclination of orbit	14° 41'	17° 18'
Long. of ascending node	51° 18'	51° 26'
Long. of perihelion	42° 24'	42° 24'
Direction of motion	Retrograde	Retrograde

Schiaparelli (1867) had also calculated the orbits of the Perseid meteors that he had recorded on 9, 10 and 11 August 1866.

It had long been known that there were more meteors in the second half of the night than in the first. Due to the orbital motion of the Earth the morning sky is at the front of the globe and the evening sky at the back. So in the morning we are pushing into the cloud of cosmic dust particles whereas in the evening they have to catch up. Schiaparelli (1866) found that the ratio between the two fluxes was $\sqrt{2}$, exactly the ratio between the parabolic and circular velocity at a specific heliocentric distance. He thus concluded that meteors in general—including the Perseids—moved in parabolic orbits.

Schiaparelli quickly realised that the Perseid meteor orbital elements bore very close resemblance to the elliptic elements of Comet 1862III. This can be seen from the table below.

	Ring of August meteors	Comet 1862III
Perihelion passage	1866, July 23.62	1862, Aug. 22.9
Ascending node passage	Aug. 10.75	—
Long. of perihelion	343° 28'	344° 41'
Long. of ascending node	138° 16'	137° 27'
Inclination of orbit	64° 3'	66° 25'
Perihelion distance	0.9643	0.9626
Direction of motion	Retrograde	Retrograde

The orbits of comets and meteor streams thus became closely associated. E Weiss (1837–1917) (Weiss 1867) published a list of 33 comets which had nodes close to the Earth's orbit. H L d'Arrest (1822–75) (d'Arrest 1867) noted that the November Andromedids had an orbit close to that of Biela's comet (1852III) and J Galle (1812–1910) (Galle 1867) associated the April Lyrids with Comet 1861I.

A S Herschel (1836–1907) (Herschel 1878) listed about 75 suggested coincidences between comets and meteor streams. J Glaisher (1809–1903) went so far as to say that 'the intimate connection now know to exist between comets and meteors is perhaps the most striking and novel discovery of a purely astronomical kind that has been made in our time'.

It must be remembered that many researchers at that time (cf Sir Norman Lockyer (1836–1920) (Lockyer 1890) regarded comets as nothing more than immense aggregations of meteors 'and naturally if they throw off anything it is meteoric particles which they throw off'. In former years panic had been caused by the prediction that a cometary collision with the Earth would be catastrophic. Now it seemed that it would only produce a brilliant display of celestial fireworks, the meteoroid stones burning up in the envelope of air that protects the Earth.

The marriage between meteoroid streams and comets has stood the test of time and was placed on an even stronger foundation by the 'dirty snow-ball' model of the cometary nucleus (see Whipple 1950). Here the fount of all cometary activity was a single kilometric sized irregular agglomeration of silicate dust and snow (predominatly of H_2O), a hypothesis that has recently been confirmed by the Giotto spacecraft's fly-by of Comet Halley. The sublimation of the cometary snows at a time when the comet was near the perihelion of its orbit provides an ample source of momentum for the detachment of meteoroid dust from the surface of the nucleus and the acceleration of these dust particles up to relative velocities of the order of 1 km s^{-1}. These velocities put the meteoroids onto orbits which are very similar to those of the parent comet but which cause the meteoroids to either progressively gain on or fall behind the comet as time passes, (see Plavec 1955, Hughes 1977). The end product is a stream of meteoroids around the cometary orbit.

1878–99: More radiants and stationary ones

Observational meteor astronomy during this period was mainly the domain of the dedicated amateur. Little apparatus was required, simply a trained eye and a considerable familiarity with the constellations. Meteor tracks were carefully recorded on star charts and this enabled W F Denning (1899) to publish his catalogue of 4367 individual radiants. According to Denning there were up to 50 meteor showers active each night, the large majority being very minor. Denning's list has been criticised because he did not realise the complexity of radiants or their rapid movement. The fact that even a skilled observer could only obtain beginning and end points of a meteor train to an accuracy of $\pm 1°$ is also a hindrance (Lovell 1954).

Denning (1878a) also insisted that the radiants of some of the major showers were stationary, being fixed in their apparent direction on the sky. More precise photographic observations have since disproved this conclusion but it did provide a considerable stumbling block to the early theoreticians in their attempts to model meteor stream characteristics. To quote C P Olivier (1925), 'From that time to the present, stationary radiation has been the most disputed point in meteoric astronomy.'

It is clear that if meteors follow, for example, a specific cometary orbit their radiant moves steadily from day to day with respect to the sky background. To quote Denning (1884)

If a radiant point were found persistent for several months ... the particles must pursue different orbits. It would consist of a series of streams arranged in a peculiar manner, but not associated in a common orbit.

Olivier's extremely thorough discussion summarises a range of theoretical approaches to the problem but essentially concludes that poor data plus a poor analysis technique meant that 'by uncritical combinations of existing material one can prove that almost anything in the way of motion of radiants or its opposite has been detected'. The fact that the speed of the movement of the radiant was *inversely* proportional to the meteoroids' geocentric velocity coupled with the fact that the observed luminosity is proportional to velocity to the fourth power clearly does not help.

1866–80: Sporadic meteors

Denning's work led to even more confusion—G F Chambers (1841–1915) (Chambers 1889) concluding that sporadic meteors do not exist, 'for it has been proved that as a rule the seemingly erratic members belong to definite systems whose radiant points are capable of being discovered by prolonged and critical observations'.

Chambers is incorrect here. Observations of sporadic meteors, i.e. meteors not associated with showers, led to two firm results. First the number of meteoroids observed per unit time varies in a quasi sinusoidal fashion through the day and maximises at about 04.00 UT (see Hughes 1974). Secondly more sporadic meteors are seen during the second part of the year, July to December.

Schiaparelli (1867a) made the first attempt to explain these observations. He started by having meteor radiants of uniform intensity, these being uniformly distributed over the celestial sphere, and he then considered the concentration of radiants around the apex of the Earth's way that would be produced by the Earth's orbital motion.

Denning (1878b) was the first to tackle the problem of meteor magnitude distribution. After extracting reports of 35 134 meteors from various published catalogues he concluded that the percentage of first, second and third magnitude meteors was 10.6, 18.4 and 26.2 respectively, 3 per cent were brighter than first magnitude and the remaining 41.8 per cent were of fourth magnitude or fainter.

1848–90: Meteoroids as solar fuel

During the 1840s, geological evidence for a solar system much older than the traditional 6000 years became well established. The source of solar energy thus became a topic of much discussion. Chemical burning was quickly ruled out as being quite inadequate to support the solar luminosity for any reasonable time. J R Mayer (1814–78) (Mayer 1848), a pioneer of the law of conservation of energy, proposed that the infall of meteorites and meteoroids transferred enough kinetic energy to the Sun to balance the radiation loss. The resulting increase in the solar mass (which would obviously change the semi-major axis and period of the Earth's orbit) was to be balanced by a loss due to the mass carried away as Newtonian light corpuscles.

In 1853 Hermann von Helmholtz (1821–94) (Helmholtz 1856, p. 505) proposed that gravitational contraction was a reasonable source (the rate of contraction is below observational sensitivity (about 50 m per year) and the process will provide sufficient energy for several million years of constant luminosity). This mechanism dominated most of nineteenth-century thought on the subject. Lockyer (1890),

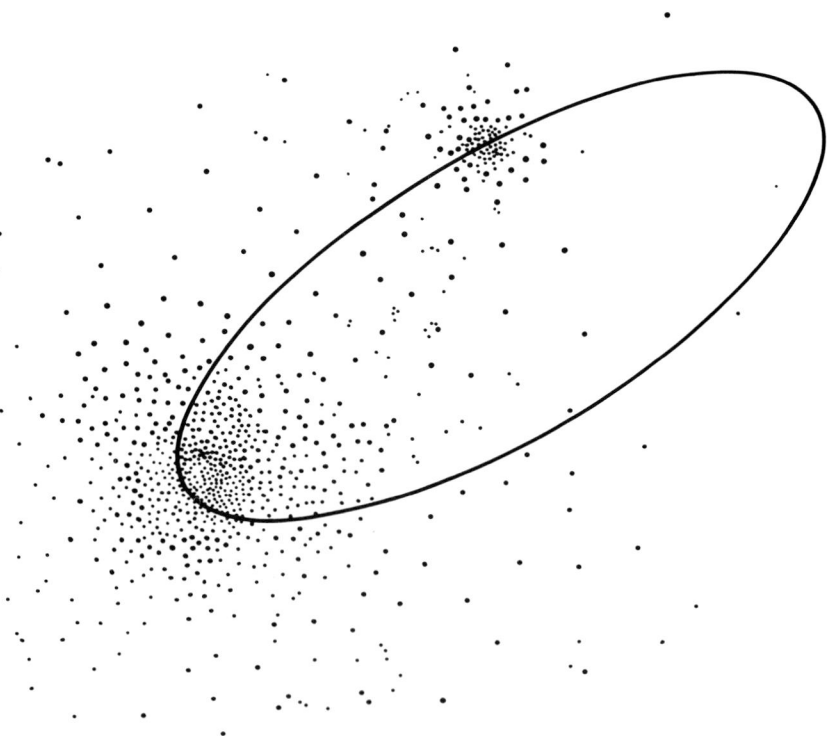

Figure 15.7 Norman Lockyer's picture of a variable star, two meteor swarms orbiting such that they periodically interpenetrate.

however, still held fast to the meteoritic hypothesis and even widened it, proposing that 'all self-luminous bodies in the celestial space are composed either of swarms of meteorites or of masses of meteoric vapour produced by heat. New stars ... are produced by the clash of meteor swarms.' Variable stars were meteor swarms orbiting in such a way that they interpenetrated. When one swarm rushed through the other, collisions and therefore brilliancy increased enormously. When they separated collisions were fewer and the light fainter (see figure 15.7).

1866: Meteor spectroscopy

The predicted profusion of Leonids led John Browning (1867) to employ his direct-vision prism with a view to obtaining as many meteor spectra as possible. After trying to 'catch' spectra by observing in different directions he finally fixed his prism pointing to the west of Leo and with its axis parallel to the horizon. All meteors thus travelled in a direction parallel to the axis. Note that the 'spectrometer' was slitless so sharply defined lines would not be shown. Four types of spectra were seen:

A. Continuous spectra with all the colours of the solar spectrum except the violet rays.
B. Spectra in which the yellow greatly predominated but in other respects resembled A.
C. Spectra of almost purely homogeneous yellow but with a faint trace of red and green continuum.
D. Spectra of purely homogeneous green light (only two seen).

Browning referred to the practice of chemists and mineralogists who inferred the presence of certain elements in a substance 'from the colour communicated by the substance to the blowpipe flame'. Yellow indicated sodium, green barium or thallium, red strontium and so on. No conclusions as to the composition of the Leonid meteoroids were drawn from his observations. This was due to the fact that nineteenth-century spectroscopy was mainly qualitative and there was still considerable disagreement over the interpretation of results (see McGucken 1969). Von Konkoly (1842–1916) (Von Konkoly 1874), however, observed a persistent meteor train for 25 minutes using a direct-vision spectroscope attached to a comet seeker telescope. He recognised sodium and magnesium lines and four bands which he identified as methane.

Meteor photography

Photograph the sky and you cannot help but pick up the occasional meteor. As far back as 1896, meteors appeared on the Harvard patrol plates. This patrol was originally set up to take routine time exposures of the sky that could subsequently be scanned for particular purposes —detecting variable stars, orbital studies of comets and asteorids and so on. The meteor flux rate is low. The 18 inch Schmidt camera on Mount Palomar was used on a supernova patrol. The field of view is 9°. After 10 663 minutes of exposure Watson (1942) found that only 81 meteors had been recorded, a rate of one every 2.2 hours. By comparing this rate with the naked-eye visual rate Watson concluded that the cameras were recording all the meteors brighter than about second or third magnitude.

In 1894 double-camera meteor photography was initiated by W F Elkin (1855–1933) (Elkin 1899). Groups of cameras were placed at two stations two miles apart. In 1900 Elkin introduced a rotating shutter. The exposures were interrupted by placing the cameras behind a rotating bicycle wheel equipped with 12 opaque screens. The screens produced a series of gaps in the photographed trail and the angular velocity of the meteor could be determined from the gap spacing. If the meteor is photographed from two stations the linear velocity (with the latest equipment) may be determined to ± 0.1 per cent. Unfortunately Elkin chose too small a baseline. Photographic work was carried out subsequently in England by F A Lindemann (1886–1957) and G M Dobson (1889–1976) (Lindemann and Dobson 1923) and in Russia by Fedynski and Stanjukowitsch (Fedynski and Stanjukowitsch 1935). The technique was, however, revolutionised in 1936 when the commerical development of a lightweight synchronous motor was coupled with the electrical power companies providing alternating current at a fairly precise frequency. The main development took place at Harvard. One camera was at Cambridge and the other 26 miles west at the Agassiz station. Typically the trails were interrupted 20 times a second and the plates were exposed for two hours or one hour. 2000 hours of exposure time produced 27 meteors the limiting magnitude being about -1 (see Millman and Hoffleit 1937, Whipple 1938, 1949).

The results were extremely impressive; the two-station camera work not only gave the beginning and end height of the meteor train but also the radiant and meteoroid velocity. These latter two results, combined with knowledge of where and when the meteor occurred, led to a complete set of orbital parameters. The deceleration of the meteoroid could also be measured as it burned up in the atmosphere and this quantity led to a value for the meteoroid density.

Hyperbolic meteors

The source of sporadic meteors generated much interest, most of which concentrated on their velocity distribution. The main question was whether their orbits were elliptical, in which case the meteoroids and their parent bodies originated in the solar system, or hyperbolic, which would point to an interstellar origin. Von Niessl and Hoffmeister (1925) published a catalogue of 611 fireball orbits. The velocity in the atmosphere had been calculated from the observed duration and the flight geometry as deduced from eyewitness reports. In 79 per cent of the cases the causative bodies had hyperbolic orbits.

E J Öpik (1893–1985) (Öpik 1932) suggested that interstellar space contained a large population of meteors and comets and he calculated that, over the lifetime of the solar system, a cloud of meteoroids out to a distance of about 200 000 AU (1 pc) was relatively stable against gravitational dissipation by passing stars. Not only did the Sun have a cloud of meteoroids in orbit around it but larger stars had larger clouds. So a significant fraction of sporadic meteors should be not only interstellar but also originate from orbits that were once closed around other stars.

The Harvard Meteor Expedition to Arizona (1931–3) was mounted to check this hypothesis. The main purpose was to measure meteor radiants and velocities. Öpik devised an ingenious technique now known as the rocking-mirror method and this is described in detail in Shapley *et al* (1932). A six-inch square plate-glass mirror was supported at three places. Two of the supports oscillated out of phase so that a meteor, seen reflected in the mirror, would appear to describe a pseudocycloidal trajectory. The shape and length of the trajectory enabled the velocity to be calculated. Öpik (1934, 1940a) concluded that 60 per cent of the observed meteors were hyperbolic. Now if this was correct one would expect nearby stars to produce meteors with specific radiants.

Whipple (1972a) talked of his calculations, in 1933, of the expected radiants of meteors associated with the star Sirius. A diligent search of the catalogues found none.

Fred Whipple and Fletcher G Watson were responsible for the two-camera meteor programme introduced at the northern stations of the Harvard Observatory. Watson never became involved in the analysis but November 1938 saw the publication of the results obtained from the first six meteor trails (see Whipple 1938). Some of the meteoroids seemed to have asteroidal-type orbits and Whipple suggested that a sequence of orbital types ranging from cometary to asteroidal might be found when sufficient data had been collected. 'Whether there exists an hyperbolic type of orbit among the bright meteors remains to be seen', was, however, the view put forward by Whipple in 1938. Öpik's work (1934) in Arizona suggested that the percentage of hyperbolic orbits decreased rapidly with increasing brightness and became very small for photographic meteors (then about -1^m†) and fireballs. The persistence of the Taurid meteor shower (it lasts for nearly a month) and the slow daily motion of the radiant had led certain researchers to suggest that the orbits of the Taurid shower meteoroids were hyperbolic. Fourteen Taurids photographed by Whipple (1940) using the twin cameras led him to conclude that this was not so and that the orbits were closed and of short period.

The subject cried out for better observational data and proved a stimulus for the development of a whole series of meteor cameras culminating in the famous Baker Super-Schmidt which would photograph 260 meteors in 100 hours of exposure, going down to about 3.5^m. Fred Whipple was responsible for this development. He conceived the programmes (see Whipple 1947), scrambled for financial support, managed the technical developments, personally selected the observing sites, administered the programmes, directed the science and was involved in much of the measurement, reduction and analysis (see, for example, Wright and Whipple (1950, 1953), Whipple (1954) and Jacchia and Whipple (1956)). Observations with the Baker Super-Schmidt cameras started in the summer of 1951 and 200 doubly photographed meteors had already been recorded by September 1952. One problem with the camera work was that there was no knowledge of exactly when the meteor occurred during the exposure period of the camera. Whipple (1972a) found that the resulting meteor orbit was hyperbolic if one assumed an instant of apparition during one part of the exposure, but elliptical in the other part. No meteor was observed that gave a hyperbolic solution over the whole exposure period.

Sinding (1948) produced telling evidence in favour of an 'all solar system' solution. Very careful calculations of the planetary perturbations on comets showed that even if they had slightly hyperbolic orbits in the inner solar system all comets with well determined orbits had elliptical orbits before they entered the planetary region.

The Baker Super-Schmidt programme showed finally that the hyperbolic component of the sporadic meteoroid influx, even if it exists at all, does not exceed one per cent. McCrosky and Posen (1961), Hawkins and Southworth (1961) and Jacchia and Whipple (1961) all

---

† -1^m means minus one in magnitude, where 'magnitude' here is a logarithmic measure of relative brightness. It has been chosen by historical accident such that the more negative the measure, the greater the brightness. For example, the bright star Sirius has a magnitude -1.5 and the much fainter Pole Star a magnitude of 2.12 (ed).

contributed to this programme. The camera scientists had, however, been pipped at the post by the radar meteor groups. By 1948–9 pulsed techniques using narrow beam aerials at Jodrell Bank (Almond *et al* 1951) had shown that only a small fraction of the faint ($\sim 7.6^m$) meteors were produced by hyperbolic meteoroids (see Lovell 1954, p. 233).

Öpik (1969) eventually retracted his rocking-mirror results and wrote 'failures are inevitable, but usually not in vain: from them we sometimes learn even more than from the successes'. The hyperbolic meteor search certainly proved this.

1907–30: Meteoroids and the Earth's atmosphere

The cosmic dust particles that produce meteors when they burn up in the atmosphere are known as meteoroids. There is a complex relationship between the mass, density, composition and velocity of the meteoroid and the resultant atmospheric phenomenon, the meteor.

One of the first relationships between meteoroid characteristics and the resultant meteor was noticed by G von Niessl (1839–1919) (von Niessl 1907). The height of disappearance of the meteor varied with mass. Perseid meteors disappeared at 88 km, meteors of magnitudes 1 to 5 at 86.3 km, large meteors at 49.7 km, fireballs at 60 km, detonating meteors at 31.0 km and meteorites at 22 km—all these values being averages over many observations. According to von Niessl

> Thus it may well be proper to explain these different types of shooting stars as due to a gradual increase in mass, since larger masses experience a relatively smaller resistance in the atmosphere and thus can penetrate deeper than the smaller masses.

Von Niessl also found that terminal altitude and geocentric velocity were directly related, such that slow meteoroids penetrated deeper into the atmosphere.

But why do meteors begin and end so precisely? Why does the brightness not change appreciably during flight? Why is the retardation so small? What masses do meteoroids have? The first person to seriously tackle the theory of meteoroid ablation was Öpik (1922). He concentrated on the Perseids and concluded that their colour was similar to the star Alpha Persei (F5 spectral type) so the effective temperature was assumed (incorrectly) to be about 7000 K. He also concluded that a Perseid of second magnitude was produced by a meteoroid of mass 10^{-3} g. Öpik showed, for the first time, that mass loss by the meteoroid, not velocity decrease, was the main source of the energy needed to produce a meteor. A Perseid meteoroid with a cosmical velocity of 56 km s^{-1} would have lost 99 per cent of its mass by the time it had retarded to a velocity of 54.4 km s^{-1}.

The second keynote paper in this field was by Lindemann and Dobson (1923b). They started with Denning's data on meteor heights (see Denning 1898) which stated that, on average, meteors first appeared at 118.5 km, disappeared at 72.9 km, had a path length of 100 km and a velocity of 43.3 km s^{-1}. Lindemann and Dobson assumed that a typical first-magnitude meteor appeared at an altitude of 100 km, ended at 80 km, and travelled a distance of 60 km at a velocity of 40 km s^{-1}; in other words, was visible for 1.5 seconds, and gave out energy at the rate of 4400 h.p. (3.3×10^{10} erg s^{-1}). The total energy ($1.5 \times 3.3 \times 10^{10}$ erg) is equated to the initial kinetic energy of the meteoroid giving a mass of 6×10^{-3} g). They gave the explanation of meteors as follows:

> A meteor appears when a cosmic particle of matter moving at a sufficiently high speed relative to the earth, becomes heated by atmospheric friction to such a temperature that it evaporates, and that it disappears when all the matter originally present evaporates. The molecules which distil off the meteor, as also the molecules of air in its path after collision, are moving at approximately the speed of the meteor and lose their kinetic energy largely in form of radiation, and by collision with other atmospheric molecules.
>
> The only tenable explanation is that the meteor becomes visible when its surface becomes hot enough to evaporate appreciably, and its fast vapor molecules collide with the air molecules. It disappears when it has evaporated practically completely. The velocity changes little in the process, also the radiation that comes to us and by which we see the meteor has its source in the vapor: the radiation from the particle being negligible.

Among other results they found the mass of the assumed typical meteor to be 6×10^{-3} g, i.e. if the meteoroid were made of iron its diameter would be about 0.1 cm. If a meteoroid had the same speed and duration as mentioned and should appear as bright as the full moon it need only be 2.5 cm in diameter.

We now know that these ideas are essentially correct but we must not forget that they were very controversial conclusions at the time. W H Pickering (1858–1938) (Pickering 1902, 1909) stated that small meteoroid masses were usually derived from 'some very doubtful assumption with regard to the amount of energy converted into light by a candle'. He deduced that if the intrinsic brilliancy of a meteor of third magnitude was about that of the incandescent portion of the carbon of an electric arc light the meteoroid must be about 17 cm in diameter. 'Fireballs might be 5 or 6 feet, or sometimes larger.'

Pickering thought that most of the meteor light came from the bright lines in its spectrum and thus that the nucleus was not very incandescent. His conclusions opposed the ideas of very small meteoroid masses.

Let us quote Olivier in 1925:

> Unfortunately the question can as yet be solved only by assumptions, and we know so little of the true conditions existing in the upper air. The solution again falls into the domain of the mathematical physicist rather than of the astronomers and the writer does not venture to say who is right. Certainly *a priori* larger values would seem more probable, but most men who have attacked the problem find the smaller.

In passing let me draw your attention to the fact that many researchers used meteors as probes of the Earth's atmosphere. Lindemann and Dobson (1923b) for example found that the density at a height of 150 km was *1000 times* that formerly calculated. Prior to the early 1920s the knowledge of atmospheric temperature had been obtained by climbing mountains and using small balloons. The height limit was 25 km. Most people had assumed that the temperature of the air at all heights above the tropopause was $-53°C$. Densities were calculated using meteor height measurements made by skilled amateur meteor observers. Values were obtained over the 35 to 160 km range. There was considerable scatter in the deduced densities and the results came as a great surprise. They were also the subject of much scepticism (see Dobson 1968). Their veracity was confirmed by photographic meteor work and later on by rocket flights.

Lindemann and Dobson suggested that the temperature at heights above 50 km must be much higher than expected. From 60 to 160 km they calculated that the temperature was about $+30\,°C$ instead of $-53\,°C$. The high-temperature region about 50 km above the ground was produced by the absorption of solar ultraviolet radiation which also created the ozone layer. Meteor observations were responsible for the discovery of this ozone region.

Atmospheric density and temperature calculations were continued by Whipple (1943) using the Harvard 1936–7 photographic meteor data. The conclusion was that there was a flat temperature maximum of about $100\,°C$ at a height of 60 km, a rapid drop to $-23\,°C$ near 80 km and a constant or slowly rising temperature at greater heights, up to about 110 km. The heights of these temperature regions were found to rise by 5.3 km between midsummer and midwinter.

In meteoroid physics Lindemann and Dobson placed far too much emphasis on the formation of an air cap around the incoming particle. At heights above 50 km the atmospheric density is so low that ablation is due to the direct impacts of single air molecules with the surface of the meteoroid. The work of C M Sparrow (1880–1941) (Sparrow 1926) and later of N Herlofson (1946) must be mentioned in this context.

Öpik (1933, 1937, 1940b) realised (by analogy with the impact of fast-moving positive ions on metal surfaces) that the impacting air molecule would not just eject a single meteoroid molecule but would become trapped in the meteoroid surface, and the kinetic energy so released would result in the 'boiling off' of a host of meteoroid atoms, these having thermal energy plus the meteoroid's geocentric velocity. For complete evaporation the air mass intercepted would be small in comparison with the meteoroid mass. So the atmosphere does not greatly decelerate the meteoroid but essentially breaks it into atoms and molecules which are then dispersed along the track.

1897–1932: Photographic meteor spectroscopy

Harvard provided the main database for this subject, as the result of a programme designed to perform another task. The Henry Draper Catalogue of stellar spectra was largely based on objective prism spectra obtained at Harvard. Inevitably the odd meteor appeared on the plates and the meteor of 11.00 PM 18 July 1897 was discussed in detail by Pickering (see *Harvard Circular* no 20). Its spectra contained six bright lines at 3954 (Hε), 4121 (Hδ), 4195, 4344 (Hγ), 4636 and 4857 (Hβ). Blajko (1907) discussed two photographic spectra obtained at the Moscow Observatory. The emission spectra of each were wholly different. No continuous spectrum could be seen.

Millman used the Harvard spectra set as the basis of his doctoral thesis (see Millman 1932a,b). By 1959 the world list of meteor spectra had reached 318, 75 per cent being from the major showers.

The spectra are predominantly emission line types with the energies of the observed multiplets being in the 2 to 14 eV range. Millman discovered that the meteor bursts were anomalous. They had lower excitation levels but higher ionisation levels than the preburst path. Millman (1950) applied a rotating shutter technique to meteor spectrometry and thus obtained spectra of the meteor wake. He found that low-excitation multiplets were particularly strong there.

Visual observations of long-enduring trails lasting from a few seconds up to several minutes have been made, but very few spectra have been obtained. The explanation of these trails has proved to be a stimulus to research in upper atmospheric chemistry with special emphasis being given to the possible catalytic release mechanism of latent energy produced by the introduction of sodium and caesium by the meteoroid.

1929–50: Meteor radar

The probing of the ionised layers in the upper atmosphere using radio-wave reflection started in the mid-1920s. Around 1929–30 a number of investigators noticed that unusual echoes were sometimes obtained at night from the E region. Observations suggested that the density of ionisation was enhanced for short periods of time. The first to attribute this to meteors was H Nagaoka (1865–1950) (Nagaoka 1929) but unfortunately he suggested that the meteoroids swept away ionisation causing a local change in refractive index. A M Skellett (b.1901) (Skellett 1931, 1935) came to the right conclusion and suggested that meteors contributed ionisation.

Typically radio meteors are thin columns some 10 km in length with electron line densities of 10^{10} to 10^{12} cm^{-1}. The radius of the column is initially of the order of the atmospheric mean free path, say 10 cm at a height of 100 km. So the electron volume density is 10^8 to 10^{10} cm^{-3}.

During the Leonid meteor shower of 1932 Skellett found that visual meteors overhead were correlated with transient increases in the E region. Unfortunately, the results did not show precisely what part the meteoric ionisation played in the production of sporadic E (see Lovell 1948), so Skellett's conclusions were not generally accepted. Conclusive evidence for the relationship was only produced in 1947 (see Appleton and Naismith 1947).

T L Eckersley (1886–1959) (Eckersley 1937) measured the height of the ionisation that produced the echoes and went on to measure the durations of the echoes and the variation of echo rate as a function of time of day. Appleton and Piddington (1938) found that the ionisation enhancement had a linear dimension of less than 30 m and that the maximum concentration was at 115 km. High-frequency forward scatter had been noted and the association of meteoric ionisation with some types of sporadic E was stressed by Pierce (1938).

Up to the time of World War II the photographic, visual and theoretical meteor astronomers seemed to have had minimal contact with the radio-wave propagation scientists and the use of radar as a tool for the investigation of meteors had not been fully recognised.

World War II spawned the detailed study of meteors by radar. J S Hey, a member of the British Army Operational Research Group was working, from 1943 onwards, on the detection of V2 long-range rockets using modified anti-aircraft radars. Radar echoes had to be received from objects which were at great heights and large distances. Spasmodic transient echoes were being picked up from objects at heights of around 95 km, at a rate of about 5 to 10 per hour. Two forms of irregular ionisation in the ionospheric E layer were known at that time, a very brief 'short-scatter' and longer lasting 'sporadic E'.

Short-scatter could be detected using wavelengths of a few metres, and this was what the V2 radars were picking up. In spring 1945 Hey's team observed a specific E layer region from three locations, Aldeburgh, Walmer and Richmond Park. Short-scatter echoes never appeared simultaneously on the three radars so they concluded that they were detecting meteor ionisation and Hey had thus discovered that backscatter echoes were only picked up from meteor trails that were perpendicular to the radar beam (see Hey and Stewart 1947). This obviously was of great significance. Radar could be used to determine the orientation of the trails and thus the direction of approach of the meteoroids and their radiant. This was of considerable interest to meteor astronomers but much less so to ionospheric physicists. The Hey and Stewart technique required three observing sites placed near the apex of an equilateral triangle. They found a diurnal variation in the numbers of echoes, the peak rate occurring first at one site and then, a few hours later, at another. So in the summer of 1945 Hey and Stewart had discovered the first daytime meteor shower activity, the successive occurrence of perpendicular aspect causing the rate variations. Daytime radar meteor shower work then switched to Jodrell Bank and scientists there quickly discovered a complex and extensive shower activity throughout June, July and August. This activity was greater than the known night-time showers and was reported by Clegg *et al* (1947), Aspinall *et al* (1949), Ellyett (1949), Aspinall and Hawkins (1951), Davies and Greenhow (1951) and Almond (1951).

A much more practical method than Hey's using a movable narrow-beam aerial at a single site was developed by Clegg (see Clegg 1948) and was used at Jodrell Bank from 1947. Probable radiant errors of $\pm 2°$ in right ascension and $\pm 3°$ in declination could be obtained. The workers at Jodrell Bank could thus clearly delineate where the radiants of the daytime showers were and so should be given the credit for discovering the great series of daytime streams.

Forward-scatter studies were seriously initiated in 1943. The Federal Communications Commission of the USA were monitoring the meteor forward-scattered signals from several high-powered, frequency modulated transmitters operating at around 45 MHz. E W Allen (1948) noted correlations between short increases in signal strength and the occurrence of bright visual meteors. The diurnal variation, maximising at 0600 hours local time and minimising at 18.00, was recognised. The recorded annual variation corresponded strikingly with that of meteor activity.

Sir Bernard Lovell has stated that the major stimulus in the post-war development of radar meteor studies was the desire to find out exactly what caused the transient echoes (Lovell private communication).

Jodrell Bank owed its origin to Lovell's initial belief that the radar echoes might have been caused by reflection from ionisation produced by large cosmic ray showers. Trailers of army radar were brought to Jodrell in 1945–6 specifically to help test this hypothesis (see Gilbert (1976); apparently Lovell paid a nominal sum of £10 for the lot although to buy it on the commercial market at that time would probably have cost £500 000). Another visual–radar collaboration also took place there at the time of the 1946 Perseid maximum. J Prentice *et al* (1947) led the meteor section of the British Astronomical Association in the visual campaign. Good correlation was found between the occurrence of visual meteors and 72.4 MHz radar echoes of duration greater than 0.5 s. Correlation with shorter-duration echoes was, however, very poor. Prentice concluded that this was due to the fact that these meteors were too faint to be seen easily. But it was only after the October 1946 Giacobinids that people became certain that *all* echoes were from meteors and that only meteors have an unambiguous specular refraction property.

When listening to the signal received from an unmodulated short-wave transmitter Chamanlal and K Venkatareman (1941) heard short-lived whistles, usually of descending pitch. This was a radio Doppler effect produced by a heterodyne beat between the transmitted and reflected waves. Unfortunately Chamanlal and Venkatareman thought that the Doppler shift was caused entirely by the retardation of the meteoroid. The correct interpretation was given by Appleton and Naismith (1947). The Doppler shift is due to the variation in the apparent radial velocity of the meteoroid as it crosses the beam of the radar transmitter at a relatively constant velocity. Measurement of this shift gives the meteoroid velocity and so radar could be applied to the pressing problem of the verification of the existence of hyperbolic meteoroids. Within about four years of the end of World War II, radar apparatus had been designed and built which measured radiant positions and meteoroid velocities with sufficient accuracy so that the hyperbolic component of the meteoroid influx was shown to be negligible. Ellyett and Davies (1948) used the Fresnel zone technique (as suggested by Herlofson (1946–7)) to measure the velocities of Geminid meteoroids and found them to be $34.4 \pm 2.8 \text{ km s}^{-1}$. Meteoroids were solar system objects.

The early history of meteor observations was closely linked with the Leonid shower and this brief history of meteors and meteor streams can fittingly end with another shower, the Giacobinids. The orbit of the Giacobinid (or Draconid) meteoroid stream coincides with that of Comet Giacobini–Zinner (1900III). In 1933 the Earth crossed the orbit about 80 days after the parent comet and the resulting meteor storm was the most spectacular of this century. (During the few minutes of peak activity the visual rate was as high as 5000 per hour.) It was predicted that the Earth would cross the orbit 15 days behind the comet in 1946 and considerable efforts were made to observe the shower both visually and by radar. The 4.2 m radar at Jodrell Bank recorded meteors at the rate of 10 000 per hour for the few minutes close to maximum activity (see Lovell 1954). A new narrow-beam directional aerial was specifically constructed for this work (Lovell *et al* 1947). Coincident visual observations were also attempted and the results agreed well with the expectations. It was found that meteoroids which produced less than about 10^{10} electrons per centimetre of path were below the limit of naked eye visibility, this result supporting Herlofson's (1946–7) theory of meteor ionisation. The fact that visual trains must be perpendicular to the aerial beam for radar detection was firmly established then (Lovell *et al* 1947) and this, according to Lovell, was the cardinal observation responsible for placing the radio echo studies on a secure footing. In the USA and UK, pulsed radars working at 27, 64, 100 and 212 MHz all recorded Giacobinid meteors. Hey's group recorded 22 head echoes in which the change in the range of the causative meteoroids was clearly shown as a function of time. Analysis of these echoes gave a mean velocity for the meteoroids of 22.9 km s^{-1}—the first meteoroid velocities ever to be measured by radar (Hey *et al* 1947).

Lovell wrote to me in May 1983—'I feel sure that at this stage and in retrospect all seems obvious, but it was not so in the immediate post-war years. The Establishment was cautious in the extreme and even Blackett remained unconvinced that we had a new method for observing meteors until that great Giacobinid shower of 1946'.

Conclusion

Over the last 300 years we have seen the true nature of meteors revealed. The meteor parents, the cosmic dust particles, are the debris of decaying comets and colliding asteroids. The meteors, those 'stars' which shoot across the sky, can be observed at night with the naked eye, telescopes, television systems and cameras. Since the 1940s they have also been observed by radar and so it is possible to monitor their influx 24 hours a day. The space age has enabled scientists not only to detect meteoroids in the near-Earth environment but also, thanks to Comet Halley, to travel to within 605 km of the cometary nucleus source of two of the major meteoroid streams.

The history of meteors has always been a story of the slow solution of problems. Aristotelian inertia, fixed radiants and hyperbolic velocities teased the early scientist. Today we worry more about the selectivity of detector systems producing bias in our data, the constancy of the flux and the physical and chemical structure of the meteoroids. It is always difficult to measure something which, at its slowest, hits the atmosphere at $11.2\,\mathrm{km\,s^{-1}}$ and also rarely gets to within 80 km of the observer.

Acknowledgments

I would like to thank Professor Stuart Malin for inviting me to give a talk on the history of meteors and meteor showers and thus encouraging me (indirectly) to write this chapter. I am also very grateful to Professor Fred Whipple and to Professor Sir Bernard Lovell for their most helpful comments.

References

Adams J C 1867 On the orbit of the November meteors *Mon. Not. R. Astron. Soc.* **27** 247

Allen E W 1948 Reflections of very high frequency radio waves from meteoric ionization *Proc. Inst. Radio Eng.* **36** 346–52

Almond M 1951 The summer daytime meteor streams of 1949 and 1950 III. Computation of the orbits *Mon. Not. R. Astron. Soc.* **111** 37–44

Almond M, Davies J G and Lovell A C B 1951 The velocity distribution of sporadic meteors *Mon. Not. R. Astron. Soc.* **111** 585–608

Appleton E V and Naismith R 1947 The radar detection of meteor trails and allied phenomena *Proc. Phys. Soc.* **59** 461–72

Appleton E V and Piddington J H 1938 Reflection coefficients of ionospheric regions *Proc. R. Soc.* A **164** 467–76

Arago F 1857 *Astronomie Populaire* vol. 4 p. 310

Aristotle 1952 *Meteorologicae* transl. H Lee (London: Heinemann)

d'Arrest H L 1867 Ueber eine Merkwurdige Meteor Falle beim durchgange der Erde durch die Bahn des Biela'schen Cometen *Astron. Nachr.* **69** (no 1633) col. 7–10

Aspinall A, Clegg J A and Lovell A C B 1949 The daytime meteor streams of 1948 I. *Mon. Not. R. Astron. Soc.* **109** 352–8

Aspinall A and Hawkins G S 1951 The summer daytime meteor streams of 1949 and 1950 *Mon. Not. R. Astron. Soc.* **111** 18

Ball R S 1886 *The Story of the Heavens* (London: Cassell)

Beech M 1990 Halley's meteoric hypothesis *The Astronomy Quarterly* in press

Bible Readings for the Home Circle 1899 (Michigan: Review and Publishing Company of Battle Creek)

Biot J B 1803 An account presented to the Institut National de France on 17 July 1803 *Mem. Classe Sci., Math. Phys. L'Inst. Natl. France* (1807) 224–66

Blajko S 1907 Spectra of two meteors *Astrophys. J.* **26** 341–8

Boguslawski G von 1867 Notiz über Kosmisches Theorie der Feuermeteore *Ann. Phys., Lpz.* **130** 165–8

Browning J 1867 On the spectra of meteors 13–14 November 1866 *Mon. Not. R. Astron. Soc.* **27** 77–9

Burke J G 1986 *Cosmic Debris* (Berkeley: University of California Press) p. 7

Chamanlal and Venkataraman K 1941 Whistling meteors—a Doppler effect produced by meteors *Electrotechnics* **14** 28 (see also Whistling meteors *Nature* **149** 416)

Chambers G F 1889 *A Handbook of Descriptive and Practical Astronomy* 4th edn vol. I (Oxford: Clarendon) p. 609

—— 1893 *Pictorial Astronomy for General Readers* (London: Whittaker)

—— 1910 *The Story of the Comets* (London: Oxford University Press)

Chladni E F F 1794 Uber den Ursprung der von Pallas gerfunden und anderer ihr ähnlicher Eisenmasse und über einig damin in Berkindug Stebende Naturerscheinungen (Riga: Harttnoch)

—— 1819 Uber Feure-Meteore und über die mit denselben harangefallen Massen (Vienna: Heuber)

Clegg J A 1948 Determination of meteor radiants by observation of radio echoes from meteor trails *Phil. Mag.* **39** (ser. 7) 577–94

Clegg J A, Hughes V A and Lovell A C B 1947 The daylight meteor streams of 1947 May–August *Mon. Not. R. Astron. Soc.* **107** 369–78

Condé J A 1854–5 *History of the Dominion of the Arabs in Spain* transl. J Foster 3 vols (London: Bohn)

Davies J G and Greenhow J S 1951 The summer daytime meteor streams of 1949 and 1950 *Mon. Not. R. Astron. Soc.* **111** 26–36

Denning W F 1878a Suspected repetition, or second outbursts from radiant points; and the long duration of meteor showers *Mon. Not. R. Astron. Soc.* **38** 111–14

—— 1878b Average per cent of the recorded magnitudes of shooting stars *The Observatory* **2** 20–1

—— 1884 The long duration of meteoric radiant points *Mon. Not. R. Astron. Soc.* **45** 93–116

—— 1899 General catalogue of the radiant points of meteoric showers and of fireballs and shooting stars observed at more than one station *Mem. R. Astron. Soc.* **53** 203–92

—— 1898 The heights of meteors *Nature* **57** 540–2

Devens R M 1876 *Our First Century* (Atlanta, GA: Nebhut)

Dobson G M B 1968 *Exploring the Atmosphere* (London: Oxford University Press) p. 38

Dunkin E 1891 *The Midnight Sky* (London: Religious Tract Society)

Eckersley T L 1937 Irregular ionic clouds in the E layer of the atmosphere *Nature* **140** 846–7

Elkin W F 1899 Results of the photographic observations of the Leonids, November 14–15, 1898 at the Yale Observatory *Astrophys. J.* **9** 20–2 and **10** 25–8

—— 1900 The velocity of meteors as deduced from photographs at the Yale Observatory *Astrophys. J.* **12** 4–7

Ellicott A 1803 *The Journal of Andrew Ellicott* (Philadelphia: Dobson)

Ellyett C D 1949 The daytime meteor streams of 1948 *Mon. Not. R. Astron. Soc.* **109** 359–64

Ellyett C D and Davies J G 1948 Velocity of meteors measured by diffraction of radio waves during formation *Nature* **161** 596–7

Erman G A 1939 Ueber einige Tatsachen, welche wahrscheinlich machen, dass die Asteroiden der Augustperiode sich im Februar etc *Ann. Phys. Chem.* **48** 585–601

Fedynski V and Stanjukowitsch K 1935 *Astron. J. USSR* **12** 440

Galle J G 1867 Ueber den muthmasslichen zusammenhang der periodischen Steenschauppen des 20 April mit dem ersten Cometen des Jahres 1861 *Astron. Nachr.* **69** (no 1635) col. 33–6

Gilbert G K 1895 The Moon's face: a study of the origin of its features *Bull. Phil. Soc. Wash.* **12** 241–92

Gilbert G N 1976 *Perspectives on the Emergence of Scientific Disciplines* (The Hauge: Mouton) p. 188

Halley E 1686 A discourse of the rule of the decrease of the height of the mercury in the barometer, according as places are elevated above the surface of the Earth, with an attempt to discover the true reason of the rising and falling of the mercury upon change of the weather *Phil. Trans. R. Soc.* **16** 104–16

—— 1705 *A Synopsis of the Astronomy of Comets* (London: Senex)

—— 1714a An account of several extraordinary meteors or lights in the sky *Phil. Trans. R. Soc.* **29** 159–68

—— 1714b An account of the late surprizing appearance of the lights seen in the air on the sixth of March last *Phil. Trans. R. Soc.* **29** 406–28

—— 1719 An account of the extraordinary meteor seen all over England on the 19th March 1718/9 with a demonstration of the uncommon height thereof *Phil. Trans. R. Soc.* **30** 978–90

Hawkins G S and Southworth R B 1961 Orbital elements of meteors *Smithsonian Contrib. Astrophys.* **4** 85–95

Helmholtz H von 1856 On the interaction of natural forces *Phil. Mag.* **2** 489–518

Herlofson N 1946–7 The theory of meteor ionisation *Rep. Prog. Phys.* **11** 444–54

Herrick E C 1841 Contributions towards a history of star-showers of former times *Am. J. Sci.* **40** 349–65

Herschel A S 1878 List of known accordances between cometary and observed meteor showers *Mon. Not. R. Astron. Soc.* **38** 369–95 (see also *Mon. Not. R. Astron. Soc.* 1876 **36** 220–3)

Herschel J F W 1866 On the meteoric shower of 1866 Nov 13–14 *Mon. Not. R. Astron. Soc.* **27** 19–21

Hey J S, Parsons S J and Stewart G S 1947 Radar observations of the Giacobinid meteors shower, 1946 *Mon. Not. R. Astron. Soc.* **107** 176–83

Hey J S and Stewart G S 1947 Radar observations of meteors *Proc. Phys. Soc.* **59** 858–83

Hind J R 1866 The meteoric shower of November 13–14 [1866] as witnessed at Mr. Bishops observatory, Twickenham *Mon. Not. R. Astron. Soc.* **27** 49–50

Hooke R 1665 *Micrographia, or Philosophical Description of Minute Bodies* (London) p. 243

Howard E C 1802 Experiments and observations on certain stony and metalline substances, which at different times are said to have fallen on the Earth *Phil. Trans. R. Soc.* **92** 168–212

Hughes D W 1974 The influx of visual sporadic meteors *Mon. Not. R. Astron. Soc.* **166** 339–43

—— 1977 Meteor stream formation after cometary decay *Space Res.* **17** 565–70

—— 1987 A mysterious woodcut *Sky and Telescope* **74** 252

Hughes D W and Emerson B 1982 The stability of the node of the perseid meteor stream *The Observatory* **102** 39–42

Humboldt A von and Bonpland A 1822 *Personal Narrative of Travels to the Equinoctial Regions* transl. H M Williams 7 vols vol. 3 (London) pp 331–3

Imoto S and Hasegawa I 1958 Historical records of meteor showers in China, Korea and Japan *Smithsonian Contrib. Astrophys.* **2** 131–44

Jacchia L G and Whipple F L 1956 The Harvard photographic meteor program *Vistas Astron.* **2** 982–94

—— 1961 Precision orbits of 413 photographic meteors *Smithsonian Contrib. Astrophys.* **4** 97–129

Kirkwood D 1861 *Danville Q. Rev.* (December)

Le Verrier U J J 1867 Sur les etoiles filantes du 13 Novembre et du 10 Aout *C. R. Acad. Sci., Paris* **64** 94

Lindemann F A and Dobson G M B 1923a Note on the photography of meteors *Mon. Not. R. Astron. Soc.* **83** 163–6

—— 1923b Theory of meteors and density and temperature of the outer atmosphere to which it leads *Proc. R. Soc.* A **102** 411–37

Lockyer J N 1890 *The Meteoric Hypothesis* (London: Macmillan)

Lovell A C B 1948 Meteoric ionization and ionospheric abnormalities *Rep. Prog. Phys.* **11** 415–44

—— 1954 *Meteor Astronomy* (Oxford: Oxford University Press)

Lovell A C B, Banwell C J and Clegg J A 1947 Radio echo observations of the Giacobinid meteors 1946 *Mon. Not. R. Astron. Soc.* **107** 164–75

McCrosky R E and Posen A 1961 Orbital elements of photographic meteors *Smithsonian Contrib. Astrophys.* **4** 15–84

McGuken W 1969 *Nineteenth Century Spectroscopy* (Baltimore: Johns Hopkins)

McKinley D W R 1951 Meteor velocities determined by radio observation *Astrophys. J.* **113** 225–67

Mayer J R 1848 *Beiträge zur Dynamik des Himmels in Populärer Darstellung* (Heilbronn)

Mechanics' Magazine 1833 (2 November) 287

Millman P M 1932a Meteor spectra *Ann. Harvard Coll. Obs.* **82** 113–46

—— 1932b An analysis of meteor spectra *Ann. Harvard Coll. Obs.* **82** 149

—— 1950 Spectrum of a meteor train *Nature* **165** 1013–14

Millman P M and Hoffleit D 1937 Meteor photographs taken through a rotating shutter *Ann. Harvard Coll. Obs.* **105** 601–21

Nagaoka H 1929 Possibility of the radio transmission being disturbed by meteoric showers *Proc. Imp. Acad. Tokyo* **5** 233

Newton H A 1863 Evidence of the cosmic origin of shooting stars derived from the dates of early star-showers *Am. J. Sci. Arts* **36** 145–9
—— 1864a The original accounts of the displays in former times of the November star shower etc. *Am. J. Sci. Arts* **37** 377–89 and **38** 53–61
—— 1864b Altitudes of shooting stars *Am. J. Sci. Arts* **38** 135–41
Newton I 1730 *Optiks* (London: William Innys) p. 355
—— 1934 *Mathematical Principles of Natural Philosophy* transl. A Motte (Berkeley: University of California Press) p. 541
von Niessl G 1907 Determination of meteor-orbits in the solar system (see *Smithsonian Miscellaneous Collections* **66** no 16 (1917) for English translation)
von Niessl G and Hoffmeister C 1926 Katalog der Bestimmungsgrossen fur 611 Bahnen grosser Meteore *Denkschr. Akad. Wiss. Math.-Naturwiss. Klasse, Wien* **100** 1–70
Olbers H W 1837 Die Sternschnuppen *Schumacher's Jahrbuch für 1837* pp 37–64
Olivier C P 1925 *Meteors* (Baltimore: Williams and Wilkins)
Olmsted D 1834 Observations of the meteors of November 13, 1833 *Am. J. Sci.* **25** 363–411 and **26** 132, 137–74
Öpik E J 1922 A statistical method of counting stars and its application to the Perseid shower of 1920 *Publ. Obs. Astron. Univ. Tartu* **25** no 1 1–56
—— 1932 Stellar perturbations of nearly parabolic orbits *Proc. Am. Acad. Arts Sci.* **6** 169–83
—— 1933 Atomic collisions and radiation of meteors *Acta Comm. Univ. Tartu* A **26** (Harvard Reprints no 100)
—— 1937 Basis of the physical theory of meteor phenomena *Publ. Obs. Astron. Univ. Tartu* **29** no 5 1–69
—— 1940a Analysis of 1436 meteor velocities *Publ. Obs. Astron. Univ. Tartu* **30** no 5 86
—— 1940b Meteors *Mon. Not. R. Astron. Soc.* **100** 315–26
—— 1969 The failures *Irish Astron. J.* **9** 156–9
Öpik E J and Boothroyd S L 1934 Arizona expedition for the study of meteors *Circ. Harvard Coll. Obs.* nos 388, 389, 390, 391
von Oppolzer T 1867 Bahnbestimmung des Cometen I *Astron. Nachr.* **68** 241–50
Peters C F W 1867 Bemerkung über Sternschruppenfall vom 13 November und 10 August 1866 *Astron. Nachr.* **68** 287
Pickering W H 1902 The meteor shower of 1877 *Ann. Harvard Coll. Obs.* **41** 133–51 (in 140–1)
—— 1909 The size of meteors *Astrophys. J.* **29** 365–80
Pierce J A 1938 Abnormal ionization in the E region of the ionosphere *Proc. Inst. Radio Eng.* **26** 892–908
Plavec M 1955 Ejection theory of the meteor shower formation. I. Orbit of an ejected meteor *Bull. Astron. Inst. Czech.* **6** 20–3
Prentice J P M, Lovell A C B and Banwell C J 1947 Radio echo observations of meteors *Mon. Not. R. Astron. Soc.* **107** 155–63
Quetelet L A J 1839 *Catalogue des Principles Apparitions d'Etoiles Filants* (Bruxelles: Hayez)
—— 1861 *Sur la Physique du Globe* (Bruxelles: Hayez)

Schaeberle J M 1895 Simultaneous meteor observations made at Mt Hamilton and Mt Diabalo on Aug 9–10, 1894 *Contrib. Lick Obs.* **5** 31–54
Schiaparelli G V 1867a Note e riflessioni intorno alla teoria astronomica delle stelle cadenti *Mem. Math. Fis. Soc. Ital. Sci.* **1** 153–284
—— 1867b Sur la relation qui existe entre les comètes et les étoiles filantes *Astron. Nachr.* **68** 331
Seneca 1971 *Naturales Quaestiones* transl. H Corcoran (London: Heinemann Loeb)
Shapley H, Öpik E J and Boothroyd S L 1932 The Arizona expedition for the study of meteors *Proc. Natl. Acad. Sci. USA* **18** 16–23
Sinding E 1948 On the systematic changes of the eccentricities of nearly parabolic orbits *K. Danske Vidensk. Selsk., Mat.-Fys. Meddr.* **24** no 16
Skellett A M 1931 The effects of meteors on radio transmission through the Kennelly–Heaviside layer *Phys. Rev.* **37** 1668
—— 1935 The ionizing effects of meteors *Proc. Inst. Radio Eng.* **23** 132–49
Smyth C Piazzi 1866 On the meteor-shower of 1866, November 13–14 *Mon. Not. R. Astron. Soc.* **27** 23–8
Sparrow C M 1926 Physical theory of meteors *Astrophys. J.* **63** 90
Strickland S 1853 *Twenty-seven Years in Canada West* ed. A Strickland (London: Bentley)
Twining A C 1834 Investigations respecting the meteors of 13 November 1833: remarks on Professor Olmsted's theory respecting the cause *Am. J. Sci. Arts* **26** 320–52 and **27** (1835) 339–40
Von Konkoly T M 1874 On the spectroscopic observation of a meteor *Astron. Reg.* **12** 3–4
Walker S C 1843 Researches concerning the periodical meteors of August and November *Trans. Am. Phil. Soc.* **8** 87–140
Watson F 1942 Meteors photographed with the 18-inch Schmidt camera on Mount Palomar *Pop. Astron.* **50**(3) 138–40
Weiss E 1867 Bemerkungen über den Zusammenhang zwischen Cometen und Sternschnuppen *Astron. Nachr.* **68** 381–4
Whipple F L 1938 Photographic meteor studies I *Proc. Am. Phil. Soc.* **79** 499
—— 1940 Photographic meteor studies III *Proc. Am. Phil. Soc.* **83** 711
—— 1943 Meteors and the Earth's upper atmosphere *Rev. Mod. Phys.* **15** 246–64
—— 1947 *Harvard College Observatory Technical Report* no 1 (see also Whipple 1972b p. 131)
—— 1949 The Harvard photographic meteor program *Sky and Telescope* **8**(4) 90
—— 1950 A comet model. I. The acceleration of the comet Encke *Astrophys. J.* **111** 375–94 and **113** (1951) 464–74
—— 1954 Photographic δ-Aquarid meteors *Astron. J.* **59** 400–6
—— 1972a The incentives of a bold hypothesis: hyperbolic meteors and comets *Ann. NY Acad. Sci.* **198** 219–24
—— 1972b *The Collected Contributions of Fred L. Whipple* (Cambridge, MA: Smithsonian Astrophysical Observatory)
Wright F W and Whipple F L 1950 The photographic Taurid meteors *Harvard*

College Observatory Technical Report no 6 (see also Whipple 1972b vol. 1 pp 180–216)

—— 1953 The photographic Perseid meteors *Harvard College Observatory Technical Report* no 11 (see also Whipple 1972b vol. 1 pp 253–81)

16

Solar Variability and Terrestrial Weather†

A J Meadows

Introduction

One interesting aspect of science is the way in which particular topics move in and out of fashion. Some proposed scientific explanations may never be totally disproved; but their degree of confirmation may also never reach an acceptable level. An excellent case study of such an 'in-between' topic is provided by research on solar variability and terrestrial weather. From the 1860s onwards, a major effort was put into the search for a link between solar changes and changes in the Earth's atmosphere, but it never really reached a satisfactory conclusion.

Solar variability

The first detailed observations of the Sun's disc were made by Galileo (1564–1642) early in the seventeenth century (North 1974). He identified the most striking feature present—the sunspots. Observers soon realised that the number of sunspots visible on the disc varied with time, but no attempt was made to look for periodicity. At the end

† Based on a presentation given at the conference on the History of Atmospheric Physics, 2 April 1986, City University, London.

of the eighteenth century, William Herschel (1738–1822) (Herschel 1801) collected sunspot data over several years. Again he did not look for periodicity, but he did make the first attempt to find a link between solar variation and terrestrial weather. He tried to correlate changes in sunspot numbers with variations in the average price of corn, arguing that the latter was an index of the average weather experienced each year. Herschel claimed to have identified a similar trend in both sets of figures, but most of his contemporaries discounted his results as being explicable by chance.

The major breakthrough with regard to sunspot numbers was made by the German amateur astronomer S H Schwabe (1789–1875) in the mid-nineteenth century. Schwabe had started counting the number of spots on the Sun's disc in 1826, initially in an attempt to identify the passage across it of a proposed new planet. He pursued his observations on every fine day for a number of years, and eventually announced (Schwabe 1844) that his results suggested the number of sunspots was varying periodically (with a period of about 10 years). His announcement caused little stir until it was taken up and publicised by one of the most famous German scientists of the day, Baron A von Humboldt (1769–1859) in his multi-volume work, *Kosmos* (Humboldt 1852). Meanwhile, a leading English geophysicist, Colonel E Sabine (1788–1883), had started analysing data on the Earth's magnetic field that had been collected at British observing stations overseas. His results indicated that disturbances to the geomagnetic field seemed to vary in a periodic way with time (Sabine 1851–6). (A similar result was obtained independently at about the same time by J von Lamont (1805–79) (Lamont 1852) at Munich on the basis of German measurements.) Sabine's wife was the English translator of Humboldt's *Kosmos*, and had just completed the volume which referred to Schwabe's results. A copy of the translation was sent to William Herschel's son, John (1792–1871), who in his reply remarked on the sunspot cycle. Sabine, now that his attention was drawn to it, responded immediately (Sabine 1852):

> With reference to Schwabe's period of 10 years having a minimum in 1843 and a maximum in 1848, it happens by a most curious coincidence (if it is nothing more than a *coincidence*) that in a paper waiting to be read at the R.S., I trace the very same years as those of minimum and maximum ... in the frequency and magnitude of the magnetic disturbances.

Very shortly afterwards A Gautier (1793–1881) (Gautier 1852) independently noted the similarity of the sunspot and geomagnetic periods, as did Rudolf Wolf (1816–93) (Wolf 1852). Further work by these two established that the period was more like 11 years than the initially suggested 10.

Auroral variability

The geomagnetic results led immediately to a search for a corresponding variability in the number of aurorae seen. It had been accepted since the eighteenth century that aurorae and magnetic storms were connected, so such a correlation was to be expected. John Dalton (1766–1844), though best remembered today as a chemist, was a keen meteorologist, and his *Meteorological Essays* (1793, 1834) ably summarised the state of current knowledge. He noted that aurorae seemed to be governed by the Earth's magnetism, and appeared high in the atmosphere (at 150 miles, he estimated). However, he also believed there could be a connection between activity in the high atmosphere and what happened to weather in the lower atmosphere. For example, aurorae were, he thought, an indication of future fine weather.

Periodicity of the aurorae proved hard to pin down, not least because many aurorae went unobserved. But, by the 1860s, there was fair agreement that aurorae, too, varied in frequency during the solar cycle. Many people, like John Dalton, expected that influences on the upper atmosphere would produce some effect in the lower atmosphere. So it was natural for them to turn next to terrestrial meteorology. This automatic connection was reinforced by a generally accepted analogy between the solar and terrestrial atmospheres. Both were believed to experience cyclonic and anti-cyclonic wind circulations, and the upper regions of the solar atmosphere were supposed to experience electrical discharges akin to those thought to cause the aurorae on Earth.

Meteorological variability

As a result of this chain of argument. Schwabe's discovery of the solar cycle generated an interest in possible periodic changes in the weather. Indeed, Schwabe, Gautier and Wolf all looked for such changes, but with no great success. Almost all the meteorological variables were analysed for periodicity by someone. Although this work was undertaken in a number of countries, there were good reasons why it should have been pressed most strongly in Britain. The responsibilities of a wide-flung Empire had led to a consequent interest in global weather changes. All the colonies depended to a great extent on agriculture for their livelihood. This was especially true in India, where famine was a recurrent disaster. In addition, the Indian subcontinent experienced a much simpler annual weather pattern than a country such as Britain. Hence, it was more important to predict the weather for India, whilst, at the same time, it should to all appearances be easier. For this reason, a number of the British attempts to look for periodicity in meteorological data looked at the general area of the Indian Ocean.

The early 1870s was a period of sunspot maxima when aurorae were particularly common. A cluster of important papers appeared at this time looking at solar–terrestrial weather links. J Baxendell (1815–87) (Baxendell 1872) published an analysis of the meteorological data collected at the Radcliffe Observatory, Oxford, since the previous solar maximum. He claimed that variations of both atmospheric pressure and temperature had occurred in step with solar activity. At about the same time, E J Stone (1831–97) (Stone 1871), at the Cape of Good Hope, examined the previous 30 years' meteorological observations there and concluded that the mean annual temperature varied with the solar cycle. C Meldrum (1821–1901) (Meldrum 1873) in Mauritius carried out an even more wide-ranging study, using data on rainfall from 18 observing stations and covering five sunspot periods. He claimed that, with one exception, all the years of maximum and minimum rainfall coincided quite well with sunspot maxima and minima. He also noted that cyclones in the Indian Ocean were more frequent and more violent at sunspot maxima than at minima.

The most impressive results, however, were published by C Piazzi Smyth (1819–1900) (Smyth 1880), the Astronomer Royal for Scotland. He had analysed data from a long series of temperature measurements started long before in 1837. These were obtained from thermometers buried deep in the rock at Edinburgh, so as to remove diurnal temperature variations and only record longer-term changes. Not long before, Piazzi Smyth had visited Lamont in Germany, and this helped stimulate him to look for periodic variations in the temperature data. He claimed to find an appreciable change with the sunspot period.

Investigations of these, and other meteorological variables continued into the 1880s. Balfour Stewart (1828–87) (Stewart 1882), for example, examined annual variations in the depths of rivers and lakes. In 1882, he analysed data relating to the Nile and the Thames, and found that the maximum depths of both rivers occurred just after sunspot maximum, with another maximum just after sunspot minimum. He soon extended this analysis to the Elbe and the Seine, finding a similar result. He interpreted his results as implying an increase in rainfall at these times.

By the latter part of the 1870s, there was a widespread feeling that correlated Sun–weather variations occurred. It even filtered down to the school syllabus in England, where pupils were required to understand:

> Explanation of the daily and yearly changes in the meteorology of the earth. Possible 11 yearly changes.†

In the early 1880s, G Stokes (1819–1903) (Stokes 1881), who was an interested, but generally cautious, observer of the scientific scene, summarised the contemporary position as follows (Stokes 1881):

> Speculations have been made as to whether there is not a decennial period, or something of the kind, traceable even in the occurrence of Indian famines. If so, there may be some very close relationship between the solar spots and these famines. At first sight, one would be disposed to say, "What possible connection can sun-spots and famines have with one another? You might as well as speak of the connection between comets and war!" But when we go deeper below the surface, and study carefully the phenomena present to our view, we see that a possible connection between such apparently remote things as sun-spots and famines may not be chimerical.

The reaction

Stokes was right to be cautious. By the latter part of the 1870s adverse criticism of the supposed links was increasing. In 1877, Lieut-General R Strachey (1817–1908) contributed a paper to the *Proceedings of the Royal Society* 'On the alleged correspondence of the rainfall at Madras with the sun-spot period and on the true criterion of periodicity in a series of variable quantities' (Strachey 1878). He was concerned with two vital questions. Are the variations sufficiently larger than the observational scatter to be accepted as real? If the variations are real, are the data adequate to identify a genuine period? Strachey concluded:

> I would explain that I do not desire to call in question the possible or actual occurrence of periodical terrestrial phenomena corresponding to the sun-spot period, but to point out in the case of the rainfall not only has no such correspondence been established, but that there has been no sufficient evidence adduced of any periodicity at all.

In 1881, the same year as Stokes' cautious summary, H F Blanford (1834–93) (Blanford 1882), the leading Indian Government meteorologist, submitted an official report to the famine Commissioners on Sun–weather variations. He explained that he believed

† Science and Art Department Syllabus for Physiography, 1877.

solar heat varied cyclically, but had been unable to find any simple correlations with terrestrial meteorology of the type that had been suggested. He concluded that there appeared to be no immediate possibility of using solar observations to predict weather in India. Criticisms along these lines grew during the rest of the nineteenth century.

Conclusion

Arguments about the reality of weather variations which followed the solar cycle continued throughout the nineteenth century. By the early years of the present century, further progress looked unlikely, and the whole question was gently shelved. Despite occasional attempts to revive it, the existence of such variations remained an unsettled problem—one which was not considered worth pursuing by most scientists. In the 1960s, however, interest in the problem began to revive. An important reason was the development of electronic computers which allowed very large quantities of meteorological data to be handled. It was hoped that extensive data analysis followed by sophisticated methods of sifting for periodicities would provide much firmer results. But it gradually became clear that something else was essential: the identification of a mechanism by which solar changes could affect the Earth's lower atmosphere. Appropriate mechanisms were by now well established for auroral and geomagnetic variations. The difficulty in identifying a new mechanism was that the important solar changes were certainly small, and yet had to be monitored from the bottom of the Earth's rapidly varying atmosphere. Satellites seemed the obvious answer, and, indeed, satellite measurements of the Sun's radiation budget do appear to be leading in the right direction. But a fully acceptable model linking solar variation to terrestrial meteorology has still to be developed.

References

Baxendell J 1872 On changes in the distribution of barometric pressure, temperature and rainfall under different winds during a solar spot period *Proc. Lit. Phil. Soc. Manchester* **11** 111–18

Blanford H F 1882 Some further results of sun-thermometer observations with reference to atmospheric absorption and the supposed variations of solar heat *J. Asiat. Soc. Bengal* **51**(2) 72–84

Dalton J 1793 and 1834 *Meteorological Observations and Essays* (Manchester)

Gautier A 1852 Note sur quelques recherches récentes astronomiques et physiques relatives aux apparences que présente le corps du soleil *Arch. Sci. Phys. Nature* **20** 177–207, 265–82

Herschel W 1801 Observations tending to investigate the nature of the Sun, in order to find the causes or symptoms of its variable emission of light and heat, with remarks on the use that may possibly be drawn from solar observations *Phil. Trans. R. Soc.* **91** 265–318, 354–62

Humboldt A von 1852 *Cosmos* transl. E Sabine vol. III (London: Longman, Brown, Green and Longmans) p. 292

Lamont J von 1852 On the ten-year period which exhibits itself in the diurnal motion of the magnetic needle *Phil. Mag.* **3** 428–35

Meldrum C 1873 On a periodicity of cyclones and rainfall in connexion with the sun spot periodicity *Rep. Br. Assoc. Adv. Sci.* **1873** 466–78

North J D 1974 Thomas Harriot and the first telescopic observations of sunspots in *Thomas Harriot* ed. J W Shirley (Oxford: Clarendon) pp 129–65

Sabine E 1851–6 On periodical laws discernable in the mean effects of the larger magnetic disturbances *Phil. Trans. R. Soc.* 1851 635–42, 1852 103–4, 1856 357–74

—— 1852 *John Herschel Letters* no 15.235 (Sabine to Herschel, 16 March 1852) (Royal Society Archives)

Schwabe H 1844 Sonnen-beobachtungen in Jahre 1843 *Astron. Nachr.* **21** 233–6

Smyth C Piazzi 1880 Notice of the completion of the new rock thermometers of the Royal Observatory, Edinburgh, and what they are for *Trans. R. Soc. Edinb.* **29** 637–56

Stewart B 1882 On a supposed connection between the heights of rivers and the number of sun-spots on the Sun *Rep. Br. Assoc. Adv. Sci.* **1882** 462–4

Stokes G G 1881 Solar physics *Nature* **24** 593–8, 613–18

Stone E J 1871 On an approximately decennial variation of temperature at the observatory at the Cape of Good Hope between the years 1841 and 1870, viewed in connexion with the variation of the solar spots *Proc. R. Soc.* **19** 389–92

Strachey R 1878 On the alleged correspondence of the rainfall at Madras with the sun-spot period, and on the true criterion of periodicity in a series of variable quantities *Proc. R. Soc.* **26** 249–61

Wolf R 1852 Sur le retour périodique de minimum des faches solaires; concordance entre ces périodes et les variations de déclination magnétique *C. R. Acad. Sci., Paris* **35** 704–5

17

The Early History of Atmospheric Ozone

C Desmond Walshaw

Introduction

This chapter is based on a talk given to a joint meeting of the Historical Groups of the Royal Astronomical Society, the Royal Meteorological Society and The Institute of Physics on 2 April 1986. I had aimed at covering the history of atmospheric ozone up to the present, but in delving into the early history found so much of interest that I decided to stop at 1921, when the publication, delayed by World War I, of the observations of C Fabry and H Buisson at Marseilles marked the start of what may be regarded as the 'modern' period. After that G M B Dobson, summoned after the war by F Lindemann to join him in starting new research at the Clarendon Laboratory, Oxford, really took over, perfecting the methods of Fabry and Buisson with extraordinary ingenuity and discovering in the course of one or two years most of the peculiar features of the behaviour of ozone in our atmosphere which make it a unique minor constituent. The historical development of Dobson's work is readily followed from his own account of it (Dobson 1966, Houghton and Walshaw 1977). Of the earlier period, however, little seems to have been written, at least in English (see Partington 1964, Fabry 1950 and especially Schmidt 1988).

Only a brief glance at the more important and interesting events and people was attempted in my talk, and here I have done little more than round off some particularly jagged edges. A strong local bias will also be noted, for which I make no apology, as Oxford has provided a congenial home for the muses of experimental philosophy from the beginning, a fact of which many now seem sadly unaware (see, however, Simcock (1984)). A more sombre and balanced account of the subject remains to be written.

Schönbein

We must begin with the history of ozone itself. The starting point seems to be the unique 'electrical odour' noticed by the early experimenters when they worked their impressive electrostatic machines (Van Marum 1756, Partington 1964, vol. III, p. 342). That this was due to a definite chemical substance was established by Schönbein, although many of his contemporaries for long remained sceptical. Christian Friedrich Schönbein (Partington 1964, vol III, pp 190–196, Schönbein 1891, 1970) was born at Metzingen, near Baden-Baden, on 18 October 1799. He became an apprentice at a chemical factory in Böblingen, and spent the early mornings and evenings toiling at his studies in chemistry, mathematics, Latin and philosophy. After taking some posts in industrial chemistry he decided to study pure science, and attended the universities of Tübingen and Erlangen. He formed a close friendship with the philosopher F von Schelling (1775–1854), to whose ideas some of his later (and to our minds whimsical) theories of ozone may be traced. After this he travelled, spending time in Paris, where he attended the lectures of such men as A M Ampère (1775–1836) and J L Gay-Lussac (1778–1850), and in London. He seems to have liked England, and taught at Dr Mayo's educational establishment at Epsom. By 1829 he had evidently acquired some scientific reputation as he was then invited to the University of Basel to stand in for the professor of physics and chemistry, Peter Merian. There he spent the rest of his life, obtaining the chair of chemistry in 1852. He was much respected and active in public affairs, being responsible for gas lighting in the streets and promoting the establishment of the Museums of Art and Science, and in 1840 was made an honorary citizen. As well as discovering ozone, he invented gun-cotton; then, by dissolving gun-cotton in ether, collodion, thus ushering in the first really practical process of photography.

Schönbein's opinion that ozone was a definite chemical substance was based on his finding that the oxygen produced in the electrolysis of water has a smell identical to the 'electrical odour'. He studied its properties extensively, and later found that it is also produced when phosphorus glows in air (Schönbein 1840, 1854). He gave ozone its name, but the conclusive evidence that it is a form of oxygen was provided by the independent investigations of J C de Marignac (1817–94) and A de la Rive (1801–73) in 1845 (Marignac 1845, de la Rive 1845).

A vivid idea of Schönbein's personality and of the development of his ideas can be obtained from his extensive correspondence with Michael Faraday (1791–1867) with whom he formed a deep friendship and who shared his ozone enthusiasms (Kahlbaum and Darbishire 1899, Mohr 1854).

> My dear Faraday,
> I take the liberty to introduce to you Professor Vischer of Basle, an intimate friend and colleague of mind, who intends to make a stay in London for some time and is kind enough as to take charge of a parcel containing voluminous letters, scientific papers and something else destined for the Sovereign of the Royal Institution. It will perhaps interest you to learn on this occasion, that my friend, being an excellent Greek scholar, acted the part of a god-father, when I christened my child "Ozone" 19 years ago...

Just before Faraday gave his Friday Evening Discourse at the Royal Institution on 'Schönbein's Ozone and Antozone' (Faraday 1851, 1859) Schönbein's daughter Emilie died. Faraday wrote (Kahlbaum and Darbishire 1899, p. 324)

> ... Instead of a glad and buoyant heart I shall go to my work, as work indeed. I was desiring to put it off, but when I began to look about for the purpose, I found so many engagements had been made contingent upon the evening, and that even the Prince Consort [†] was coming, that I could not properly change the date. I only hope that I shall not break down. I know I shall not be able to forget for the hour, and an overpowering thought may break in...

Schönbein gave instructions for making ozone test papers from a mixture of starch and potassium iodide. Ozone in the air liberated iodine from the iodide and this turned the starch blue. The papers were usually exposed for 12 hours and then compared with a scale of blue tints numbered from 0 to 10. Schönbein was aware that other oxidising agents in the air could produce the same effect, and tried several variations (indigo, litmus), but the original starch–iodide papers seemed the most popular. Perhaps inspired by Schönbein's enthusiasm, an ozone fever now spread rapidly in scientific circles. Negretti

† HRH Prince Albert was Vice-President of the Royal Institution.

and Zambra did a brisk trade in ozone papers and supplied a neat little cage to protect them from birds etc (Chambers 1909).

Ozone, medicine and meteorology: the early period

At Oxford the subject was taken up by Charles Daubeny (1785–1867), who succeeded J Kidd (1775–1851) as Professor of Chemistry in 1822 and became Sherardian Professor of Botany in 1834. He made critical experiments on the accuracy of the Schönbein papers, particularly the effect of light on them. The date of his earliest regular observations on ozone in the atmosphere is uncertain, but his biographer, J Phillips (1869) informs us that

> During a few late winters Dr Daubeny found it desirable to exchange his residence in Oxford for the milder climate of Torquay. Here his activity of mind was equally manifested by public lectures on the temperature and other atmospheric conditions of that salubrious resort, and by experiments on ozone and the usual meteorological elements, in comparison with another series in Oxford. By this connection with Devonshire he was induced to join the Association in that county for the advancement of Science, Literature and Art; and one of his latest public addresses was delivered to that body, as President, in 1865.

Ozone seems to have become an addiction for Daubeny in his later years. Details of his Torquay series, which started in 1864, are recorded in his posthumously published paper (Daubeny 1867); this also contains an account of his elaborate (and unfortunately unrewarding) experiments on the production of ozone by the leaves of plants. I cannot resist quoting the final paragraphs:

> And this leads me to the last question to be considered, namely, the uses which ozone subserves in the economy of nature. When we consider its remarkable oxidising properties and the rapidity with which any organic matter, dead or living, undergoes a slow combustion in its presence, it seems reasonable to conclude, that this principle is an important agent for destroying putrescent animal and vegetable matter by oxidation, and thus for restoring to the atmosphere its purity.
> However, it may be urged by some, that it is premature to speculate as to the uses of a principle, until we are fully assured of its existence; and although it would seem extremely unaccountable, that air which had been in contact with a growing plant, should, when introduced into a darkened tube, affect Schönbein's paper in the manner described, unless ozone were present in it, still it would be more satisfactory if some other test, not liable to be affected by light, could be appealed to in confirmation of its presence. And if the suggestion of some new and

improved method of determining ozone should flow from the reading of this paper, I shall deem myself amply repaid for the labour it has cost me, even though the conclusions I have sought to deduce should eventually be overthrown by other more precise methods of observation.

It is time indeed that the questions as to the existence of ozone in the atmosphere should once for all be set at rest; for at present, whilst it is disputed by some eminent chemists, it is taken for granted by meteorologists, and thus the weather tables constructed in this and most other countries are swelled by an additional column devoted to ozone observations, which, if fallacious, should be swept away as fostering the popular belief in a non-entity; and if real, should be turned to some account by those whose business it is to investigate the unexplored properties of the atmosphere in which we live.

The atmospheric ozone observations referred to by Daubeny in the first of the above quotations as 'another series in Oxford' seem to be those made at the Radcliffe Observatory. Concerning these I am indebted to C G Smith, Director of the Radcliffe Meteorological Station from 1952 to 1985, for the following information:

> They were printed for the first time in Radcliffe Observations Vol. 15 for the year 1854 and appear to have commenced in January that year. They appear on p. 79 and there is a brief note on p. xxiv of the Introduction to the Meteorological Part of the volume. They were printed for the year 1856 (pp. 79, xxxvi) but not in the volumes for 1857 and 1858. After that they appear to have been published each year up to 1920. They are usually described as by Schönbein's method, but in some later years as by Moffatt's method. [†] They ceased in 1920 and there is a note in the Volume for 1921–1925 (Vol. 54, p. v) which reads "Observations of the amount of Ozone by means of Schönbein's Ozonometer were discontinued, the method being generally considered to furnish results of little value".

The possible influence of ozone on health, for better or worse, was clearly one of the chief causes of the ozone fever. Sir Henry Acland (1815–1900) who made a pioneering study of the possible effects of the meteorological conditions on the outbreak of cholera which started in Oxford Prison in the autumn of 1854 (Acland 1861), used data from the Radcliffe Observatory, confirming that 'observations with Schönbein's ozonometer were commenced during the last days of January 1854'. It seems likely that Daubeny was responsible for the

---

† I have not found a proper description of Moffat's papers, nor of James's (or De James) which are mentioned in Acland (1861). Both were included in the comparisons with Schönbein papers by Berigny (1857) and Moffat's papers are also mentioned by Daubeny (1867).

addition of ozone to the meteorological elements recorded at the Radcliffe Observatory, although there seems to be no direct evidence for this. Daubeny certainly corresponded with Schönbein, and Schönbein visited Daubeny in Oxford, signing the visitors' book on some date between August 1843 and May 1847.†

Quite a number of these early series of observations of atmospheric ozone survive.‡ A particularly vigorous observer was A Houzeau (1829–1911), Director of the Station Agronomique de la Seine-Inférieure. The series he started at the Parc de Montsouris in Paris (using the indigo variant) was only discontinued after World War II, and he made substantial contributions to the study of ozone chemistry (Houzeau 1858a, 1872a, b).

In the United States the medical aspects of ozone were taken very seriously. The Schönbein papers were exposed at some 20 stations in Michigan in a programme of meteorological observations started by R C Kedzie (1823–1902) with the aim of studying the spread of certain diseases. A daily series taken at the Michigan Agricultural College near Lansing running from 1871 to 1882 has recently been examined by Linville, Hooker and Olson (Linville *et al* 1980). In careful laboratory experiments they showed that the sensitivity of the Schönbein papers to ozone depends strongly on the relative humidity of the air. Applying their results to the Kedzie series they found similar day-to-day changes in ozone to those given by modern methods. Their calibration of the test papers has also been used to reinterpret a series of measurements made at Athens from 1901 to 1940 (Mariolopoulos *et al* 1984).

Ozone continued to attract the attention of some of the most distinguished chemists of the century,§ who found it difficult to convince themselves that it really was O_3. This was largely due to the fact that the substance had not yet been isolated in a pure state. Only rather weak mixtures with oxygen were available, usually produced either in the silent electrical discharge or by the electrolyis of water with a high current density at the anode. Conviction grew from experiments designed to measure the change in volume which occurs

---

† Formerly in the Bodleian Library (MS Sherard.265, MS A.41), this is now in the library of the Botany School, Oxford University.

‡ The series at Vienna seems to be the longest with uninterrupted use of Schönbein papers and has been reviewed by F Lauscher (1984). For others see, in Bern, R Wolf (1854, 1855) and in Cracow, F Karlinski (1854). For some early work on the reliability of the ozone papers see Berigny (1857).

§ See, for example, Odling (1872) and Palmieri (1872). We may also note in passing the first investigation, at the Royal Institution, of the infrared spectrum of ozone (Tyndall 1861).

when the ozone in such a mixture is destroyed by heat. Such work was done by T Andrews (Andrews 1855, Andrews and Tait 1857) at Dundee, confirmed later in Oxford by B Brodie (1872). In his monumental work R T Gunther (1937) tells us

> Sir Benjamin Collins Brodie, 1870–1880, ... was appointed to the Professorship of Chemistry in 1885, and finding the Old Ashmolean unsuited for his purpose was a strong advocate of the Glastonbury Kitchen and other conveniences at the New Museum in the Parks [Simcock 1984]. The apparatus used for some of his more important research relating to the production and properties of Ozone was still in existence there when I was an undergraduate, but it has now disappeared.

Ozone consolidated: 1879–1921

Returning to ozone in the atmosphere, we find the year 1879 to mark a convenient division between the 'early' history and a 'middle' period into which we now move on.† In this year the French astronomer A Cornu (1841–1902) published his paper on the ultraviolet limit of the solar spectrum (Cornu 1879, 1881). It is not clear who first noticed that all stellar spectra were cut off at about the same ultraviolet wavelength. W Hartley (1881b) quotes W Miller (1863) but, although it is obvious from his spectra as far as the solar spectrum is concerned, I cannot find any mention of the matter in his text.

The problem was 'Why does the sun's spectrum show no wavelengths shorter than about 300 nm?'. Was it due to the sun, or was it due to absorption in the earth's atmosphere? Cornu studied the variation of the wavelength of cut-off with the height of the sun above the horizon. He decided that the cause of the cut-off must be absorption in the earth's atmosphere, and that the gas responsible was neither water vapour nor carbon dioxide. To confirm his conclusion he observed the solar spectrum from the peak of the Riffelberg (2570 m),‡

---

† There was apparently a lack of interest in this subject by British meteorologists. All I could find in the *Quarterly Journal of the Royal Meteorological Society* was Twemlow (1875) and Mulvaney (1879) and ensuing discussion.

‡ The essentials of mountain spectroscopy are nicely described by Cornu (1881, p. 12): 'Mettant a profit les indications données par plusieurs savants habitués aux expériences dans les montaignes, je me suis installé d'abord au Riffelberg, dans le massif de mont Rose, à une altitude de 2570 m. Cette station, ou se trouve un hôtel convenable, est l'une de celles ou les probabilités de temps clair sont les plus grandes ...'

expecting that the reduction of the amount of air in the light path would push the cut-off further into the ultraviolet. But no such extension was found. The experiment was repeated from the Gornergrat (3136 m). This also gave a negative result. Cornu's work was thus inconclusive, and it was not until some 35 years later that his idea of an atmospheric absorber was accepted and the failure of his mountaineering experiments understood.

At almost the same moment P Hautefeuille and J Chappuis (Hautefeuille and Chappuis 1880a,b) succeeded in obtaining one drop of liquid ozone. Considering the difficulty at that time of reaching the necessary low temperature, and the tendency for liquid ozone to explode violently, this was a remarkable achievement. At last the isolation of ozone in an almost pure state had been achieved, and its existence as a real substance became undeniable.

The boiling point of ozone at atmospheric pressure is $-112\,°C$. In their first paper they compressed ozonised oxygen in a capillary tube to a pressure of 200 atm at $-23\,°C$, using methyl chloride as refrigerant. A fog of liquid ozone droplets was produced. They then (1880b) reduced the temperature to $-88\,°C$ by using boiling liquid nitric oxide. The ozone became darker blue in colour, but again no liquid separated. When carbon dioxide was added, however, a meniscus appeared over the liquid, the deep blue colour of which they noted. Once liquid nitrogen became available the liquefaction of ozone became a routine operation apart from the possibility of a violent explosion.†

In the same year, 1880, appeared the paper by J Chappuis on the absorption bands in the orange region of the spectrum which give ozone its characteristic blue colour. Then in the following year W N Hartley (1846–1913) (Hartley 1881a) obtained in his laboratory in Dublin spectra of the great ultraviolet absorption band, using a spark source.‡ He did not know the amount of ozone in his absorption cell, but studied the wavelength limits of the band as the amount was increased. Taking Houzeau's results into consideration, Hartley was able to propose that (i) the ultraviolet cut-off of the spectra of all heavenly bodies is definitely due to ozone in the earth's atmosphere, (ii) the concentration of ozone is greater in the upper atmosphere than

---

† By attending to the precautions given by M E Vigroux (1949) however I carried out this process many times when a research student at the Cavendish Laboratory, Cambridge, without ever having an explosion.
‡ Hartley's review of earlier work is of great interest. He discusses very clearly the problems of measuring atmospheric ozone by chemical methods, as well as its variation with wind direction and the limit of detection by the nose.

near the ground, and (iii) the blue colour of the sky is due to ozone.†
The following remarks surely earn for Hartley the title of 'father of
spectroscopic sounding of the atmosphere' (Hartley 1881b, p. 128):

> I am also of opinion that a spectroscopic camera, such as I have been
> using, may become a valuable meteorological instrument, because the
> ultra-violet rays are so exceedingly sensitive to the absorptive action of
> very minute traces of substances and notably to certain traces of ozone.

After this surge of activity, the next milestone seems to be 1890, when
William Huggins (1824–1910) and Margaret Huggins discovered six
new lines in the spectrum of Sirius, close to the ultraviolet limit
(Huggins and Huggins 1890).‡ There is no hint in their paper that
these lines might not originate from the star.

By 1906 the technique for obtaining high concentrations of ozone by
evaporation from the liquid had been much improved by E Ladenburg
and E Lehmann (Ladenburg and Lehmann 1906). They photographed
the spectrum in the ultraviolet using a source of light with a continuous
spectrum (magnesium ribbon). They found a series of absorption
bands which appeared to match the Sirius 'lines' discovered by
Huggins. They measured the wavelengths of the bands, but not with
great accuracy, so that the attribution of the Sirius 'lines' to ozone
could not be made with confidence.

Attempts continued to be made, following Houzeau, to detect the
decrease of ozone in the absorbing path with increased height of
observation of the solar spectrum. For example A Miethe and E
Lehmann (Miethe and Lehmann 1909) climbed Monte Rosa in 1909
and A Wigand (1913) made a balloon ascent at Halle in 1913, but all
these efforts were attended by failure.

The year 1913 is our last milestone. In it Charles Fabry (1867–
1945) (Lecomte *et al* 1973),§ his famous work on optical inter-

---

† Here, of course, Hartley went a bit too far. Chappuis (1880) has been more
discreet: 'sa couleur bleue joue donc certainement un rôle dans la coloration
du ciel'. We now know of course that generally speaking the blue of the sky is
due to the scattering by air molecules beautifully investigated by Lord
Rayleigh. To see the role of the ozone we must look at the twilight sky in the
zenith; without the ozone its colour would be not blue, but a pale blue-
greenish grey (see Hulbert 1953).
‡ They comment on the difficulty of measuring the wavelengths but give no
hint of them arising anywhere but in Sirius. The reader will be rewarded by
consulting Sir William's earlier monumental paper (Huggins 1880) with its
beautiful plate. Wavelengths were based on measurements by Mascart and
Cornu.
§ These three (separate) papers were given at the Ozone Symposium held in
Monaco in 1968 as a centenary tribute to Fabry: they contain much
interesting information based on personal acquaintance.

ference behind him, published with Henri Buisson his first paper on ozone (Fabry and Buisson 1913). In this they reported their measurements on the absorption coefficients in the Hartley band, made with extreme care over the very wide range of absorption coefficients (the absorption by ozone at its peak around 256 nm is, mass for mass, as strong as the absorption by metals in the visible spectrum). These measurements served for many years as the quantitative basis for ozone measurement. Fabry and Buisson went on to discuss the spectroscopic measurement of ozone in the atmosphere. In a good plate they compare the spectra of the Sun near the zenith at Marseilles, an iron arc, and the same arc with about 5 mm (at NTP) of ozone in the path. The publication of their continued work was delayed by World War I (see below).

For the tying up of the remaining loose ends we turn to the careful work of R W Strutt (1842–1919) (later the Fourth Baron Rayleigh) soon after the war. In his first paper with A Fowler (Fowler and Strutt 1917) accurate measurements of the wavelengths of the weaker absorption bands revealed the errors in the data of Ladenburg and Lehmann and led to a clear assignment of the Huggins bands to ozone.† The following year Strutt (1918) reported measurements of the ultraviolet absorption of the lower atmosphere over a path of 6.5 km. His conclusion that the lower atmosphere contains very little ozone explained why the mountain and balloon measurements had found no decrease—most of the ozone lay well above the highest attempted ascents.

The delayed paper of Fabry and Buisson (1921) mentioned above brings to a close this period of ozone history. Their method of measuring the amount of ozone in the vertical column of air above the observer (now usually referred to as the total amount of atmospheric ozone) is simple in principle but beset by great practical difficulties. The method of photographic photometry developed by astronomers was used to measure the ratio of the intensities of sunlight at two wavelengths in the region of the ultraviolet cut-off as the sun rose or set. The logarithm of this ratio was then plotted against the secant of the Sun's zenith distance, a quantity proportional to the mass of ozone in the line of sight. Provided that the ozone amount did not change during the observations, a straight line should be obtained. From the

---

† The plate accompanying this paper is of great interest. Sir Alfred Egerton (1949) tells how Fowler, armed with 'some photographs of the spectrum of Sirius taken by R. A. Sampson at Edinburgh Observatory with a quartz primatic camera' secured the attention of the busy Lord Rayleigh in this matter. See also Howard (1964) which is entirely devoted to the Third and Fourth Barons Rayleigh.

slope of this line and the laboratory measurements of the absorption coefficients at the two wavelengths the amount of ozone could then be determined. The first major difficulty is the extremely feeble intensity of the ultraviolet sunlight compared with the visible part of the spectrum; Fabry and Buisson overcame this by the use of a double spectrograph. Secondly, the reduction of intensity due to the scattering of light out of the beam by air molecules (Rayleigh scattering) had to be taken into account. Thirdly there was the difficulty caused by other atmospheric constituents with absorption bands or scattering properties in the ultraviolet, the most obvious being mists, smoke and dust. Fourthly, the method would obviously not work if changes in the ozone amount took place during the period of observation (as it often does). Finally, the accuracy demanded of the photographic photometry was much higher than was normally acceptable to astronomers.

Notwithstanding all these obstacles, Fabry and Buisson found the total ozone amount, when reduced to standard temperature and pressure, to be about 3 mm. They were able to observe significant changes from day to day during 14 days between 21 May 1920 and 23 June 1920, taking all spectra at noon on the same plate to reduce photometric errors. They had brought the study of atmospheric ozone, a natural phenomenon now established beyond all reasonable doubt, into the region of practical observation, and had demonstrated its tantalising and unexplained day-to-day variations. They had laid the foundations on which Dobson could now build. They had brought us closer to attempting Daubeny's 'last question ... namely, the uses which ozone subserves in the economy of nature'. We are still candidates in that examination.

Acknowledgments

In assembling this chapter I am most grateful for the help and encouragement I have received from Dr Tony Cox, Lady Dalrymple-Champneys, Dr H C Harley and Dr Michael Shortland.

References

Acland H W 1861 *Memoir on the Cholera at Oxford* (London: Churchill and Parker)
Andrews T 1855 On the constitution and physical properties of ozone *Proc. R. Soc.* **7** 475–7
Andrews T and Tait P G 1857 Note on the density of ozone *Proc. R. Soc.* **8** 498–500

Berigny A 1857 Recherches et observations pratiques sur le papier ozonomètrique *C. R. Acad. Sci., Paris* **44** 1104–8

Brodie B C 1872 An experimental enquiry on the action of electricity on gases—I. On the action of electricity on oxygen *Phil. Trans. R. Soc.* **162** 435–84

Chambers G F 1909 *Story of the Weather* (London: Newnes p. 108)

Chappuis J 1880 Sur le spectre d'absorption de l'ozone *C.R. Acad. Sci., Paris* **91** 985–6

Cornu A 1879 Observations de la limite ultra-violette du spectre solaire à diverses altitudes *C. R. Acad. Sci., Paris* **89** 808–17

—— 1881 Sur l'absorption atmosphèrique des radiations ultra-violets *J. Physique* **10** 5–17

Daubeny C 1867 On ozone *J. Chem. Soc.* **20** (new ser. 5) 1–28

Dobson G M B 1966 *Forty Years' Research on Atmospheric Ozone at Oxford—a History* (Atmospheric Physics Memorandum no 66.1 Clarendon Laboratory, Oxford). (Reprinted in *Appl. Opt.* **7** 387–405, 1968)

Egerton A 1949 Lord Rayleigh 1875–1947 *Obit. Not. Fell. R. Soc.* **6** 503–38

Fabry C 1950 *L'ozone Atmosphèrique* (Paris: CNRS) ch. 1

Fabry C and Buisson H 1913 L'absorption de l'ultra-violet par l'ozone et la limite du spectre solaire *J. Physique* (5me ser.) **3** 196–206

—— 1921 Etude de l'extremité ultra-violette du spectre solaire *J. Physique* (6me ser.) **2** 197–226

Faraday M 1851 On Schönbein's ozone *Proc. R. Inst.* **1** 94–97

—— 1859 On Schönbein's ozone and antozone *Proc. R. Inst.* **3** 70–1

Fowler A and Strutt R J 1917 Absorption bands of atmospheric ozone in the spectra of sun and stars *Proc. R. Soc.* A **93** 577–86

Gunther R T 1937 *Early Science in Oxford* vol. XI (Oxford) p. 74

Hartley W N 1881a On the absorption spectrum of ozone *J. Chem. Soc.* **39** 57–60

—— 1881b On the absorption of solar rays by atmospheric ozone *J. Chem. Soc.* **39** 111–28

Hautefeuille P and Chappuis J 1880a Sur la liquéfaction de l'ozone et sur sa couleur a l'état gazeux *C. R. Acad. Sci., Paris* **91** 522–5

—— 1880b Sur la liquéfaction de l'ozone en présence de l'acide carbonique et sur sa couleur a l'état liquide *C. R. Acad. Sci., Paris* **91** 815–17.

Houghton J T and Walshaw C D 1977 Gordon Miller Bourne Dobson 1889–1976 *Biog. Mem. Fell. R. Soc.* **23** 41–57

Houzeau A 1858a Preuve de la présence dans l'atmosphère d'un nouveau principe gazeux, l'oxygène naissant *C. R. Acad. Sci., Paris* **46** 89–91

—— 1858b Rapport sur plusieurs Mémoires de M. Houzeau, relatifs à l'ozygène odorant (l'ozone) *C. R. Acad. Sci., Paris* **46** 670–3

—— 1872a Sur la preparation de l'ozone à l'état concentré *C. R. Acad. Sci., Paris* **74** 256–7

—— 1872b Sur la proportion d'ozone contenue dans l'air de la campagne et sur son origine *C. R. Acad. Sci., Paris* **74** 712–15

Howard J N (ed.) 1964 John William Strutt 'Rayleigh' *Appl. Opt.* **3** 1091–115

Huggins W 1880 On the photographic spectra of stars *Phil. Trans. R. Soc.* **171** 669–90

Huggins W and Huggins M 1890 On a new group of lines in the photographic spectrum of Sirius *Proc. R. Soc.* **48** 216–17

Hulbert E O 1953 Explanation of the brightness and color of the sky, particularly the twilight sky *J. Opt. Soc. Am.* **43** 113–18 (and subsequent letter from E V Ashburn *ibid.* 805–6)

Kahlbaum G W A and Darbishire F V (ed.) 1899 *The Letters of Faraday and Schönbein 1836–1862* (Basel: B Schwabe and London: Williams and Nergate) p. 307

Karlinski F 1854 Erste Resultate ozonometrische Beobachtungen in Krakau *Ann. Phys., Lpz.* **91** 627–8.

Ladenburg E and Lehmann E 1906 Über Versuche mit hochprozentigen Ozon *Ann. Phys., Lpz.* **21** 305–18

Lauscher F 1984 *Ozonbeobachtungen in Wien von 1853 bis 1981. Zusammenhänge zwischen Ozon und Wetterlagen* (Wein: Zentralanstaldt für Meteorologie und Geodynamik) publikation nr 284

Lecomte J, Arnulf A and Vassy E 1973 Charles Fabry *Appl. Opt.* **12** 1117–29

Linville D E, Hooker W J and Olson B 1980 Ozone in Michigan's environment 1876–1880 *Mon. Weather Rev.* **108** 1883–91

Marignac J C de 1845 Sur la production et la nature de l'ozone *C. R. Acad. Sci., Paris* **20** 808–11

Mariolopoulos E G, Repapis C C, Zerefos C S, Varotsos C, Ziomas I and Bais A 1984 A neglected long-term series of ground-level ozone in *Atmospheric Ozone* (ed. C S Zerefos and A Ghazi) (Dordrecht: Reidel) pp 788–90

Miethe A and Lehmann E 1909 [Ultra-violet region of the solar spectrum] *Sitzungsber. König. Preuss. Akad. Wiss. Berlin* **8** 268–77

Miller W A 1863 On the photographic transparency of various bodies, and on the photographic effects of metallic and other spectra obtained by means of the electric spark *Phil. Trans. R. Soc.* **152** 861–87

Mohr F 1854 Aelteste Nachricht über Ozon und seine Benennung *Ann. Phys., Lpz.* **91** 625–7

Mulvaney J 1879 Ozone in nature: its relations, sources and influences etc. From fifteen years' observations ashore and afloat, under all conditions of climate *Q. J. R. Meteorol. Soc.* **6** 184–90

Odling W 1871 On the history of ozone *Proc. R. Inst.* **5** 546–55

Palmieri L 1872 Sur l'ozone atmosphèrique *C. R. Acad. Sci., Paris* **74** 1266–7

Partington J R 1964 *A History of Chemistry* vol. IV (London: Macmillan)

Phillips J 1869 *Proc. R. Soc.* **17** lxxiv–lxxx

de la Rive A 1845 Sur les movements vibratoires que determinent dans le corps, soit la transmission des courants electriques, soit leur influence extérieuse *C. R. Acad. Sci., Paris* **20**, 1287

Schmidt M 1988 *Pioneers of Ozone Research: a Historical Survey* (Göttingen: Max-Planck Institut für Aeronomie)

Schönbein C F 1840 Recherches sur la nature de l'odeur qui se manifeste dans certaines actions chimiques *C. R. Acad. Sci., Paris* **10** 706–10

—— 1854 Über verschiedene Zustände des Sauerstoffes *Ann. Chem. Pharm.* **89** 257–300

—— 1891 *Allgemeine Deutsche Biographie* (Leipzig: Bayrische Akademie der Wissenschaften) pp 256–9

—— 1970 *Dictionary of Scientific Biography* 1970 ed. C C Gillispie vol. 12, (New York: Scribner) pp 196–9

Simcock A V 1984 *The Ashmolean Museum and Oxford Science* (Oxford: Museum of the History of Science)

Strutt R J 1918 Ultra-violet transparency of the lower atmosphere, and its relative poverty in ozone *Proc. R. Soc.* A **94** 260–8

Twemlow F E 1875 Ozone *Q. J. R. Meteorol. Soc.* **2** 383–90

Tyndall J 1861 On the action of gases and vapours on radiant heat *Proc. R. Inst.* **3** 295–8

Van Marum M 1756 *Beschreibung einer Elektrisiermaschine* (Leipzig) (summary in Partington (1964, vol. 3, p. 342))

Vigroux M E 1949 Remarques sur l'ozone *Bull. Soc. Chim. Fr.* 402–3

Wigand A 1913 Das ultra-violette Ende des Sonnenspektrums in verschiedenen Höhen bis 9000 m *Ber. Deutsch. Phys. Gesell.* **15** 1090–9

Wolf R 1854 Ueber Beobachtungen mit dem Schönbein'schen Ozonometer *Ann. Phys., Lpz.* **91** 314–15

—— 1855 Resultate der Ozonometer-Beobachtungen in Bern vom Dezember 1853 bis Ende November 1854 *Ann. Phys., Lpz.* **94** 335–6

18

Ionospheric Research—Some Personal Recollections†

W J Granville Beynon

Introduction

My involvement with studies of the earth's upper atmosphere started towards the end of 1937 when I joined the staff of the Radio Division of the National Physical Laboratory. At that time part of the division was located at Ditton Park near Slough and I was assigned to a small group working on various radio research problems. The total station complement (scientific, technical and administrative) at Ditton Park was about 20, with about six or so on the scientific staff. Small though it was, this Radio Research Station was well known in the radio world—particularly in the ionospheric world—mainly because of the work of Professor E V (later Sir Edward) Appleton (1892–1965) (Ratcliffe 1966). Later the 'RRS' was to become famous (although it was at the time a well guarded secret) as the place at which the early pioneer experiments were carried out by R A Watson-Watt (later Sir Robert Watson-Watt (1892–1973)) and his colleagues on the 'radio location of aircraft' or, as it was later termed, 'radar'.

The research programme of the RRS was under the overall direction of the Radio Research Board and in the late 1930s the principal lines of research were radio direction-finding (RDF), radio sounding of the

† Based on a lecture given at University College, Cardiff, 17 November 1986, to the IOP History of Physics Group.

ionosphere and radio atmospherics. Part of the research programme called for regular routine soundings of the ionosphere at vertical incidence, i.e. pulsed radio signals were radiated upwards to the ionosphere and received back again at the same site. RRS proudly laid claim to the fact that it was the first site in the world at which routine radio soundings of the ionosphere were made. Beginning in 1931 these daily ionospheric observations have continued at Slough for more than 57 years. The main programme of radio research activities at the Ditton Park site formally terminated in 1979, but a programme of ionospheric sounding, using automated remotely controlled equipment, still continues.

For studying the vertical structure of the ionosphere, radio soundings carried out at vertical incidence have obvious advantages, but the practical problems of radio communication over long distances, using high-frequency signals, involves oblique incidence reflections from the ionosphere. In the late 1930s the problem of how to apply information about the ionosphere derived from vertical sounding experiments to actual radio communication problems was one of considerable practical interest and this was the research area in which I worked soon after joining the station. In 1938 a memorandum which I wrote on calculating 'maximum usable frequencies' for ionospheric layers with parabolic distributions of electron density came to the notice of Professor E V Appleton (then at the Cavendish Laboratory Cambridge and a member of the Radio Research Board). He contacted me about this work and thus began a collaboration and a friendship between us which was to continue until his death 27 years later.

So much for my personal initiation, just 50 years ago, into the ionospheric research field.

Early milestones in ionospheric history

The origins of ionospheric studies really date to the early years of this century and some of the milestones along the road are briefly summarised below.

In 1901 G Marconi (1874–1937) (Marconi 1902) succeeded in sending a wireless signal from Cornwall to Newfoundland—across a few thousand miles of ocean and around a fair portion of the curved earth. In theory of course it was 'impossible' to do this but, as has often been said, the young Marconi did not know much about the theory of long distance radio propagation and not knowing that it was 'impossible' to send a radio signal from England to Canada he just went and did it! At first some of the theoreticians suggested that diffraction around the curved surface of the earth could explain Marconi's success but others concluded otherwise. (Incidentally, the signal which Marconi sent across the Atlantic was the single letter 'S' which in Morse code is simply three dots. It would appear that Marconi and his assistants chose this simplest of code messages, because of the output power limitations of his transmitting equipment.)

In 1902, A E Kennelly (1861–1939) in America and Oliver Heaviside (1850–1925) (Ratcliffe 1970) in Britain, almost simultaneously suggested that the existence of an electrically conducting layer in the upper atmosphere could explain transatlantic radio propagation and Heaviside went further and suggested that this layer might consist of ions and electrons produced most probably by the ionising action of solar radiation. As evidence for such a reflecting layer grew, it became known as the 'Kennelly–Heaviside layer' and this layer was later identified with what is now called the E layer of the ionosphere at a height of about 100 km.

Although over the years, the ionosphere has come to be associated with the subject of radio science, it is to be remembered that the first suggestion for electrically conducting layers high in the earth's atmosphere, came not from radio experiments but from studies of the earth's magnetic field. In fact in 1878, some 24 years before the Kennelly–Heaviside hypothesis, and some 10 years before radio waves had been discovered, Balfour Stewart (1828–1887) (Stewart 1883), then Director of the Kew magnetic observatory, in an article in the *Encyclopedia Britannica*, suggested that the well known minute but regular semidiurnal changes in the earth's magnetic field could be caused by electric currents flowing at some level high in the earth's atmosphere.

After the experiment by Marconi in 1901 and the Kennelly–Heaviside hypothesis of 1902 probably the next significant date in ionospheric history was the publication in 1912 by W H Eccles (1875–1966) (Eccles 1912) of the theory of the refraction of radio waves by an ionised medium. Then in December 1924, E Appleton and M Barnett (Appleton and Barnett 1925) provided the first direct experimental evidence for the existence of a layer in the high atmosphere which reflected and refracted radio waves. A year or so later two Americans, G Breit and M Tuve (Breit and Tuve 1926), using short radio pulses, provided further very conclusive evidence for radio-echo signals from the upper atmosphere. This ability to generate and record short pulsed radio signals was a considerable technical step forward and one which was rapidly exploited for investigating the newly discovered ionosphere. Later of course, this pulse sounding technique was to form the whole basis of 'radar'.

In 1932 Appleton (Appleton 1932) developed and published the theory of the refraction of radio waves in the ionosphere in the presence of a magnetic field—the so-called magneto-ionic theory, a theory which provided the basis for the interpretation of the observed properties of radio signals reflected and refracted by the ionosphere.

These then are some of the principal milestones in early ionospheric history. Then in the 1930s there came a further important development when Sydney Chapman (Chapman 1931) published his theory of the formation of an ionospheric layer by solar ultraviolet radiation.

In the 1920s and 1930s two of the leading figures in the UK radio field were Edward Appleton and Robert Watson-Watt. In 1923 they collaborated in some joint work on radio atmospherics, (Appleton and Watson-Watt 1923), both were associated with early ionospheric work at Slough and both played important roles in the planning and execution of the Second International Polar Year of 1932–3 (see the later section on ionospheric research and international years). Today Watson-Watt is remembered in Britain as the man mainly responsible for the development of radar but in ionospheric history he deserves

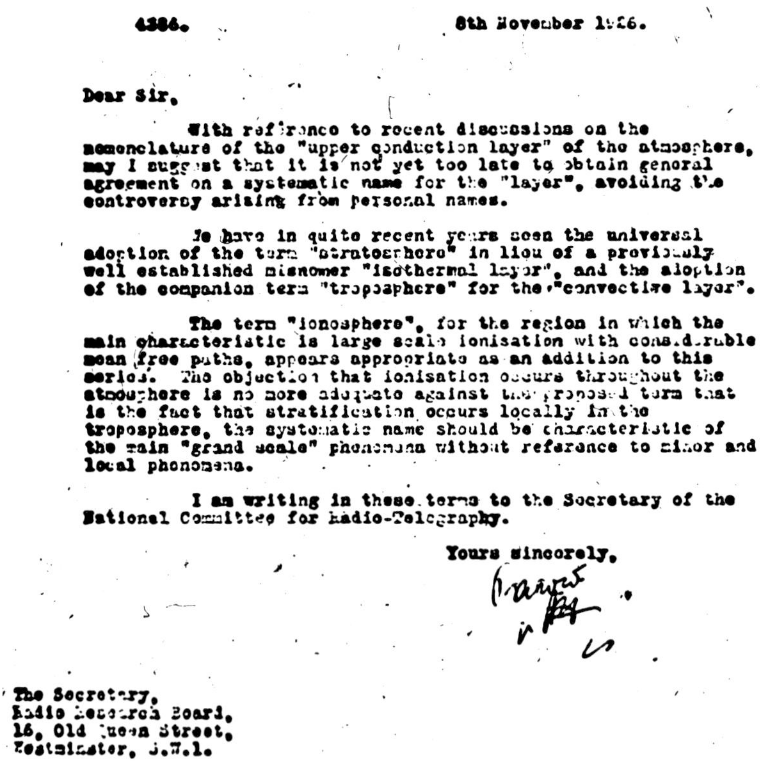

4386. 8th November 1926.

Dear Sir,

With reference to recent discussions on the nomenclature of the "upper conduction layer" of the atmosphere, may I suggest that it is not yet too late to obtain general agreement on a systematic name for the "layer", avoiding the controversy arising from personal names.

We have in quite recent years seen the universal adoption of the term "stratosphere" in lieu of a previously well established misnomer "isothermal layer", and the adoption of the companion term "troposphere" for the "convective layer".

The term "ionosphere", for the region in which the main characteristic is large scale ionisation with considerable mean free paths, appears appropriate as an addition to this series. The objection that ionisation occurs throughout the atmosphere is no more adequate against the proposed term than is the fact that stratification occurs locally in the troposphere, the systematic name should be characteristic of the main "grand scale" phenomena without reference to minor and local phenomena.

I am writing in these terms to the Secretary of the National Committee for Radio-Telegraphy.

Yours sincerely,

The Secretary,
Radio Research Board,
16, Old Queen Street,
Westminster, S.W.1.

Figure 18.1 Letter from Watson-Watt (November 1926) in which he proposes the new word 'ionosphere'.

mention for the fact that he coined the word 'ionosphere'. Figure 18.1 is a copy of a letter which Watson-Watt wrote in November 1926 to the Secretary of the Radio Research Board in which he suggested the new word 'ionosphere' for what he called the 'upper conduction layer of the atmosphere'. Around that time the term 'Appleton layer' was beginning to be used for that 'upper conducting layer' with the term 'Heaviside layer' reserved for the lower (E) layer. For one reason or another Watson-Watt's proposed new term 'ionosphere' did not catch on very quickly and in July 1929 (nearly three years after his original proposal) he published an article in the Quarterly Journal of the Royal Meteorogolocial Society (Watson-Watt 1929) which contained the paragraph shown in figure 18.2. As Watson-Watt rightly suggested in this article, his term 'ionosphere' fitted logically into a systematic group of terms which aptly described different parts of the atmosphere and it was to be expected that sooner or later it would be universally adopted by the scientific community and this indeed proved to be the case.

> combined in electrically neutral molecules. So we conclude that the indirect ray reaches our receiver because it is returned from the upper air by a mirage effect, and that the study of the indirect ray, besides explaining the vagaries of wireless reception, may offer new and invaluable data on the ionisation in the upper atmosphere. The region of the atmosphere in which this mirage effect originates has been called the Heaviside layer, an admirable title which is defective in only two particulars, first that the ascription to Heaviside is inexact and second that the region is not a layer. It might be permissible to call it the Balfour - Stewart - Fitzgerald - Heaviside - Kennelly - Zenneck - Schuster - Eccles - Larmor - Appleton space, but something less indirect is desirable. I have
> ⟶ suggested the name ionosphere to make a systematic group troposphere, stratosphere, ionosphere, but meanwhile the term " upper conducting layers " seems to hold the field.
>
> **Watson-Watt. Q.J.R.Met.Soc. July 1929.**

Figure 18.2 Extract from letter of Watson-Watt to the *Quarterly Journal of the Royal Meteorological Society* (July 1929).

Radio sounding of the ionosphere 1930–48

Sixty years ago when Appleton carried out his pioneer experiments which provided the first direct experimental evidence for a reflecting layer in the upper atmosphere his recording instrument was a string galvanometer of the type used in physics laboratory experiments of that time. What Appleton effectively did in that historic pioneer experiment of 1924–5 (Appleton and Barnett 1925) was a large-scale radio version of the Lloyd's mirror experiment in optics. A simple radio receiver located at a suitable distance from a transmitter received two signals—one arriving directly along the ground and the other after reflection/refraction from a layer in the upper atmosphere. When he varied the frequency of the transmitter by a known amount he observed interference 'beats' between the two signals, a count of which provided a measure of the path difference between them and hence the height of the reflecting layer. Figure 18.3 shows some of the actual records obtained in that 1924 experiment—the first positive records of radio signals reflected by the ionosphere.

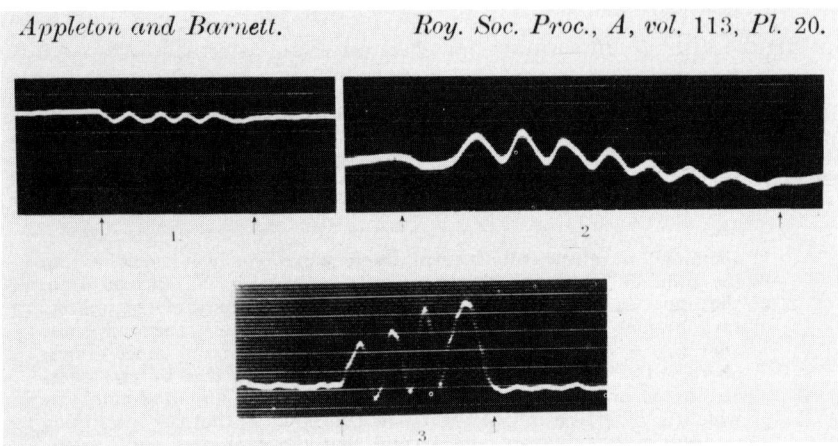

Figure 18.3 Records obtained by Appleton and Barnett in their frequency-change method for measuring reflection height of signals reflected by the ionosphere (Appleton and Barnett 1925).

Just a few years later the radio technique for studying the ionosphere made a great leap forward when it became possible to generate, radiate and record short pulse signals of duration 100 μs and less. Around the same time the cathode ray (CR) tube was being developed on which these pulse signals could be clearly displayed (figure 18.4). In the mid-1930s the recording of ionospheric echo signals was developed further, and figure 18.5 is a record obtained at Slough on 27 December 1933. In this record the frequency of the transmitter has been continuously increased slowly from 1.5 Mc s^{-1} (now MHz) up to beyond 7 Mc s^{-1}. The broad white line running the length of the record is the ground signal and the traces above it show the echoes from different levels of the ionosphere. From records of this sort (later to be called 'ionograms') one can readily deduce the levels of reflection of the signals and also deduce the variation of electron density with

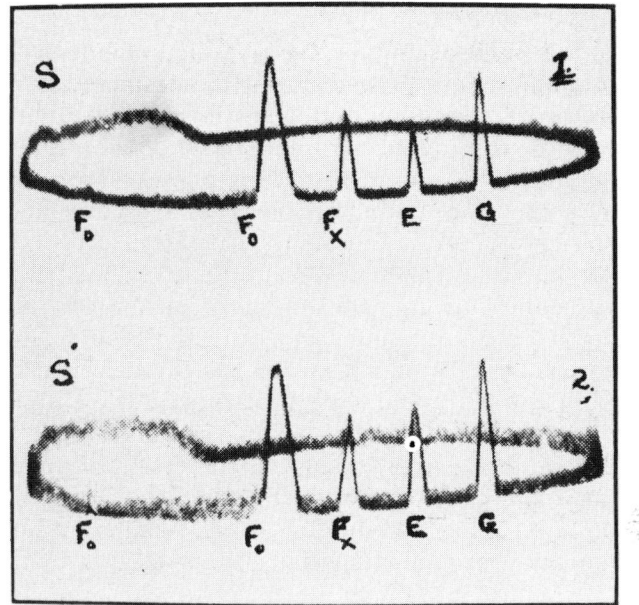

Figure 18.4 Display of pulse-pattern on CR tube with elliptical time-base. (Courtesy of the Rutherford–Appleton Laboratory).

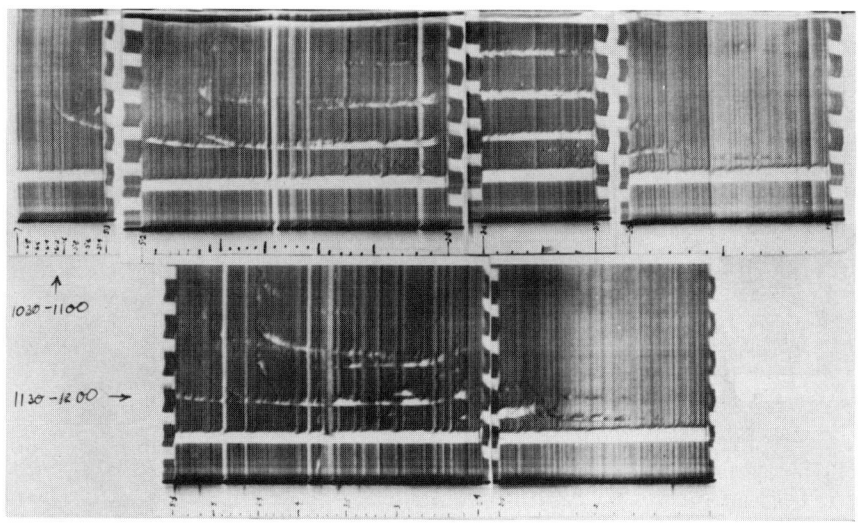

Figure 18.5 Swept-frequency (P′, f) records ('ionograms') for Slough 27 December 1933. (Courtesy of the Rutherford–Appleton Laboratory.)

height in the ionospheric layers, i.e. deduce the vertical structure of the ionosphere.

At first records such as these were obtained 'manually', i.e. the transmitter frequency was changed slowly, the tuning condenser being driven forward by a small motor and the receiver was kept in tune with the transmitter by continuous manual adjustment of the receiving tuning condensers. These were the kinds of record we obtained at Slough when I joined the staff nearly 50 years ago and since we often had to make observations right round the clock, especially on so-called 'International Days', we spent many all-night sessions with our eyes intently fixed on the changing pulse pattern on the CR tube screen. Figure 18.6 gives an overall view of this the first Slough 'ionosonde'. The receiver was contained in the large mumetal screened boxes seen in the centre of the picture and I recall that it had three tuned circuits with three separate tuning condensers coupled together with string and Meccano pulley wheels. Later an automated version of this ionosonde was developed which covered a limited frequency range, the receiver and transmitter were kept in tune by a series of pulleys and strings linked mechanically through a specially shaped cam. Unfortunately

Figure 18.6 Overall view of first Slough 'ionosonde', developed by L H Bainbridge-Bell in the early thirties.

temperature changes and other causes resulted in the receiver continually going off tune and when this happened we had to adjust the shape of the cam by sticking pieces of paper on to some part of it (this was in the days before Sellotape!) and taking bits off in other places. In my early days at Slough this was one of the first jobs I had to do first thing every morning, readjust the shape of the cam by the 'sticky paper technique'.

These then are some of my recollections of life with one of the first ionosondes at Slough.

The post-war period—some unsolved ionospheric problems in 1948

In 1948 Appleton and I had occasion to examine the current state of ionospheric knowledge and we published a paper outlining what we considered to be 12 major unsolved ionospheric problems. It is interesting now—some 40 years later—to recall that list (Ratcliffe 1963):

(i) What is the precise solar source of the ultraviolet light which causes the E and F1 layers?

(ii) How far do we need to invoke the additional action of corpuscles in the formation of the F2 layer?

(iii) What is the detailed explanation of the stratification of the ionosphere into several layers?

(iv) What are the nature of the atomic processes by which electrons disappear in the various layers?

(v) Why does the scale height of the atmosphere increase with height above 100 km?

(vi) Why is the F2 layer exceptionally subject to geomagnetic control?

(vii) What is the cause of the changes in the F2 layer during ionospheric and geomagnetic storms?

(viii) What is the influence of atmospheric oscillations in the ionosphere?

(ix) How far do ionospheric currents give rise to Hall effect phenomena due to the influence of the earth's permanent magnetic field?

(x) What is the evidence suggesting the existence of ionospheric winds?

(xi) What are the ionospheric levels of the currents responsible for the quiet-day magnetic variations?

(xii) What are the ionising agencies responsible for the production of sporadic or abnormal E-layer ionisation in different lattitudes?

These were the major problems we considered to be unsolved in 1948.

Fifteen years after Appleton and I prepared that list, J A Ratcliffe, in the URSI Golden Jubilee Volume, again referred to it and commented that, at the time of Appleton's death in 1965, five of these twelve problems had been solved and four others had been partly solved (Ratcliffe 1963).

The period which has elapsed since 1948 has been marked by two major events which between them have profoundly changed the whole scene. I refer to the International Geophysical Year of 1957–8 (and to the complementary worldwide enterprise at the sunspot minimum period of 1964–5—the 'IQSY'), and secondly to the development of rocket and satellite techniques in the 'space age'. It is not surprising then that a list of the major unsolved ionospheric problems at the end of the 1980s would look very different from that in the pre-IGY and pre-space era. Until the early 1950s, when rocket soundings first started, our knowledge of the ionosphere was necessarily almost entirely limited to studies of the electron component—little or nothing could be learned about the much more numerous ion or neutral constituents. It was recognised, of course, that in the ionosphere (at least up to the peak of the F layer) the electrons only formed a small part of the whole atmosphere, but since our radio waves were only affected by electrons the radio techniques then available to us were not of much assistance. However, all this started to change with the development of rockets and satellites, and in the past few decades studies of the ions and the neutral air at all levels have steadily grown until they now form a large part, if not indeed the major part, of the whole ionospheric research effort. This new and extended interest in the ions (and in the complex chemistry and dynamics of this ion and neutral atmosphere) is, I suppose, one of the main features which distinguishes modern ionospheric studies from those of 20 and 40 years ago.

Ionospheric research and the 'International Years'

A scientific understanding of such geophysical phenomena as the atmosphere, the geomagnetic field and the aurora calls for simultaneous observations at many locations and over extended periods of time. Progress in these scientific disciplines thus benefits greatly from experiments coordinated internationally and from time to time workers in these fields have planned and carried out many large-scale cooperative projects. One of the earliest (but not by any means the first) was the so-called 'International Polar Year of 1882–3'. In that year geomagneticians and meteorologists from some 11 countries joined in an international cooperative study at polar latitudes. Fifty years later, in 1932–3 the anniversary was marked by a renewed venture of a similar kind. The 1882–3 enterprise was of course some five years before Hertz had generated and detected radio waves and man-made radio waves thus played no part in the scientific programmes of that year. Fifty years later radio waves were part of the common scene and routine soundings of the ionosphere by the so-called 'critical frequency' method were getting under way. In that Second Polar Year there were about 10 ionospheric stations in the whole world but only one or two made more or less regular daily soundings. The agreed programme of soundings for the whole Polar Year in 1932–3 consisted only in noon measurements of electron density on a total of 31 days.

A measure of the expansion in the scale of radio soundings of the ionosphere in the two or three decades which followed 1932 is emphasised when this modest Second Polar year programme is compared with that recommended 25 years later for the International Geophysical Year (IGY) 1957–8 when each of the 170 sounding stations distributed across the globe from the poles to the equator were recommended to make quarter-hourly observations on a normal day and at five minute intervals on World Days and Special World Intervals. The expanded ionospheric sounding programme in the 25 year period 1932–57 was of course an enormous step forward but within the matter of another five years even the output of the entire IGY ionosphere sounding programme was to be dwarfed by that of the first satellite designed for studying the top-side of the ionosphere. Alouette-1 launched in September 1962 produced more ionograms in one year than all the 170 or so ground-based stations put together produce in 30 years. Such is the progress of change that top-side sounding of the ionosphere has now almost become something of the past—something that already belongs to history. In just a few decades, well within our own lifetimes, ionospheric research has changed out of all recognition.

I sometimes look back and recall a remark Appleton once made to me in 1939. I had been trying to get some quantitive measurements of scale heights in the E and F regions from our rather crude ionospheric records of the time—trying to see whether we could deduce temperatures in the ionosphere and in particular to see whether we might detect any seasonal changes in these temperatures. When I showed my results to Appleton his comment was that my values seemed rather high to him but then he added 'But never mind, whatever we say we can be quite safe—no one will ever get up there to say whether we are right or not'!

The two decades between the discovery of the ionosphere in 1924/5 to the end of World War II, were marked by a systematic unravelling of the structure of the ionosphere, its short- and longer-term variations and of its role in short-wave radio communication problems. The period of the war itself saw the development of methods for forecasting the behaviour of the ionosphere to meet the various requirements of the many different groups concerned with radio communication.

The International Geophysical Year 1957-8

Credit for suggesting that the Polar Years of 1882 and 1932 should be followed by another major international cooperative project in 1957-8, that is after an interval of 25 years, goes to an American Lloyd V Berkner. He made the suggestion on 5 April 1950 at a small private social gathering in the home of Professor J van Allen. One of those present at that small party was Sydney Chapman and four months later, in September 1950, he and Berkner put the idea to a small ICSU Commission—the 'Mixed Commission on the Ionosphere'. My personal involvement with what was then called the 'Third International Polar Year' (later renamed the 'International Geophysical Year 1957-8') began at that meeting in Brussels on 4 September 1950.

Originally an officer in the US Navy, Berkner had had first-hand experience of polar exploration—he had been with Admiral Byrd in Antarctica. A man with breadth of vision and a flair for thinking and planning on a large scale, he applied these qualities in many directions. The Mixed Commission on the Ionosphere, to which he and Chapman brought the proposal for a Third International Polar Year, was a small body of about 12 scientists nominated by the four Scientific Unions with a direct interest in the ionosphere. The Unions concerned were URSI (Union Radio Scientifique Internationale) (radio science), IUGG (International Union of Geodesy and Geophysics) (geodesy and geophysics), IAU (International Astronomical Union) (astronomy) and IUPAP (International Union of Pure and Applied Physics). Appleton was Chairman of the Commission and between 1948 and 1957 we met at intervals of two or three years. They were always three-day meetings, the discussions were always informal and leisurely with the last half day always devoted to the formulation of resolutions—which were later transmitted to the four International Scientific Unions concerned, to ICSU (The International Council of Scientific Unions) and, when relevant, to national organisations

Ionospheric Research

Recommendations

Great advances in our understanding of the physics of the earth's atmosphere are to be expected by combining special observations in the north and south polar regions with observations of a similar nature carried out at lower latitudes. It is therefore recommended that :

(i) The year 1957-58 be designated an International Polar Year.

(ii) A Commission be set up by I. C. S. U. similar to that established for previous polar years to encourage, through the various Unions and their National Committees, the establishment of a proper network of observing stations.

(iii) In view of the complexities of the apparatus needed to exploit the potentialities of modern technique, the above Commission be established in 1951, so as to give at least five full years of preparation and trial.

(iv) A permanent secretariat should be formed to operate during the most active period of the Commission's work, from about two years prior to the polar year until about three years after the polar year.

Figure 18.7 Proposal for a 'Third International Polar Year' adopted by the Mixed Commission on the Ionosphere 6 September 1950.

Figure 18.8 Group of members of CSAGI—the International Special Committee for the IGY. From left to right: M Nicolet, J Coulomb, L V Berkner, V Laursen, V V Beloussov, S Chapman, A H Shapley, J Van Mieghem, W J G Beynon, G Laclavere.

concerned with the funding of ionospheric studies. In the course of five meetings, spread over nine years, the Mixed Commission on the Ionosphere formulated some 76 resolutions and there can be little doubt that these resolutions greatly influenced the course of ionospheric research during this period. Figure 18.7 shows the text of a resolution which the Commission adopted on 6 September 1950. This is the original proposal that 'the Third International Polar Year be nominated for 1957–8'. The document which was attached to the resolution briefly outlined the types of observation which it was thought should be carried out in various fields such as radio, geomagnetism, aurorae, cosmic rays, solar astronomy, rocket experiments, etc. Thus, what ultimately became the IGY of 1957–8 was set on its way within the ICSU framework. In 1952 ICSU set up a Special Committee for the IGY—which became known as CSAGI (the acronym of the title in French—Comité Speciale de l'Année Geophysique Internationale) under the Chairmanship of Sydney Chapman. Figure 18.8 shows a group of CSAGI members.

The first artificial earth satellite

In August 1954 URSI held its Eleventh General Assembly at The Hague and during a meeting of Commission III (on Ionospheric Radio) Drs Fred Singer and Lloyd Berkner handed me, as Secretary of the Commission, a scrap of paper on which was a draft of a resolution which the US delegation wished to put before the Commission. After some discussion the Commission (and later the whole General Assembly) adopted the resolution a facsimile of which is shown in figure 18.9. This was I believe the first occasion on which an international scientific organisation made a proposal to put an artificial earth satellite into orbit. Just over three years later, and three months after the start of the IGY, on 4 October 1957, the USSR successfully launched Sputnik 1—an historic event which was soon to change the whole geophysical scene and particularly the solar–terrestrial part of it. In the years immediately following the IGY, the rapid development of space research techniques opened up vast new areas of geophysical, solar and interplanetary study. New topics, new terms and new concepts flooded on to the stage—the solar wind, trapped particles, the topside ionosphere, the magnetosphere, the plasmapause, etc, etc—many completely new exciting discoveries all of direct concern to the solar–terrestrial physics community. If the IGY had greatly expanded our vistas and had brought vast numbers of new enthusiastic workers into the field, as it certainly did, then the space era, which quickly followed it, continued and amplified those processes to a degree which few could have imagined.

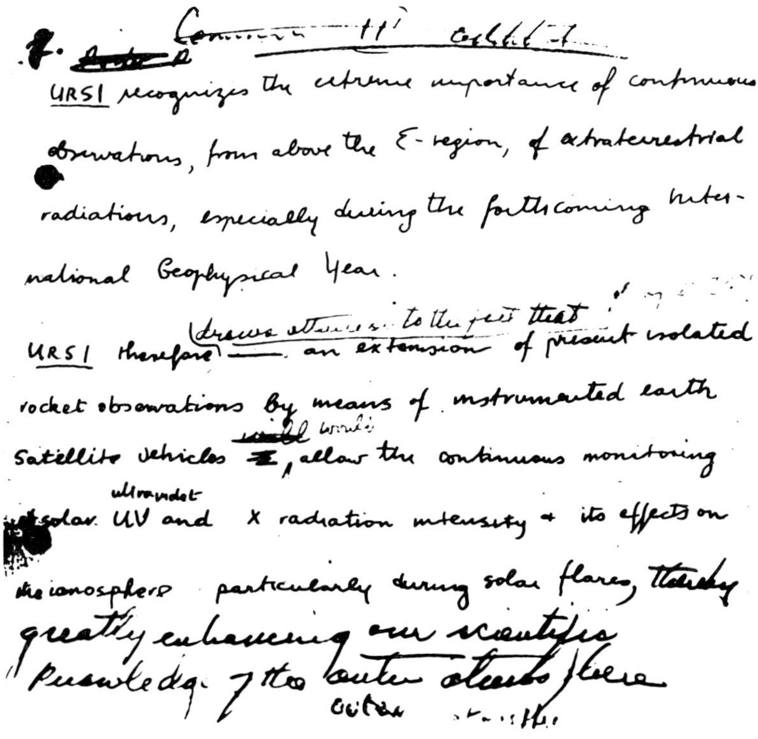
URSI recognizes the extreme importance of continuous observations, from above the E-region, of extraterrestrial radiations, especially during the forthcoming International Geophysical Year.

URSI therefore draws attention to the fact that an extension of present isolated rocket observations by means of instrumented earth satellite vehicles would allow the continuous monitoring of solar ultraviolet and X radiation intensity and its effects on the ionosphere particularly during solar flares, thereby greatly enhancing our scientific knowledge of the outer atmosphere.

FIRST DRAFT OF ORIGINAL RESOLUTION PROPOSING THE USE OF 'INSTRUMENTED EARTH SATELLITE VEHICLES' FOR THE STUDY OF SOLAR RADIATION.

Submitted at XI Assembly of U.R.S.I. at The Hague
30 August 1954

Figure 18.9 Resolution adopted by the Eleventh General Assembly of URSI on 30 August 1954, proposing the launch of an 'artificial earth-satellite' to study solar UV radiation.

Incoherent scatter soundings of the ionosphere

In 1929 during the earliest days of radio sounding of the ionosphere the suggestion was made by C Fabry (1867–1945) (Fabry 1928) that it might be possible to detect at the ground 'Thomson scattering' from electrons in the ionosphere and that the received signal might show a Doppler broadening corresponding to the thermal velocity of the electrons. When, soon afterwards, strong reflections were obtained from the ionosphere by ordinary refraction processes using quite small transmitter powers this suggestion of Fabry—which would clearly call for very powerful transmitters and/or very sensitive receivers—was not followed up. However, in the 1950s, following the wartime development of very powerful radar detection systems, interest in the possibility of detecting scattering at the ground from electrons in the ionosphere was renewed and in 1958 W E Gordon (Gordon 1958) predicted that with the large antennae then available it should be possible to detect this Thomson scattering from electrons in the ionosphere and permit quantitive measurements of both the numbers of electrons and the electron temperature over a range of ionospheric heights. Soon afterwards another American scientist K L Bowles (Bowles 1958) using a powerful radar at 1.8 m wavelength, observed scattered signals from the ionosphere of intensity roughly equal to that predicted by Gordon but the bandwidth of the received spectrum was only a small fraction of that calculated by Gordon.

A fuller analysis of the problem then showed that the scattering process was rather more complicated than had been predicted by Gordon—whilst the power scattered back corresponded to the cross section of electrons the frequency spectrum of the scattered signals corresponded to the thermal velocity of the ions in the ionosphere. (Because of the random nature of the motions of the electrons the scattered signals have random phases and the scatter is termed 'incoherent scatter'.)

This unexpected complexity of the scattering process has, in the event, proved to be a large bonus, since we find that this incoherent scatter technique yields accurate information, not merely on the two parameters originally predicted by Gordon, namely electron density and electron temperature, but also on a number of other parameters such as ion temperature, ion mass, plasma drift velocity, ion–neutral collision frequency and electric current. Several of these parameters are interdependent and when some are measured, others can be deduced—just as in completing a jigsaw puzzle the more pieces that there are in place the easier it becomes to solve the whole problem.

Figures 18.10 and 18.11 show values of electron temperature and electron density for all heights between 150 and 650 km obtained with an incoherent scatter system.

Several of the outstanding current problems of ionospheric physics relate to phenomena at high latitudes—to the auroral zones where charged particle streams from the sun enter the atmosphere and are a major source of the overall energy input. Such is the scientific interest in high-latitude atmospheric phenomena that in December 1975 a group of six European nations, Finland, France, Germany, Norway, Sweden and the UK, agreed to establish a very powerful incoherent

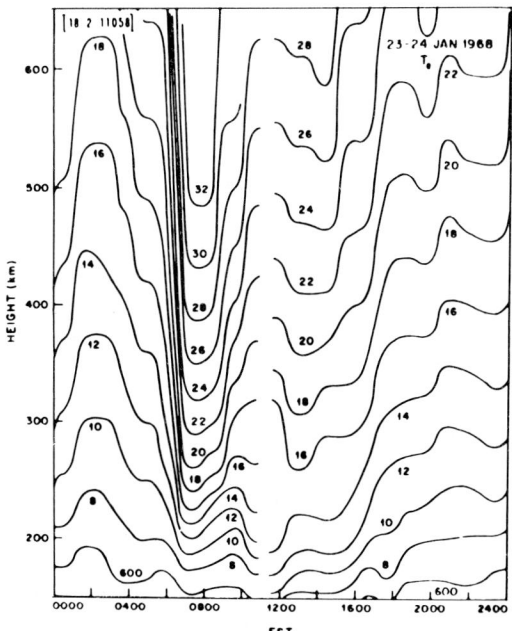

Figure 18.10 Incoherent scatter measurements of electron temperature (T_e) between 150 and 650 km.

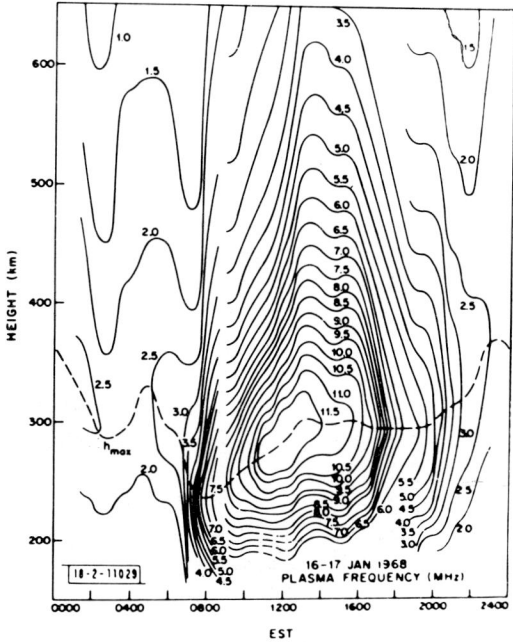

Figure 18.11 Contours of 'plasma frequency' ($\propto$(electron density)$^{1/2}$) over height range 150–650 km.

Figure 18.12 EISCAT VHF transmitter antenna at Tromso (1978). (Courtesy of the Rutherford–Appleton Laboratory).

Figure 18.13 Appleton and Naismith digging a hole at Tromso in 1932 for wooden poles to support a dipole antenna. (Courtesy of the Rutherford–Appleton Laboratory.)

scatter radar system in northern Scandinavia to study these problems. The project known by the acronym EISCAT has been built up over a number of years and consists in two very high-power radars operating on 930 Mhz (UHF) and 270 Mhz (VHF). These are located in Norway (Tromso), Sweden (Kiruna) and Finland (Sodankyla). Figure 18.12 shows a photograph of the VHF transmitting antenna at Tromso—a parabolic cylinder design—taken during the construction phase in 1978. A measure of the change in years in the design of a radio antenna for sounding the ionosphere is well illustrated by comparing figure 18.12 with another picture also taken at Tromso but in 1932. Figure 18.13 shows Appleton and his assistant Naismith valiantly struggling to dig a hole in the rocky ground at Tromso for their 30 ft wooden poles needed to support a 12 m dipole—poles which stayed up until 1949 when a great winter gale blew them down!

Post-IGY international cooperation in solar–terrestrial physics

One can point to several remarkable direct consequences of the IGY. The imminent rapid expansion around that time in space, in oceanography and in Antarctica was foreseen even before the IGY ended and steps were quickly taken in 1958 to set up permanent international organisations to continue the cooperation first established in the IGY. Another significant feature of the IGY planning was the decision to establish international centres in which all the scientific data collected by the 67 participating nations would be stored, catalogued, processed and generally made available to scientific workers the world over. Originally known as the 'IGY World Data Centres' (the 'IGY' was later dropped from the title) some 25 of these vast repositories of geophysical data are, over 30 years later, still in full active operation. Funded and managed entirely by the individual countries in which they are located, the World Data Centres make a vast range of geophysical data available to research workers, and authors of innumerable papers published in these past 30 years have only been too willing to acknowledge the help they have received from them.

The World Data Centres, like a number of other IGY initiatives, have lived on long after the project ended, but there was another important part of the IGY exercise which had a planned fixed life and one which merits a brief mention here—the *Annals of the International Geophysical Year*. Preparation and publication started in 1956 and continued over a period of some 14 years, and the 48 volumes of the series constitute a comprehensive record of the project.

The IGY had been planned for a period of solar maximum activity and, as it turned out, the highest sunspot numbers ever were recorded during the IGY. However, even in the planning stages of the project, it became clear that the full fruits of the enterprise would only be gained if data, comparable in quantity and quality, became available for a period of low solar activity. Furthermore such an enterprise would be able to make full use of the powerful new space-research techniques developed after the IGY. In 1962 ICSU set up a Special Committee, under my chairmanship, to organise what, in due time, became the 'International Years of the Quiet Sun 1964–65' (IQSY). The plans which we drew up for the 1964–5 sunspot minimum years received enthusiastic scientific and financial support from some 71 nations and in the disciplines concerned which included a large-scale programme of rocket and satellite work, the overall effort was as large, if not larger, than that during the IGY. Following the pattern of the IGY a summary record of the IQSY activity was prepared in a seven-volume series, the *Annals of the IQSY*. Published in 1967–9, these *Annals* describe measurement techniques, observation schedules, treatment of data, details of solar activity and geophysical events together with critical reviews summarising the advances achieved in each discipline.

The IGY and the IQSY, without doubt, left a permanent imprint on the whole field of solar–terrestrial physics and as a result of these two world-wide projects a remarkable new degree of international cooperation was achieved. Before the end of the IQSY repeated pleas were made from many quarters that the close international cooperation which the IGY initiated and which the IQSY enhanced should not be allowed to lapse, and ICSU was urged to establish some long-term mechanism for maintaining this cooperation in the solar–terrestrial physics area. In 1966 at the General Assembly of ICSU I presented our final report on the IQSY and urged the Council to consider setting up some permanent organisation in the solar–terrestrial physics area as they had already done for space, oceanography, etc. The Council responded favourably and agreed to establish an Inter-Union Commission for solar–terrestrial physics. Later, in 1972, this Commission became a Special Committee of ICSU (SCOSTEP, Special Committee on Solar–Terrestrial Physics) and in 1979 this body was given permanent status as a Scientific Committee of ICSU.

Acknowledgments

The author is indebted to various colleagues for permission to reproduce some of the figures and to the Rutherford–Appleton Laboratory, Chilton, for figures 18.4, 5, 6, 12 and 13.

References

Appleton E 1932 Wireless studies of the ionosphere *J. Inst. Electr. Eng.* **71** 642

Appleton E and Barnett M 1925 On some direct evidence for downward atmospheric reflection of electric rays *Proc. R. Soc.* A **109** 621

Appleton E and Watson-Watt R 1923 On the nature of atmospherics. I. *Proc. R. Soc.* A **103** 84

Bowles K L 1958 Observation of vertical-incidence scatter from the ionosphere at 41 Mc/sec *Phys. Rev. Lett.* **1** 454–5

Breit G and Tuve M 1926 *A Test of the Existence of the Conducting Layer* (Baltimore)

Chapman S 1931 Some phenomena of the upper atmosphere *Proc. R. Soc.* A **132** 353–74

Eccles W H 1912 On the diurnal variations of the electric waves occurring in nature, and on the propagation of electric waves round the bend of the earth *Proc. R. Soc.* A **87** 79–99

Fabry C 1928 Remarques sur la diffusion de la lumière et des ondes Hertziennes par les electrons libres *C. R. Acad. Sci., Paris* **187** 777–81

Gordon W E 1958 Incoherent scattering of radio waves by free electrons with applications to space exploration by radar *Proc. Inst. Radio Eng.* **46** 1824–9

Marconi G 1902 The progress of electric space telegraphy *Proc. R. Inst.* **17** 195–210

Ratcliffe J A 1963 Ionospheric radio in *U.R.S.I.—Union Radio Scientifique Internationale Golden Jubilee Volume* (Brussels: URSI) pp 60–1

—— 1966 Edward Victor Appleton 1892–1965 *Biog. Mem. Fell. R. Soc.* **12** 1–15

—— 1970 *Sun, Earth and Radio* (London: Weidenfeld and Nicolson) pp 45–6

Stewart B 1883 Terrestrial magnetism (under Meteorology) *Encyclopaedia Britannica* 9th edn vol. XVI (Edinburgh: A and C Black) p. 181

Watson-Watt R 1929 Weather and wireless: Symond memorial lecture *Q. J. R. Meteorol. Soc.* **55** 273

19

The Origins of Ionisation and Plasma Physics†

F Llewellyn-Jones

Introduction

A plasma is composed of a volume of ionised gas which satisfies certain conditions. The dimensions must be larger than a certain approximate minimum depending upon the concentrations of the constituent particles of electrons, positive ions and neutral gas molecules, and on their mean energies of agitation; also, the assembly must be macroscopically electrically neutral, so that the concentration of the electrons is approximately everywhere equal to that of the positive ions.

Common forms in which plasmas are produced and observed in the laboratory are found in the uniform glows of high-frequency or direct current electrical discharges in gases at low gas pressure, as well as in arcs and sparks. Indeed, it was to the uniform glow region (previously known as the uniform positive column) that the term *plasma* was first applied in 1923 by Irving Langmuir (1881–1957), and later research established the existence of plasmas over an extremely wide range of conditions. On the one hand, plasmas exist in the upper reaches of the terrestrial atmosphere, and further out still, where there exist important reactions of plasmas and magnetic fields (Alfvén 1950), while, on the other hand, practically invisibly small microplasmas occur when an

† Based on a lecture given at University College, Cardiff, 12 November 1986 to the IOP History of Physics Group.

ordinary low-voltage (about 5 V) low-current (about 5 A) circuit is opened (Llewellyn-Jones 1979). In fact, the sun and stars themselves are now regarded as gigantic concentrations of plasma at high (thermal) temperatures, and the notion is now extended to nebulae and even to interstellar space. In this sense, a plasma is today regarded as a fourth state of matter in its most widespread general form (Frank-Kamenetski 1972). Over this range, the particle density concerned extends from about 10^6 to 10^{24} cm^{-3}. The forces between the particles vary from short-range forces of the so-called elastic 'gas-kinetic' collisions to long-range Coulomb forces between charged particles.

Practical interest in the physics of plasmas can be related to two approximate separate areas designated 'cold' plasmas in which the mean electron energies are not greater than about 5 eV with the gas assembly at about room temperature, and 'hot' plasmas in which the charged particle mean energies are of the order of keV and the actual thermodynamic temperatures in the region of 10^9 K. There is then a physical meaning to the expressions *electron* or *ion temperatures*.

Developments in the fields of magneto-hydrodynamics and of thermonuclear fusion research have been responsible for considerably increased interest in plasma physics in recent years. To investigate and understand the characteristics and properties of all types of plasmas, it is clearly necessary to understand the motions and interaction of the fundamental particles composing assemblies of electrons, positive ions, neutral gas atoms and radiation quanta in the presence of applied electric and magnetic fields, so that in this sense plasma physics is a natural development of ionisation or collisional physics. The origin of plasma physics thus lies in the origin of the study of the nature and properties of the fundamental particles of matter.

Although the general development of science is broadly continuous, there are certain fairly short periods of years during which the progress and development of new understanding of phenomena appear to have been more than usually rapid. Such a period was the decade following the discovery of x-rays by W C Röntgen (1845–1923) in Germany in 1895. The years of that decade saw an almost explosive advance in establishing the atomic nature of electricity in gases, the motion and properties of atomic particles, including their transport coefficients, the important theory of ionisation by collision, the fundamental principles of the spatio-temporal growth of ionisation leading to the electrical breakdown of dielectrics, and to electrical discharges in general. This field covers the basis of the study of electron and ion swarms both in equilibrium and non-equilibrium states, i.e. of plasmas.

Before the turn of the century, Cambridge University started to encourage young graduates from elsewhere to enter the Cavendish Laboratory to undertake research work without working formally for a degree, and the first two who arrived there in 1895 to work with J J Thomson (1856–1940) were Ernest Rutherford (1871–1937) from New Zealand and John Townsend (1868–1957) from Trinity College, Dublin. Others who later came to work there in the same field included F W Aston (1877–1945), O W Richardson (1879–1959), C T R Wilson (1869–1959) and H A Wilson (1906–64) from England, Paul Langevin (1872–1946) from Paris and J Zeleny (1872–1951), later of Yale. Langevin afterwards consolidated the French connection in having a considerable influence in the Paris work of Jean Perrin (1870–1942) on cathode rays and on ionised gases, on which Eugene Bloch was also working. In Germany, the work of E Wiechert (1861–1928) was prominent.

Detailed accounts of the experimental investigations of this period are to be found in papers published in *Proceedings of the Cambridge Philosophical Society*, *Philosophical Transactions*, *Proceedings of the Royal Society* and the *Philosophical Magazine*, while many original papers were republished in the volume *Ions, Electrons, Corpuscles*, edited by H Abraham and P Langevin and published by the Société de Physique of Paris about 1905. Valuable discussion and comment on the work are given in the treatises of J J Thomson (1903, 1906).

The development of research in ionisation physics, as presumably in other areas, falls approximately into two phases separated by World War II. Limitations of available materials and techniques meant that, in general, early investigations had to be restricted to steady state, or quasi steady state, conditions. However, the collisional processes involving normal or metastable atoms, electrons, ions and photons covered a range of time intervals over a factor of 10^8, from the long diffusion times of metastable molecules in gases down to short interelectrode photon transit times. Today, as a result of advances in electronics developed from wartime radar experiments and later solid state research, there are available circuits capable of producing high-electric-field delta function pulses, oscilloscopes capable of recording time intervals much less than nanoseconds, laser sources of short radiation pulses of very high energy, as well as the introduction of fast computers for complicated analyses. In fact, computer simulation of extremely fast electrical discharges is now commonplace. Before World War II no such techniques were available, and a particularly serious deficiency was the absence of time-resolution techniques in the early work, although in the 1930s, time intervals in some experiments were controlled by megahertz oscillators using thermionic radio valves. In the early work with steady state conditions, electrode potentials were maintained by using banks of secondary cells, but this apparent measurement limitation proved later to be of crucial importance and advantage in the measurement of small ionisation currents when maintenance of the necessary constant electric field was essential. However, many of these diagnostic difficulties were overcome by brilliant design coupled with outstanding practical skill, and it is with this basic early period from 1895 to 1939 that this chapter is concerned.

Early concepts of basic particles

In the last decade of the nineteenth century, views on the fundamental nature of electricity were brought to a head by experimental studies of the discharge of electricity through rarefied gases, and particularly by the work of W Hittorf (1824–1914) in Germany and of William Crookes (1832–1919) in England. They concluded that their 'cathode rays' consisted of minute negatively charged particles, but it was a view so contested by strong supporters of Maxwell's electromagnetic field theory that A Schuster (1851–1934) later commented that 'the view that a current of electricity was only a flow of aether appealed generally to the scientific world and was held almost universally' (Thomson 1933). However, experiments on the deflection of cathode rays by electric or magnetic fields, especially those by E Wiechert (1861–1928), Walter Kaufmann (1871–1947) and J J Thomson, established the particle nature of the rays and produced reliable values for the ratio e/m of the particle charge to its mass; but their individual values were still unknown.

Now, Faraday's work on electrolysis with the concept of charged ions in salt solutions supported the idea of elementary ionic charges. Then the analysis by W Nernst (1864–1941) in 1889 of electrolytic phenomena, using methods not strictly applicable for gases, led to an expression for the product Ne, of Avogadro's number N to the monovalent ionic charge e, in terms of the ratio of the diffusion coefficient of the ions to their mobility; but neither N nor e was accurately known separately. An analogous concept for the properties of electricity in gases was not generally accepted; and, as was previously the case with cathode rays, strong supporters of Maxwell's theory of the electromagnetic field formed the view that concentrations of charges like ions were really of electromagnetic origin to which 'gas statistics' did not apply, and that consequently the concept of a material particle was unnecessary.

Nevertheless, Thomson and Rutherford set about examining the consequences of the electrification (i.e. ionisation) of gases by Röntgen rays, by the emanations from radioactive substances, or by the light from spark discharges. They investigated how the resulting electric current depended upon the applied field intensity, or, when not swept away by a field, how the electrification decayed spontaneously (due to recombination). These important processes of mobility and recombination to form neutral particles were more carefully analysed, and Rutherford's experiments in 1897 were probably the first specifically made to measure their coefficients. These proved to be the forerunners of numerous experiments designed to measure ion mobilities with increasing accuracy, because of their significance in collision theory, right up to modern times.

In the absence of means of recording very short times, Rutherford (1901) employed long path distances using a pendulum to record corresponding time intervals; while Zeleny (1901) used a quasi steady state technique in which ions were drawn longitudinally down a tube by a streaming ambient gas as they were also driven radially to the walls by a weak field. Later, Langevin devised a method in which field reversal was obtained by operating successive switches by fast-falling weights in order to apply and reverse the voltage from secondary batteries, so producing voltage pulses. Experimental data were normally analysed on the basis of Langevin's classical theoretical treatment; the wave-mechanical calculations were first carried out in 1934 by H S W Massey and C B O Mohr (Massey and Mohr 1934).

Now, Townsend held the same view of the material nature of ions in gases as did Thomson and Rutherford; but he also considered that the excess charge on a gaseous molecule (which thus constituted an ion) in fact did little to interfere with the ordinary intermolecular collisional process, other than to produce a unidirectional mechanical force which deviated each ionic free path, thus causing a general drift along the direction of the external electric field. In effect, the cloud of charged ions could be regarded as another gas mixed with the ambient gas of normal molecules and taking part in all their usual collisional activity. In other words, this 'ion gas' would be subject to the Maxwell–Boltzmann statistics applicable to ideal gas molecular collisions, and so exhibit the macroscopic transport phenomena of diffusion, drift and, when ions of opposite signs are present, of recombination as well. The adoption of this view was a step forward of fundamental importance, as it finally led to the detailed mathematical treatment of the collisions of electrons, ions and molecules in weak fields.

In a Bakerian Lecture, Rutherford had discussed the identity of gaseous ions, and he pointed out that their material existence could be established decisively if their individual electric charge and their coefficient of diffusion in gases were measured, and added that this would be difficult to do. It was not long, however, before both these quantities were determined experimentally by Townsend, who took the view that the ions contained in a gas could be regarded as themselves forming a gas which obeyed Maxwell–Boltzmann statistics. An interesting discussion of the almost contemporary treatment of kinetic theory was given by H W Watson (1827–1903) (Watson 1876).

In his experiments, Townsend used the fact that gases liberated by electrolysis carried electric charges and possessed the property of forming a cloud in passing through aqueous vapours without any cooling or expansion; and also that the weight of the cloud was proportional to the charge in the gas. He then measured the rate of fall of the cloud, its weight and total charge, and, using Stokes' law, deduced the individual drop size and so their total number, finally giving the value of 5×10^{-10} e.s.u. for the ionic charge (Townsend 1897, 1898). This brilliant work in 1897 and 1898 contained, in the view of R A Millikan (1868–1953), essential elements of subsequent improved determinations by J J Thomson, H A Wilson and of Millikan himself (Millikan 1917, 1937, 1947).

In order to tackle the problem of diffusion experimentally, Townsend passed ionised gas through a metal tube which then acquired some of the ionic charge from the radial diffusion of ions to the tube walls. By treating the process theoretically on the basis of the kinetic theory of gaseous diffusion, the diffusion coefficient could be found in terms of the measured radial loss of charge of the ion stream. Further, in a general analysis of ion motion in an electric field using Maxwell's transport equations, Townsend deduced that the ratio of the diffusion coefficient to the ionic mobility was proportional to the product Ne where e is the ionic charge. This was of the same form as the equation which Nernst had deduced for electrolytes, the form being known later, for no discernible reason, as the Einstein equation. Then, using Rutherford's value for the mobility with his own determination of the diffusion coefficient, Townsend, in 1899, obtained a value of Ne for ions which was sufficiently close to the value accepted for monovalent electrolytic ions as to establish the equality of electric charge on the two types of ions (Townsend 1900, 1901).† Consequently this was taken to be the elementary atom of electricity, which was also accepted as the charge on the negative particles—the electrons of cathode rays. Another important result followed immediately, since the electronic mass could be calculated from the measured bending of cathode rays in a magnetic field.

Important experimental diagnostic methods arose from the studies of cathode and the corresponding anode rays; these were the cathode ray oscilloscope and the mass spectrometer of Aston. Also, the observations on gases evolved in electrolysis showed that clouds could

---

† These contain the first published application of Maxwell–Boltzmann statistics to the motions of ions in gas

be formed by condensation on charge centres, and this led to the development by C T R Wilson of the cloud chamber.

Atomic collision processes

The next important step in the elucidation of ionisation phenomena was the discovery of the generation of electrons and ions by collisional processes in gases. Thomson and Rutherford and others had studied the increase of ionisation produced in a gas in an electric field between parallel plates when irradiated from external sources. They had observed that these increases depended on gas pressure, electric field and even linear dimensions of the apparatus. A current view was that these confusing phenomena were produced by the existence on electrodes of the so-called 'double-layers' of electricity, a view which had been suggested in the eighteenth century in connection with electrostatic machines; an applied electric field was thought to be able to 'peel off' layers and so liberate new charges. Indeed, it was also considered that the charges thought to exist in all molecules were held there partly by the bombardment of adjacent molecules. Thus, if the bombardment rate is reduced, as at reduced pressure, then charges could become detached or 'leak' off; the lower the pressure, the greater was the leak and so the electrification.

In considering this problem, again treating gaseous ions as a small gas additive to the ambient gas, Townsend directed attention to the action of the electric field on the gaseous ions themselves. He realised that in the free flight of an ion between collisions with gas molecules, the ion must gain energy from the field, while the average loss of energy in ideal gas collisions with molecules remained largely unaffected. Energy gained in this way could eventually attain values sufficient to produce damage to a gas molecule in collision, causing the break-off of an electron to produce a new free positive ion and with an additional free electron. In the same way, these two new particles would attain more energy and themselves repeat this disruptive process and general amplification would thus continue. Townsend soon realised that the most effective ionising agent must be the electron and not the heavy ion so his assumptions were valid. Thus the vital electric field intensity E and the mean free path λ, which in any given gas at constant temperature is inversely proportional to the gas pressure p, give E/p as the essential energy parameter.

Consequently, in his own experiments on ionisation growth, he maintained E/p constant while he varied the distance travelled by an electron cloud, thus obtaining the well known curves of exponential growth. From such curves, Townsend and collaborators, working at

Oxford since 1900, disclosed additional processes of ionisation resulting from the primary electron process, and he specified coefficients corresponding to these collisional processes involving the positive ions and photons as well as the nature of the electrode surfaces. In this way the important concept of an electron avalanche was evolved.

Further, his theoretical deduction of an expression for the primary coefficient α for ionisation of neutral molecules in electron collisions involved specification of a critical atomic potential, different for different gases, and the magnitudes of these ionisation potentials were obtained by comparing theoretical expressions with measured values of α. This appears to be the first example of a critical potential being attributed to atoms and molecules, many years before the introduction of the Bohr atomic theory of which a critical energy level was an essential feature (Townsend 1910, 1915).

Another result obtained from this theory of the α coefficient was that the atomic cross section operative in an ionising collision could be considerably smaller than that effective in a gas-kinetic elastic collision; but the full significance of this concept was not generally appreciated at the time. In Göttingen, J Franck and G Hertz, using a beam of electrons from a hot filament in a gas at low pressure, found that large energy losses attributed to ionisation of atoms appeared to occur at or very near a critical ionising potential. Some controversy followed concerned with the values of mean fractional energies lost by electrons in collisions with gas atoms as well as with the values of the effective cross sections in collisions involving high critical energy losses. Finally, it was realised that well focused monoenergetic electron beams, especially from heated cathodes, were not easy to obtain, and the nature and magnitude of the gas scattering were not then fully appreciated, so that the required accuracy was not so readily obtainable as was first thought (Franck and Hertz 1913, 1916, Hertz 1917, Townsend 1948).

The discovery of ionisation by collision proved the death-knell of any 'double-layer' theory of electrification, because increase of ionisation was obtainable merely by increasing interelectrode distance when the electric field at the electrode surface was kept constant and any supposed 'peeling-off' of layers should then be constant.

The theory of ionisation by collision and the statistics of collisional phenomena lie at the basis of our understanding of the electrical discharge through gases, and, of course, of plasma physics. The theory of growth by primary and secondary ionising processes at once led to an explanation of the measured values of the sharply defined static breakdown, or sparking, potentials of gases, and of Paschen's law and the general similarity principle in discharges (Llewellyn-Jones 1966a). This subject is today of great practical importance, especially in the field of high-voltage power technology using compressed gases, as well as in space science (Llewellyn-Jones 1957, 1966b). The modern form of the energy parameter E/p is E/n, where n is the molecular gas concentration, and its unit is called the Townsend. Extensive data on ionisation coefficients for most gases are used today with numerical methods and computers to produce simulation of extremely rapid ionisation and radiation growth, such as that which occurs with the sudden voltage collapse of an over-volted spark gap (Davies et al 1964, Davies and Evans 1973, Davies 1986). These calculations cannot be done with formal analytical methods owing to the rapid spatio-temporal field variation due to predominating space charges at high current densities. The early work on low-pressure discharges between coaxial cylinders elucidated the action of Geiger counters, and ionisation theory also produced the spark chamber as another diagnostic tool in nuclear physics. Such high speeds of ionisation development can now be recorded with streak photographs. It is interesting to recall that Schuster in 1900 tried to record radiation development from a spark using rotating mirror photography. These methods, though valid in principle, were too slow by a factor of about 10^7 to record radiation development in the spark (Schuster and Hemsalech 1900).

Accurate measurement of cross sections

In all this work Townsend realised the need for accurate data on collisional processes involving electrons, ions, photons and electrode surfaces; in particular, he was concerned with mobilities, free paths (i.e. cross sections), losses of energy in collisions and mean energies of agitation in electric fields. Consequently, he developed at Oxford a successful technique, which in refined and more sophisticated forms is in use today throughout the world and known as the 'swarm' method.

In essence, this procedure is a development of his original method of measuring the product Ne for ions based on the Nernst–Townsend equation. A stream of electrons (or ions) in equilibrium is driven through a gas by an electron field, and its lateral diffusion measured; with electrons a transverse magnetic field deflects the beam to appropriate electrodes. The theory of the ionic or electronic motion was based on a solution of Maxwell's transport equation from which the required atomic data are derived as functions of the parameter E/p. The method proved extraordinarily comprehensive and accurate; and particularly helpful data were obtained for the average fractional energy loss of electrons in atomic collisions, and their mean energies of agitation as a function of E/p, and many of his determinations are used today. In 1924 Townsend was invited to give an account of this work at the centenary celebrations of the Franklin Institute in Philadelphia, and the paper was published by the Clarendon Press (Townsend 1925). This was the first book explicitly dealing with the motions of electrons in gases; he elaborated the treatment based on the free path.

An important result, which followed from determination of the mean electron energy of a swarm as a function of E/p, was that in a helium glow discharge, for instance, where atomic ionisation and excitation requiring energies exceeding about 20 eV clearly took place, the mean electron energy could be as low as about 4 eV. Unfortunately, this conclusion was the source of some controversy at the time. Nevertheless, the result drew Townsend's attention to the energy distribution function which was then investigated by his colleague, F B Pidduck (1885–1952) (Pidduck 1913, 1916, 1936) and later by M J Druyvesteyn, Townsend and many others (Morse *et al* 1935, Chapman and Cowling 1952). Pidduck's treatment of electronic collisions and the distribution function was not generally appreciated at the time; in fact it forestalled much of the later work. Today, it is appreciated that this function is still of great importance in analyses of collisional processes and of plasma physics in general. The modern approach is through solution of the Boltzmann equation using the latest data on cross sections, as, indeed, the comprehensive discussions at the NATO ASI Meeting at Bourges-St Maurice, France, in 1981 showed (Kunhardt and Luessen 1983).

Another result obtained in 1913 by Townsend and Henry Tizard (1885–1959) with this swarm method was that the atomic cross section in collisions with slow electrons appeared to depend upon the electron velocity; a result quite inexplicable at the time from the point of view of classical mechanics (Townsend and Tizard 1913). However, this research was closed down by the outbreak of World War I, but, fortunately, was resumed by Townsend and V A Bailey (1895–1964) after the war. About the same time, C Ramsauer (1921, 1928) and R Kollath in Germany were investigating electron scattering by using collimated beams of electrons passing through a gas at pressures low enough adequately to reduce the number of collisions; electron energies lay over a wide range above about 30 eV. Variation of atomic cross section with electron velocity was also found, but of forms somewhat different from those obtained by swarm methods with slow electrons, the results of which were not then considered to be as accurate as those obtained by beam methods. However, differences were cleared up when it was realised that the two methods were, in fact, complementary, one being more accurate for the higher enegies and the other for slow electrons; in fact, the swarm method, when based on a full examination of the electron energy distribution function, is today the only one available for use with very low and even thermal electron energies. The phenomenon of the variation of atomic cross section in electron collisions is accordingly now known as the Townsend–Ramsauer effect; and its explanation had to await the advent of wave mechanics. One may remark here that the controversies which arose at the time over comparisons of data obtained by statistical swarm methods with those found by the apparently more direct beam methods, were both unnecessary and unfortunate. On reflection today, it seems that they were based upon a lack of appreciation of the limits of experimental accuracy of the two very different approaches. Criticism of the swarm method, allegedly based on the quantum theory, tended to encourage Townsend to feel somewhat sceptical about that theory as presented to him by some of the critics, but the whole matter really illustrated Eddington's comment, 'that a doctrine is not to be judged by the follies that have been committed in its name'!

The years from 1915 to 1930 proved to be of considerable interest in the development of the theory of plasma. The basic characteristics of the uniform positive column of an electrical discharge in, say, a long Geissler tube had long been recognised, and in his treatise of 1915 Townsend had set out the equations expressing the conditions of maintenance of a discharge based on the basic processes of ionisation, drift, diffusion and recombination. At the lower gas pressures recombination is negligible in the uniform glow, and the higher drift and diffusion rates of electrons set up a positive charge due to residual positive ions, but the resulting electric field reduces dispersal rates of electrons while increasing those of positive ions, so equalising the two rates. In 1924 W Schottky gave a formal expression for this ambipolar diffusion (Schottky 1924a, b). Further, the similarity of such conditions to those in an electrolyte was realised, and the Debye analysis was then applied to gas discharge conditions; this Debye screening is of great importance in plasma physics.

With the development of radio-frequency oscillators, interest in high-frequency glow discharges grew, and considerable experimental investigations of them were carried out by A Gutton (1872–1963) and others in Paris and by the Townsend school (Townsend 1915, p. 44, 1928). In 1930, Townsend produced a theory of maintenance of a high-frequency discharge which involved ambipolar diffusion and the equivalence of the glows of direct current and high-frequency discharges in certain circumstances of gas pressure and oscillation frequency f. In fact, the factor f/p is involved in the analysis just as in the energy parameter E/p (Llewellyn-Jones 1931).

The interaction of radio waves with a glow discharge or plasma remains a subject of considerable scientific interest and is used as a diagnostic tool. In very general terms it can be seen that when the wave oscillation frequency exceeds the plasma frequency the wave penetrates the plasma, but when the plasma frequency exceeds the wave frequency, the wave is absorbed or rejected, and the plasma can then be treated on magneto-hydrodynamic concepts.

When a magnetic field is applied to a plasma the resulting force on the ions and electrons can set up different modes of plasma oscillations, depending upon the direction of the applied field and the polarisation of the incident waves.

A property of a Bohr atom which, however, had a profound effect on our understanding of ionisation phenomena was that of the long-life metastable state. The long life time is due to the general non-occurrence of spontaneous radiative transitions to return the atom to a lower or ground state; destruction requires the action of another body. Striking experiments by F M Penning in Holland showed how high-energy metastable atoms could ionise molecules of a different gas of lower ionisation potential, especially when that value was near the metastable energy of the colliding atom (Penning 1931). The importance of the effect, known as a collision of the second kind, was due to the fact that during a long life time of some milliseconds, a metastable atom consequently had a high probability of colliding with an impurity molecule even at relative concentrations less than 10^{-5}. The effect emphasised the necessity of using the highest possible gas purity in, for example, determination of ionisation coefficients, particularly with the noble gases of high ionisation potential.

This then, is a suitable place briefly to describe the general experimental technique of gas purification considered necessary by Townsend and S P MacCallum at Oxford for ionisation measurements in the late 1920s and early 1930s.

Experimental techniques

Ionisation currents of about 10^{-13} A, which avoided space charge problems, were measured with a Dolezalek quadrant electrometer in conjunction with the Townsend electrostatic balance which maintained electrodes at constant potential. Internal electrodes were usually mounted on quartz, while all capacitors connected to electrodes were mounted on amber or ebonite using split insulation and appropriate guard rings to avoid leakage. Where possible, electrodes themselves were made of pure nickel, but in some cases gold, silver and copper were employed; they were usually pre-heated but were finally outgassed by heating *in vacuo* by radiant heat from external gas burners or electric heating coils wound around the chamber. Copper and molybdenum gave trouble during such out-gassing owing to sputtering which could reduce insulation. External connection to internal electrodes were usually made through tungsten-hard glass seals or by copper stranded wire sealed by molten lead into long quartz side tubes to a quartz ionisation chamber. All electrical connections, condensers, etc were completely screened by earthed metal covers.

Evacuation was carried out using Langmuir diffusion pumps, with mercury or, later, with the then newly introduced Burch low-vapour-pressure oil, and backed by Gaede or oil rotary pumps. The diffusion pumps were suitably isolated by liquid-air traps and charcoal traps. Noble gases required most careful purification, especially helium owing to its high ionisation and metastable potentials. This gas was sometimes actually prepared in the laboratory from thorianite, which was first dried and freed from many adsorbed gases by heating in a quartz vessel connected to traps containing phosphorus pentoxide and activated charcoal under evacuation. On raising the temperature to red heat the thorianite released occluded helium, which was continuously circulated over heated copper oxide (to remove hydrogen) and stored over charcoal cooled by liquid air.

A difficult problem at that time was isolation of the ionisation chamber from the final stop-cock of the gas system. This was necessary because, before the introduction of the Alpert all-metal bakeable tap, the stop-cock was made of vacuum grease-lubricated glass, and so could not be out-gassed. The solution adopted therefore lay in provision of an adequate gettering system. Initial diffusion of grease vapour was prevented first by liquid-air traps and then by a succeeding getter system, consisting of a long narrow quartz tube leading into a pair of quartz liquid-air traps, one of which contained pure calcium. When this was heated while its companion cooled, calcium vapour from the heated trap condensed into the cooler trap, so producing a getter action which adsorbed impurities; this action was also assisted by the cooling of the previously out-gassed long lead-in tube. This calcium transfer process was, of course, reversible. The system thus presented some of the properties of modern vacuum techniques.

Traces of gas impurity possibly evolved by electrodes under local bombardment of ions and electrons during measurements were removed, or certainly reduced, by using the 'clean-up' effect of a 20 MHz high-frequency electrodeless discharge through the ambient gas in an adjacent wide side tube. Accurate measurement of the sparking potential of the ambient gas proved to be a good indication of the degree of purity attained. To judge from accounts in contemporary published papers, it seems safe to conclude that the gas purity attained in the measurement of ionisation coefficients by Townsend and MacCallum at that time, was at least as good as that used in similar work anywhere else. Nevertheless, some contemporary criticism, even in textbooks, of their ionisation data was based on supposed grounds of inadequate gas purity! Traces of impurities could produce instabilities in the glow discharges and the same effect was found in more recent years with high-energy hot plasmas. Incidence of a plasma on any containing vessel is a strong source of impurity gases.

Recent work and conclusions

In concluding this brief survey of the period 1895 to 1939, in which the study of the basis of plasma physics was consolidated, it may be of interest to recall the achievement of Robert J Van de Graaff (1901–67). While his main work was of greatest relevance to nuclear physics, his first experimental work on ionic transport was not without interest in plasma physics.

In the early 1920s, Van de Graaff became fascinated by accounts in textbooks in his father's library in Birmingham, Alabama, of gaseous ions and atomic particles. Wishing to learn more of this subject, in 1924 he relinquished his appointment as engineer in the local power company to spend a year in the University of Paris where Marie Curie (1867–1934) and others were actually carrying out research work on atomic particles and the emanations from radioactive substances. This experience apparently decided him henceforth to devote his energies to work in this field in general, and in the motions of ions in particular. Consequently, on his return home to the USA, he applied for and was awarded a Rhodes Scholarship to work with Townsend in Oxford, and in 1925 started to undertake research work there as a graduate student, later obtaining the research degrees of BSc and DPhil for his work on the mobilities of positive ions. Ideas on atomic disintegration were never far from him, and he appeared to keep in mind the use of positive ions as controlled high-speed atomic projectiles.

However, for Van de Graaff's doctoral research project, Townsend proposed the more accurate determinations of ionic mobilities, as he considered that this was still an outstanding experimental problem which had important theoretical implications concerning our understanding of collisional phenomena. In this matter, it did seem that history was repeating itself since, as was stated above, more than a quarter of a century earlier J J Thomson in the Cavendish Laboratory had made a similar decision in explaining to Rutherford, to Zeleny and to Langevin, the need at that time for more accurate determinations of ionic mobilities. Again, the work of Townsend and Tizard on electron mobilities, cut short by the outbreak of World War I in 1914, was the forerunner of extensive post-war investigation of the variation of atomic collisional cross sections with slow incident electrons. Finally, about 1926, the need for still more accurate data on positive ion mobilities was also seen, quite independently, by A M Tyndall (1881–1961) (Tyndall 1950) in Bristol, as well as by Townsend in Oxford. Actually, Van de Graaff was able to publish his time-of-flight method (Van de Graaff 1928) just prior to the paper by Tyndall, L H Starr and C F Powell (1903–69) (Tyndall *et al* 1928) in the same year describing a similar technique of mobility measurement, but using different ion sources; this Bristol work became the basis of their distinguished research on precision measurements of the mobilities of alkali and gaseous ions moving in the noble gases, data from which could be compared with the theoretical values calculated with wave mechanics, as well as those calculated from Langevin's theory and its development by H R Hassé (1884–1955) (Langevin 1950, Hassé 1926, Hassé and Cook 1931).

In 1929, Van de Graaff left Oxford to join K T Compton (1887–1954) at Princeton, and his concept of a high-voltage electrostatic generator had been so clearly formulated that within only five months of his leaving Oxford, Van de Graaff had himself designed, constructed and successfully operated his 80 kV generator, and this was the forerunner of reliable massive machines used throughout the world to the present day.

It is interesting to note that essential constituent principles of the original Princeton design had been considered and tried at one time or another during the previous two centuries of experimentations in the field of electrostatics, but it was Van de Graaff who first put the ideas together in a successful design. It may be noted that his own experimental work on ionic mobilities had particularly involved aspects such as the generation, control and motions of gaseous ions from point electrical discharges in air, the screening of conductors and insulators necessary for the establishment and appropriate location of electric fields, and the measurements of small (of the order of 10^{-12} A) ionisation currents with electrostatic balances. This clearly constituted a background of experience in ionisation physics which was relevant and helpful in consideration of the design of the electrostatic generator which he actually developed; further, his own appreciation of basic ionisation phenonema could be seen from the reasons underlying his introduction, some years later, of the inclined electric field arrangement in order to overcome breakdown difficulties met at the highest voltages obtained with a single generator.

Reference has been made above to the remarkable advances in experimental techniques of plasma physics research after World War II. Early measurements of electron and ion currents by means of electrometers could take many minutes, and complete experiments require many hours, owing to the need to monitor the constancy of electrode potentials; and in steady state conditions effects of fast electron currents could not be distinguished readily from those of ion currents, still less from those of the very slow diffusion of metastable atoms. Today, with fast pulse techniques giving high time-resolution, and current recording with computer data gathering and processing, complicated measurements involving simultaneous processes of ionisation and de-ionisation can be completed in seconds. This is well illustrated in, for example, the recent work of D K Davies at Pittsburgh on plasma parameters (Davies 1987).

Although modern methods of producing practically monoenergetic focused beams of charged particles are a considerable advance on the early techniques, the necessary calculations of the atomic collisional frequency in the beam interactions are still not easy and are best carried out with fast rather than slow electrons. To obtain reliable data for very slow electrons, especially near thermal values, the most accurate method is still that based on the swarm technique involving diffusion and drift originally due to Townsend. Such data are increasingly required for investigations of collisional processes, for example, in the upper atmosphere and gas lasers.

An account of modern theoretical and experimental developments of swarm methods is given by Huxley and Crompton of Canberra (Huxley and Crompton 1974) and other treatises also deal with the areas of slow electrons (Gilardini 1972) and of electrons and ions (Lindinger *et al* 1982). These areas are important in the general study of plasma, afterglows, recombination processes and diagnostic probes, as well as in the engineering application of gaseous electronic and drift tubes, and reactor and magnetohydrodynamic processes.

References

Alfvén H 1950 *Cosmical Electrodynamics* (Oxford: Clarendon)
Chapman S and Cowling T G 1952 *Mathematical Theory of Non-uniform Gases* 2nd edn (Cambridge: Cambridge University Press) p. 350
Davies A J 1986 Discharge simulation *Proc. IEE* **133**A 217–40
Davies A J and Evans C J 1973 *The Theory of Ionization Growth in Gases under Pulsed and Static Fields* (Geneva: CERN)
Davies A J, Evans C J and Llewellyn-Jones F 1964 Electrical breakdown of gases *Proc. R. Soc.* A **281** 164–83
Davies D K 1987 Measurements of swarm parameters in dry and humid air in *Proc. 18th Int. Conf. Phenomena in Ionized Gases, Swansea* vol. 1, ed. W T Williams (Bristol: Adam Hilger) pp 2–3
Franck J and Hertz G 1913 Über einen zusammenhang zwischen Stossionisation und Elektronenaffinität *Ber. Deutsch. Phys. Ges.* **15** 929–34

—— 1916 Über die relative Intensität der Gasspektra bei der Glimmentladung in Gasgemischen *Ber. Deutsch. Phys. Ges.* **18** 213–20
Frank-Kamenetski D A 1972 *Plasma, the Fourth State of Matter* (New York: Macmillan)
Gilardini A L 1972 *Low Energy Electron Collisions in Gases* (New York: Wiley)
Hassé H R 1926 Langevin's theory of ionic mobility *Phil. Mag.* **1** 139–60
Hassé H R and Cook W R 1931 The calculation of the mobility of monomolecular ions *Phil. Mag.* **12** 554–66
Hertz G 1917 Über den Energie austausch bei zusammenstössen zwischen langsamen Elektronen und Gasmolekülen *Ber. Deutsch. Phys. Ges.* **19** 268–88
Huxley L G H and Crompton R W 1974 *The Diffusion and Drift of Electrons in Gases* (New York: Wiley)
Kunhardt E E and Luessen L H 1983 *Electrical Breakdown and Discharges in Gases Part A: Fundamental Processes and Breakdown* (New York: Plenum)
Langevin P 1950 *Oevres Scientifiques* (Paris: CNRS)
Lindinger W, Murk T D and Howorka F 1982 *Swarms of Ions and Electrons in Gases* (Wien: Springer)
Llewellyn-Jones F 1931 High frequency and direct current discharges in helium *Phil Mag.* **11** 163–73
—— 1957, 1966b *Ionization and Breakdown in Gases* (London: Methuen)
—— 1966a *The Glow Discharge and an Introduction to Plasma Physics* (London: Methuen) pp 84–9, 113–15
—— 1979 Contact electrode processes and microplasma diagnostics *J. Physique Suppl.* **40** PE7–47
Massey H S and Mohr C B 1934 Free paths and transport phenomena in gases and the quantum theory of collisions II *Proc. R. Soc.* A **144** 188–205
Millikan R A 1917 *The Electrons* (Chicago: Chicago University Press)
—— 1937, 1947 *Electrons (+ and −), Protons, Neutrons, Mesotrons and Cosmic Rays* (Chicago: Chicago University Press)
Morse P M, Allis W P and Lamar E S 1935 Velocity distribution for elastically colliding electrons *Phys. Rev.* **48** 412–19
Penning F M 1931 The starting potentials of the corona discharge in neon *Phil. Mag.* **11** 961–80
Pidduck F 1913 The abnormal kinetic energy of an ion in a gas *Proc. R. Soc.* A **88** 296–302
—— 1916 The kinetic theory of the motion of ions in gases *Proc. Lond. Math. Soc.* **15** 89–127
—— 1936 Electrical notes *Q. J. Math.* **7** 199–209
Ramsauer C 1921 Über den Wirkungsquerschnitt der Gasmoleküle gegenüber langsamen Elektronen. I. Fortsetzung *Ann. Phys., Lpz.* **66** 546–58
—— 1928 Über den Wirkungsquerschnitt neutraler Gasmoleküle gegenüber langsamen Elektronen *Phys. Z.* **29** 823–30
Rutherford E 1901 Dependence of the current through conducting gases on the direction of the electric field *Phil Mag.* **2** 210–28
Schottky W 1924a Wandströme und theorie der positiven Säule *Phys. Z.* **25** 342–8

—— 1924b Diffusionstheorie der positiven Säule *Phys. Z.* **25** 635–40
Schuster A and Hemsalech G 1900 On the constitution of the electric spark *Phil. Trans. R. Soc.* A **193** 189–213 and plates.
Thomson J J 1903, 1906 *Conduction of Electricity through Gases* (Cambridge: Cambridge University Press)
—— 1933 *Conduction of Electricity through Gases* vol. II (Cambridge: Cambridge University Press) p. 2
Townsend J S 1897 On electricity in gases and the formation of clouds in charged gases *Proc. Camb. Phil. Soc.* **9** 244–58
—— 1898 Electrical properties of newly prepared gases *Phil. Mag.* **45** 125–31
—— 1900 The diffusion of ions in gases *Phil. Trans. R. Soc.* A **193** 129–58 (This contains the first published application of Maxwell–Boltzmann statistics to the motions of ions in gases)
—— 1901 The diffusion of ions produced in air by the action of a radio-active substance, ultra-violet light and point discharges *Phil. Trans. R. Soc.* A **195** 259–78.
—— 1910 *The Theory of Ionization of Gases* (London: Constable)
—— 1915 *Electricity in Gases* (Oxford: Clarendon)
—— 1925 *Motions of Electrons in Gases* (Oxford: Clarendon)
—— 1928 Théorie des courants de haute fréquence dans les gaz *C. R. Acad. Sci., Paris* **186** 55–8
—— 1948 *Electrons in Gases* (London: Hutchinson) chs III, IV, VI
Townsend J S and Tizard H T 1913 The motion of electrons in gases *Proc. R. Soc.* A **88** 336–47
Tyndall A M 1950 *The Mobilities of Positive Ions in Gases* (Cambridge: Cambridge University Press)
Tyndall A M, Starr L H and Powell C F 1928 The mobility of ions in air *Proc. R. Soc.* A **121** 172–84
Van de Graaff R J 1928 A new method of determining the mobility of ions or electrons in gases *Phil. Mag.* **6** 210–17
Watson H W 1876 *Kinetic Theory of Gases* (Oxford: Clarendon)
Zeleny J 1901 The velocity of ions produced in gases by Röntgen rays *Phil. Trans. R. Soc.* A **195** 193–234

20

The History of the Discovery of Weakly Coupled Superconductors†

B D Josephson

The background

In response to a request of the History of Physics Group of The Institute of Physics I have chosen to write on the historical background of the work I did on in 1962 on weakly coupled superconductors (Josephson 1962a,b). I'm afraid it involves rather more sophisticated physics than most of the other chapters in this book. Also I very much regret to say that, despite my personal involvement, the amount of historical accuracy I can give is not as high as I would wish, as although in the course of my searches I have found a certain amount of material relating to that period of time, a lot of the ideas were worked out in my head, and seem not to have been written down on paper at all.

I should perhaps, as some readers may not be familiar with this work, say a little bit about what it is. At very low temperatures some materials become perfect conductors of electricity—this is called 'superconductivity'—and we now understand that the superconducting state is a state of order which is very similar to that found in a laser,

† Based on a talk delivered to the IOP History of Physics Group, University College, Cardiff, 12 November 1986.

there being coherent electron waves where in a laser you have coherent electromagnetic waves, and the phenomenon involves something which is closely connected with the waves.

The basic equations which apply to superconducting barriers are shown below. A barrier between superconductors can be regarded as a two-dimensional system, whose behaviour (if dissipative processes are neglected) is governed by the following set of equations

$$\frac{\partial H_y}{\partial x} - \frac{\partial H_x}{\partial y} = \frac{4\pi}{c} j_z + \frac{4\pi C}{c} \frac{\partial V}{\partial t} \qquad (20.1)$$

$$\frac{\partial \varphi}{\partial x} = \frac{2ed}{\hbar c} H_y \qquad (20.2a)$$

$$\frac{\partial \varphi}{\partial y} = -\frac{2ed}{\hbar c} H_x \qquad (20.2b)$$

$$\frac{\partial \varphi}{\partial t} = \frac{2e}{\hbar} V \qquad (20.3)$$

$$j_z = j_1 \sin \varphi. \qquad (20.4)$$

These equations apply for a barrier which occupies the xy plane. φ is the difference between the values of the phase of the superconducting order parameter on the two sides of the barrier, j_1 the critical current density of the barrier, C its capacitance per unit area, d its thickness plus the sum of the penetration depths in the superconductors on the two sides, and V the potential difference across the barrier. Gaussian units have been used. Equation (20.1) (essentially one of Maxwell's equations) describes the effect of the supercurrent through the barrier on the electromagnetic field, (20.2) and (20.3) the effects of the electromagnetic field on the superconducting order parameter and finally (20.4) relates the barrier current to the order parameter (Josephson 1966).

To cut a long story short, equation (20.4) is the main result I discovered. Appearing on the right is a phase difference. The system which my work was involved with was where you have two superconductors that can exchange electrons, and the calculation I did showed that under those conditions you get a current given by that equation (20.4), which depends on the sine of the phase difference. So that was the effect, and now I will describe the background to it.

Surprisingly, perhaps, my official PhD research project was not a theoretical one but an experimental one. My view at that time in my life was that I wouldn't find sitting at a desk spending most of my time thinking too congenial, although it's what I do now more or less, and so I decided I would do an experiment instead. I ultimately got my experiment to work, though it wasn't until 1974, about 10 years afterwards, that I got round to writing it up (Josephson 1974). The experiment dealt with a phenomenon that had been of particular interest to my supervisor Professor Pippard (now Sir Brian Pippard). He believed that second sound,† a phenomenon that was well known in liquid helium, might also be found in superconductors (however, during the course of my PhD work I managed to convince myself that a published calculation claiming to demonstrate the existence of second sound in superconductors (Thouless and Tilley 1961) must be wrong, and at the present time no one seems to believe that second sound can actually occur in superconductors). There were anomalous results in the non-linear electrodynamics of superconductors which Pippard thought might be indicative of a resonance associated with second sound, and the proposed experiment involved looking in a new frequency range to see what happened there. It was because I was doing this experiment that I started looking into the theories of superconductivity.

As a consequence of the contacts between David Shoenberg, the director of the Mond Laboratory, and Soviet scientists, my first acquaintance with the theory of superconductivity was with a formidable book *A New Method in the Theory of Superconductivity*, by N Bogoliubov, V Tolmachev and D Shirkov (Bogoliubov *et al* 1959), which it was suggested I try to understand. I found this book almost totally incomprehensible, which was hardly surprising in view of the way it assumed knowledge of topics such as many-body theory, second quantisation and diagram expansions‡ which neither I nor anyone else at the Cavendish at that time knew anything at all about. In the course of time through my browsings in the literature I came across explanations of these matters, and the mysteries of superconductivity theory started to become a little clearer. I found that there were a number of approaches to the problem, all of which were in some sense equivalent since they all implicitly assumed that a superconducting state was a state of pairing of electrons of equal and opposite momentum and spin like that hypothesised by J Bardeen, L N Cooper and J R Schrieffer (Bardeen *et al* 1957), who worked more or less directly with the actual

---

† Second sound is a temperature wave in liquid helium II which results in the cold superfluid component collecting at a point of low temperature while the normal component collects at a point of 'high' temperature, half a wavelength away.

‡ For an explanation of diagram expansions see, for example, Thouless D J (1972) *The Quantum Mechanics of Many-Body Systems* (New York and London: Academic Press) chs iv, vii.

wavefunction. The Bogoliubov theory (Bogoliubov *et al* 1959) used a transformation to convert a situation that couldn't be treated by perturbation theory into one that could, while the Gor'kov theory (Gor'kov 1958, 1959) used the very powerful Green function technique. Finally, there was the approach of P Anderson (1958), who described the superconducting state in terms of interacting pseudospins.†

The phase of a superconducting current

Both the Gor'kov and the Anderson theories contain something corresponding to a phase (in the Anderson theory the pseudospins lie in a plane through the relevant axis of symmetry whose orientation is otherwise arbitrary, and which is related to the phase in the Gor'kov picture), and these theories started me thinking about phases in superconductors. I have some scribblings in a notebook which suggested that I was attempting at that time to understand the structure of the superconducting state and the nature of its long-range order in terms of the kind of correlation functions that appeared in the Gor'kov theory. And there was something in the Gor'kov theory—the so-called anomalous Green function—which did have a phase attached to it. This phase allows one to understand certain properties of superconductors and in particular implied the phenomenon of flux quantisation, a phenomenon with a somewhat curious history. It is manifested first as a suggestion in the famous book of F London (1950), and later as a prediction in a paper by A Abrikosov (1957) based on the 1951 Ginzburg–Landau theory (Ginzburg and Landau 1965), which prediction Landau himself had forbidden to be published at the time on account of the result being 'clearly unphysical', although he later changed his mind about this. The Ginzburg–Landau theory itself, despite its ability to give a simple picture of many of the features of the superconducting state, was virtually unknown in the West at that period because of the language problem and the poor communications between Eastern and Western scientists. Thus the experimental demonstration of the existence of flux quantisation (Deaver and Fairbank 1961, Doll and Näbauer 1961) came as somewhat of a surprise.

In Gor'kov's theory the phase comes over as something more mathematical than physical: it is related to the phase of a matrix

---

† Pseudospin is explained by Anderson (1958) pp 1904–5 as a set of Pauli spin matrices which are not to be confused with real physical spin operators, since they act in an imaginary space (ed).

element between two states with different numbers of electron pairs, and is essentially arbitrary because the two states could be chosen with arbitrary phases. Anderson made use of diagonal matrix elements instead, computed in states which did not have a definite number of electrons. In this approach the phase is a property of the specific state of the system and seems less of a mathematical thing, although it could well be argued that such states involving coherent combinations of states with different numbers of electrons are themselves unphysical. It would take some further analysis to sort this problem out, but in the meantime I found it very fruitful to think of the phase as something that was 'in principle real', real enough in fact to produce the phenomenon of flux quantisation.

If the phase was real, this implied that there might be a way of exploring its existence more directly. The conditions under which an effect could occur were regulated by the existence of symmetry principles. From these you could show that only phase differences could be observed, not the phases themselves. Furthermore, even phase differences could not be physically significant unless a process of electron exchange, which would have to be one that did not lose any phase information, occurred between the two regions concerned. The ultimate answer to the conundrum about the reality of the phase (discussed in some detail in my Trinity College fellowship dissertation (Josephson 1962b) was that, whatever may be the case for the phases themselves, phase differences are perfectly respectable entities (i.e. entities related to operators with non-trivial expectation values even for perfectly reasonable and physical kinds of quantum state).

It was not clear at that time that the possibility of setting up the kind of situation envisaged could be more than a dream, although in fact at around that time experimenters were actually busy creating such situations in experiments using thin films.

Boundaries and junctions

A line of investigation that led me towards the actual discovery sprung from Pippard's interest in the question of what happened at the boundary between a normal metal and a superconductor. This was interesting because superconductors can be put into what is known as an 'intermediate state' where there are fairly macroscopic regions of normal metal and superconductor that alternate with each other. If you measure the electrical resistance and thermal resistance in such materials you find it is quite low, suggesting that there is not the amount of scattering at the boundaries that you would expect. That was a little bit mysterious. In fact it wasn't too clear even how the electric current could flow at all through such materials at low temperatures, because the current flow carried by the individual particles in a normal metal has somehow got to turn into current in a superconductor, and at very low temperatures, because of the energy gap, there are hardly any particles around to carry the current, unless there is a mechanism by which the ordinary electrons in a normal metal can be transformed into superconducting electrons.

Thinking about this and trying to apply some of the concepts that were around, I succeeded in producing a qualitative understanding of what went on at a normal-to-superconducting boundary. The detailed calculations would have been very complex and I did not pursue them, but subsequently Andreev (1964) was to publish calculations based on exactly such concepts. He had enough energy to go through the calculations, which were somewhat messy, and I didn't. In fact, I don't think I've ever done any very messy calculations. I tend to go for problems that won't involve such calculations. In any event, I was as a result accustomed to using the self-consistent field method† to think about this kind of problem. A notebook in which I wrote a bird's eye view of my whole research project contained the following entry:

> Meanwhile we turned our minds to the question of transport properties of superconducting boundaries. A self-consistent field equation for quasi-particles, treating them as two-component electron-plus-hole wave functions, was developed and the consequences explored. However, curious difficulties arise when treating SIS junctions and SNS junctions. An approximate treatment was thought up and seems to imply oscillatory effects. It seems to be rather difficult to persuade experimentalists that these are more than a mathematical fiction, but we hope to publish a paper on the effects, and then someone may try the experiments ... And so it came to pass that the cautiously entitled paper 'Possible new effects in superconductive tunnelling' was published in *Physics Letters* of July 1st 1962. [Josephson 1962a]

Brian Pippard had a somewhat sceptical attitude to theoretical manipulations, and insisted on the incorporation of the word 'possible' into the title of the paper.

---

† The self-consistent field method is an approximate method due to Hartree which dealt originally with central field problems in atomic physics of finding the wavefunction of an electron in the presence of other electrons (from Gray H and Isaacs A 1975 *A New Dictionary of Physics* (London: Longman) p. 481) (ed).

Detailed origin of the theory of weakly coupled superconductors

Let me now go a little into the details. The most important thing was that the various approximations that people introduced had the feature that not only could electrons scatter into electrons, but they could also scatter into holes (a hole is the absence of an electron). The apparent violation of charge conservation involved here is dealt with by treating the background of Cooper pairs as a reservoir which can supply or absorb pairs of electrons as needed without having to be put explicitly into the equations. You then find that to a first approximation all the electron–electron interactions that are important have become miraculously incorporated into the processes of the scattering of single particles already described. To calculate anything you then have to find out what the normal modes of this kind of system are, and these then correspond to mixtures of electrons and holes which you can then regard as being independent of each other.

There was, however, an awkward technical problem involved in a situation where you had two superconductors. The approximations that people introduced had the feature that not only could electrons be scattered from one state to another, but also electrons could be scattered into holes (a hole is the antiparticle to an electron in a metal) and vice versa, i.e. charge was apparently not conserved. The requirement that charge really should be conserved was taken into account by invoking the reservoir of superconducting electrons (a situation that was in due course clarified by Anderson (1958)). Now in order to get this to work properly you had to take the energy of the electrons in the reservoir, which is the Fermi energy, as your origin of energy (otherwise you would have problems with energy conservation, since the energy of the electrons in the reservoir would not have been taken into account). This creates no problem if you have only one region of superconductor, but what do you do if there are two regions and their Fermi levels are different? Fortunately for me, other people who were thinking about two-superconductor systems regarded this difficulty as something not really significant and failed to follow it up. In particular, at about this time Cohen, Falicov and Phillips (Cohen *et al* 1962) had applied perturbation theory to treat tunnelling involving superconductors by perturbation methods. They treated the case where only one side was superconducting correctly but omitted discussion of the two-superconductor case from their paper. I learnt afterwards that one of them had been asked to examine the two-superconductor case and had found terms equivalent to those I had found later, but couldn't understand their significance and had decided that they probably had no effect.

Because they had not dealt with the problem of the energy origin their formalism was inadequate to deal with the problem in any case. I managed to invent a rather esoteric formalism to handle the situation (which has rightly gone into oblivion now), involving putting in explicitly into the Bogoliubov operators the transfer of electron pairs from the background reservoir. Although that was a rather awkward method, I did it in preference to a Green function method because I wasn't familiar with Green functions and I thought I'd have more chance of getting it right if I kept to the Bogoliubov operator method instead.

The interest in doing these calculations in the first place stemmed from the fact that Ivar Giaevar, inventor of the tunnelling method for studying the properties of superconductors, had given a very simple-minded theory of the tunnel junctions that had the merit of agreeing with experiment (Giaevar 1960, Fisher and Giaevar 1961). But it was not at all clear (as had been observed by Pippard) why Giaevar's calculation should give the right answer, because it ignored the 'coherence factors' that normally have to be taken into account in a situation involving superconductors.

The Cohen *et al* (1962) calculation justified Giaevar's formula for the single-superconductor case, which was, however, the case where the coherence factors turn out to cancel out anyway. What I expected would happen in the two-superconductor case was that there would be a non-zero coherence contribution which would have no DC component (so that Cohen *et al* were in a sense correct to believe that the extra terms didn't contribute) because it would be multiplied by a factor depending on the phase difference which would average out to zero. The physical reality of this component was linked to the physical reality of the phase, and my physical intuition told me that both were real, although this was a controversial idea at the time.

My calculations confirmed my expectations, but they predicted an additional term that I had not expected. This term was the one that came from the delta-function contribution to the energy denominators, and had the unexpected feature of not vanishing when the applied voltage was set equal to zero. I checked my calculation several times trying to find a sign error whose existence would have prevented the terms from cancelling each other out, but had to conclude in the end that the contribution that did not vanish at zero voltage really did exist.

The presence of such a term at some order of approximation was not surprising, since it merely indicated the presence of a supercurrent. I had not expected it to occur in my calculation to second order because of the following argument due to Pippard (1961): a supercurrent will involve pairs of electrons tunnelling, so that the matrix element for the process will be proportional to the square of the matrix element for the tunnelling of a single electron. The rate at which the pairs will tunnel will be proportional to the square of this again, and so to the fourth power of the single-electron tunnelling matrix element. In due course I realised where this argument was in error: the tunnelling supercurrents are a coherent flow process, and so the usual rule that its rate is proportional to the square of the matrix element does not apply. I was then satisfied that my calculation was correct and went ahead and submitted the paper for publication. (Incidentally, my original calculation was not quite correct. In it I had managed to introduce two extra factors of two, through taking two integrals over the whole real line instead of just for positive values of the energy as should have been done. Thus the result given in my first paper was too large by a factor of four. I discovered the error when writing up the calculation for my Trinity College fellowship dissertation a couple of months later, while a correct value was obtained also by Ambegaokar and Baratoff (1963) who repeated the calculation using Green functions.)

If you look at my paper you will see that there was considerable concern on my part with making a convincing case that the concepts I was using were valid ones. This raises an interesting issue. I think this is something that's liable to happen in the history of science when a different way of looking at things comes about. *Now* everyone accepts that you can just write down the phase of the wave in a superconductor because this is what all the experiments confirm, but *then* the experimenters were inclined just to think of it as just a bit of mathematics, and were very doubtful about it. Fortunately Phil Anderson was spending a year's sabbatical in Cambridge and he said: 'Yes, this is certainly right and an important and interesting result'. So I had some back-up.

Testing for predictions

Although I found the theoretical arguments for the effects convincing it was rather worrying that the theory made definite predictions of how much supercurrent there should be, and these predictions didn't seem to fit the experiments. Theory predicted that for an average sort of junction you should get something like some hundreds of microamperes of supercurrent; indeed the supercurrent should be of the same order of magnitude as typical normal currents. I noted with interest a published paper by Nicol *et al* (1960) that showed a clearly visible zero-voltage blip, but this was not typical and such events when they did occur were normally ascribed to superconducting shorts across the insulating barrier. I tried to think up coherence-destroying mechanisms for reduction of the supercurrent, such as the effects of magnetic impurities, so as to bring theory and experiment into line. If such effects were significant, it might mean that the effects would be unobservable.

One possible source of non-observation of the effects that could be investigated, which had been proposed by Anderson, was that of magnetic field, since the theory indicated that the junctions should be highly sensitive to such fields. An amount of flux in the junction corresponding to one flux quantum would have a drastic effect on the magnitude of the critical supercurrent. Anderson's suggestion was that only the flux in the insulating region would be effective, but using the Ginzburg–Landau theory to calculate the effects of the field I convinced myself that the flux penetrating into the superconducting regions also contributed, and this implied that the earth's field could be important. So I set up compensating coils to reduce the local fields to a few milligauss, and looked for zero-voltage currents using an electronic ammeter sensitive to a nanoampere, smaller by a factor of around 10^5 than the supercurrents predicted by the theory. Disappointingly, no supercurrents, even at this level, showed up, even though I was able to obtain junctions with good normal characteristics.

The predictions in my paper were quite controversial, and John Bardeen in particular asserted that my predictions were in error (Bardeen 1962) because, according to him 'the pairing does not extend into the barrier'. I did not believe in this criticism, since it seemed to involve the assumption that the situation could be treated on the basis of local equations, which did not seem to me to be correct, but the fact that the supercurrents had not been observed argued in favour of Bardeen's theories. He and I debated the issue at a special session (not published in the proceedings) during the International Low Temperature Conference held at Queen Mary College later that year. My view was supported by a theoretician there, de Gennes I believe.

It took nine months from the publication of my theoretical paper before convincing evidence for the reality of the phenomenon was obtained by Phil Anderson and John Rowell at Bell Labs (Anderson and Rowell 1963) (figure 20.1). They used lower resistance junctions made from lead instead of the aluminium that I had used, and observed zero-voltage currents whose maximum value before a voltage appeared was highly sensitive to magnetic fields. Such low-resistance junctions were liable to have metallic shorts through the barrier, which was why I had used aluminium instead, but the way the currents depended on magnetic fields provided very good evidence for the existence of a genuine effect.

Why had the Anderson–Rowell junctions showed an effect whereas mine had not? The explanation involved something that I had clearly seen, but there was an unexpected twist as well. The coupling energy across the barrier is given by $-(\hbar/2e) I_c \cos (\Delta\phi)$, and I had worked out that this was large compared with kT if T is taken to be the sample temperature and concluded that thermal fluctuations would not be large enough to destroy the phase coupling process. But what I had failed to realise is that room-temperature electrical noise which is two orders of magnitude greater than the noise at helium temperature, comes down the leads to the specimen, and that for my samples with their low I_c this was enough to destroy the phase coupling and with it the zero-voltage supercurrents.

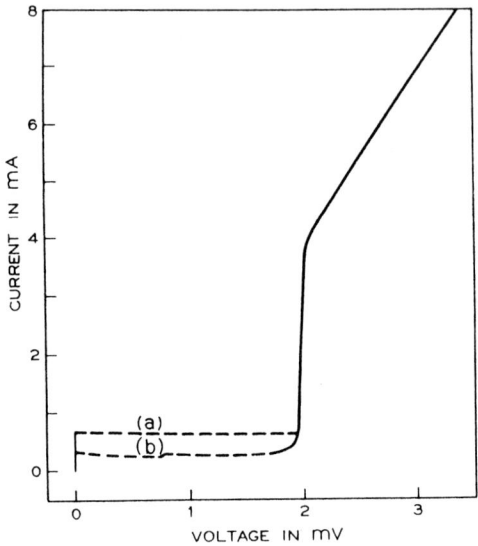

Figure 20.1 Experimental verification of the Josephson effect. Current–voltage characteristics for a tin–tin oxide–lead tunnel structure at ~1.5 K (a) for a field of 6×10^{-3} gauss and (b) for a field of 0.4 gauss (Anderson and Rowell 1963, p. 230). [Editor's insertion.]

The Anderson–Rowell paper meant that the effect was real and its prediction could really be celebrated at last. It was only a matter of time before the other effects predicted in my original paper were confirmed experimentally, and practical applications of the effects were developed over the years as well.

For a comprehensive review of subsequent developments see A Barone and G Paterno (1982) *Physics and Applications of the Josephson Effect* (New York: Wiley).

References

Abrikosov A 1957 The magnetic properties of superconductors of the second group *Sov. Phys.—JETP* **5** 1174–82

Ambegaokar V and Baratoff A 1963 Tunneling between superconductors *Phys. Rev. Lett.* **10** 486–9 (and errata *Phys. Rev. Lett.* **11** 104)
Anderson P W 1958 Random-phase approximation in the theory of superconductivity *Phys. Rev.* **112** 1900
Anderson P W and Rowell J M 1963 Probable observation of the Josephson superconducting tunnel effect *Phys. Rev. Lett.* **10** 230–2
Andreev A F 1964 The thermal conductivity of the intermediate state in superconductors. *Sov. Phys.—JETP* **19** 1228–9
Bardeen J 1962 Tunneling into Superconductors *Phys. Rev. Lett.* **9** 147–9
Bardeen J, Cooper L N and Schrieffer J R 1957 Theory of superconductivity *Phys. Rev.* **108** 1175–204
Bogoliubov N N, Tolmachev V V A and Shirkov D V 1959 *A New Method in the Theory of Superconductivity* (New York: Consultants Bureau)
Cohen M H, Falicov L M and Phillips J C 1962 Superconductive tunneling *Phys. Rev. Lett.* **8** 316–18
Deaver B S Jr and Fairbank W M 1961 Experimental evidence for quantized flux in superconducting cylinders *Phys. Rev. Lett.* **7** 43–6
Doll R and Näbauer M 1961 Experimental proof of magnetic flux quantization in a superconducting ring *Phys. Rev. Lett.* **7** 51–2
Fisher C and Giaevar I 1961 Tunneling through thin insulating layers *J. Appl. Phys.* **32** 172–7
Giaever I 1960 Energy gap in superconductors measured by electron tunneling *Phys. Rev. Lett.* **5** 147–8
Ginzburg V L and Landau L D 1965 *On the Theory of Superconductivity. Collected Papers of L D Landau* ed. D Ter Haar (New York: Gordon and Breach) pp 546–68
Gor'kov L P 1958 On the energy spectrum of superconductors *Sov. Phys.—JETP* **34** 505–8
—— 1959 Microscopic derivation of the Ginzburg–Landau equations in the theory of superconductivity *Sov. Phys.—JETP* **36** 364–7
Josephson B D 1962a Possible new effects in superconductive tunneling *Phys. Lett.* **1** 251–3
—— 1962b The relativistic shift in the Mössbauer effect and coupled superconductors *Dissertation for the Annual Election of Fellows, Trinity College, Cambridge*
—— 1966 Superconducting barriers: the plasma resonance and related properties *Quantum Fluids. Proceedings of the Sussex University Symposium 16–20 August 1965* ed. D F Brewer (Amsterdam: North-Holland) pp 174–5
—— 1974 Magnetic field dependence of the surface reactance of superconducting tin at 174 MHz *J. Phys. F: Met. Phys.* **4** 751–6
London F 1950 *Superfluids* (New York: Wiley)
Nicol J, Shapiro S and Smith P H 1960 Direct measurement of the superconducting energy gap *Phys. Rev. Lett.* **5** 461–4
Pippard A B 1961 Recent experiments on the electrical properties of superconductors *Proc. VIIth Int. Conf. Low Temp. Phys.* ed. G M Graham and A C Hollis Hallett (Amsterdam: North-Holland) pp 320–7
Thouless D J and Tilley D R 1961 Collective modes in the theory of superconductivity *Proc. Phys. Soc.* **77** 1175–81

Notes on Contributors

Dr Dennis Frederick Shaw is Director of the Radcliffe Science Library at the University of Oxford. During the years 1946–75 he taught physics at Keble College, Oxford, where he now holds a professorial fellowship, and lectured and researched in electronics and nuclear physics at the Clarendon Laboratory and later at the newly established Nuclear Physics Laboratory. His latest work is (as editor) *Information Sources in Physics* (1985) 2nd edn (London: Butterworths). He was appointed CBE in 1974. Address: Radcliffe Science Library, Parks Road, Oxford OX1 3QP, UK.

Dr Spencer Weart is Director of the Center for History of Physics at the American Institute of Physics. Among a number of works he has written or edited are *Nuclear Fear* (1988) (Harvard University Press), *Scientists in Power* (1979) (Harvard University Press), *Leo Szilard: His Version of the Facts* (1978) (Massachusetts Institute of Technology). Address: Center for the History of Physics, American Institute of Physics, 335 East 45th Street, New York, NY 10017-3483, USA.

Dr David H DeVorkin is Curator of the History of Astronomy at the National Air and Space Museum, Smithsonian Institution, Washington. His chief interests lie in the origins of modern astrophysics and in the early years of space science. He has recently published *Race to the Stratosphere:Manned Scientific Balooning in America* (1989) (Springer) and has partly completed a sequel *Science with a Vengeance: the Origins of Space Science in the V-2 Era*. Address: National Air and Space Museum, Smithsonian Institution, Washington DC 20560, USA.

John Krige has doctorates in physical chemistry and in the philosophy of science. After teaching at Sussex University for several years he moved to CERN in 1982. He is currently the leader of a new History of CERN project. Address: Study Team for CERN History, Building 54, CERN CH-1211 Geneva 23, Switzerland.

Professor Nicholas Kurti was born in Budapest in 1908. He studied in Budapest, Paris, Berlin and Breslau. He arrived in Oxford in 1933 where he carried out research at the Clarendon Laboratory in low-temperature physics. From 1940–5 he was involved with the United Kingdom Atomic Energy Project. From 1965 to 1967 he was Vice-President of the Royal Society. In 1967 he became Professor of Physics at Oxford. He has been honoured by many international bodies. His publications are mainly in low-temperature physics but he is also well known for his writings in culinary physics. Address: Department of Engineering Science, Parks Road, Oxford OX1 3PJ, UK.

Mr Ronald Grubb Stansfield was born in Southampton in 1915. From 1933 he read natural sciences at Clare College, Cambridge. During 1936–9 he was a research student in experimental physics at the Cavendish Laboratory. His war work was in Operational Research with the Royal Air Force. This prompted an interest in the human and social sciences which he has pursued ever since. Address: 62 Warwick Road, Bishop's Stortford, Herts CM23 5NW, UK.

Professor A P French was educated at Cambridge and carried out research and taught nuclear physics at the Cavendish laboratory from 1948 to 1955. Since 1962 he has been at the Massachusetts Institute of Technology where his chief interests lie in physics education and in the history of physics. He is the author of a much admired series of physics textbooks. Address: Department of Physics, Massachusetts Institute of Technology, MA 02139, USA.

Mr Brian Davies is the Regional and Public Affairs Officer of The Institute of Physics, London. He studied undergraduate physics and postgraduate history and philosophy of science at London University and education at Cambridge. After 10 years as a secondary school teacher he was appointed Senior Lecturer in Physics at the University of London Goldsmith's College, where he was responsible for postgraduate teacher training and for intercollegiate courses in the history and sociology of science. Address: The Institute of Physics, 47 Belgrave Square, London SW1X 8QX, UK.

Dr John Roche is an Associate Fellow of Manchester College, Oxford. He studied for undergraduate and postgraduate degrees in physics at

University College, Galway. From 1961 to 1972 he was head of the physics department at Strathmore College, Nairobi. From 1973 to 1977 he carried out doctoral research in the history of astronomy at Linacre College, Oxford. He now teaches history of science at Oxford University and physics at Oxford Polytechnic. Address: Linacre College, Oxford OX1 3JA, UK.

Mr Stuart Leadstone is Head of Physics at Atlantic College in South Wales. For most of his teaching career, which began in 1962, he has taught physics in schools and sixth-form colleges up to university entrance level. The difficulty of advancing one's own understanding of physics beyond a certain level has led him to the belief that a new generation of textbooks is needed in which specialist rigour joins forces with historical research in order to clarify phenomena and concepts. He is a founding member of the History of Physics Group of The Institute of Physics. Address: Atlantic College, Llantwit Major, South Glamorgan CF6 9WF, UK.

Mr Charles Boyle is a former physics lecturer who now teaches science, technology and society to science and engineering students at Nottingham Polytechnic. He is co-author of *People Science and Technology* (1984) (Wheatsheaf/Harvester). His current interest is in a comparative study of the general education of scientists and engineers in different countries. Address: Department of Chemistry and Physics, Nottingham Polytechnic, Clifton Lane, Nottingham NG11 8NS, UK.

Dr Brian Gee, until recently, was senior lecturer in physics and science education at the College of St Mark & St John in Plymouth. His research interests in the history of science began early in his career with studies in the development of stellar astronomy at Harvard College Observatory. His publications on history of physics in Institute of Physics journals bore fruit in 1984 with the setting up of the History of Physics Group in the Institute. Currently he is working on the London electrical developments of the 1830s. Address: 18 Barton Close, Landrake-Saltash, Cornwall PL12 5BA, UK.

Professor Henry Solomon Lipson is Emeritus Professor of Physics at the University of Manchester Institute of Science and Technology. He obtained a degree in physics at Liverpool University in 1930 and went on to specialise in crystallography in Liverpool, Manchester, the National Physical Laboratory and Cambridge. In 1945 he became Head of the Physics Department at Manchester College of Technology and in 1975 he was appointed Dean of the Faculty of Technology at

Manchester University. He has published numerous books and specialist papers, mainly on x-rays and crystallography. He became FRS in 1957 and was appointed CBE in 1976. Address: The University of Manchester Institute of Science and Technology, PO Box 88, Manchester M60 1QD, UK.

Mr D J Unwin is an engineer/physicist who was head of the Mechanical Laboratory and Mechanical Design at the Cambridge Instrument Company Research Department from the mid-1950s to the 1970s. His career has covered every facet of design, development and manufacture. He was also much involved in training and was chairman of the company apprentices committee for many years. His particular specialisation, both in teaching and application, has been in the principles of kinematic design. Address: 8 Hampden, Kimpton, Hitchin, Herts SG4 8QH, UK.

Dr David W Hughes is a senior lecturer in astronomy and physics at the University of Sheffield. He has spent the last 25 years researching into the origins and evolution of the minor bodies in the solar system. This class of objects includes meteoroids, cosmic dust, comets and asteroids. Address: Department of Physics, The University of Sheffield, Sheffield S3 7RH, UK.

Professor A J Meadows was formerly head of the Departments of Astronomy and History of Science and of the Primary Communications Centre at Leicester University. He is currently Head of the Department of Library and Information Studies at Loughborough University. He has published many books and specialist papers both in science and in the history of science. Address: Department of Library and Information Studies, Loughborough University, Loughborough, Leicestershire LE11 3TU, UK.

Dr C Desmond Walshaw was born in 1925 and educated at Charterhouse and Clare College, Cambridge and in the Royal Navy. His Cambridge doctoral research was in the infrared spectrum of ozone. In 1956 he became the Travelling Physicist of the International Ozone Commission, based at Professor Dobson's laboratory, Oxford. He was university lecturer in Atmospheric Physics in Oxford from 1963 to 1986 and is Fellow Emeritus at St Cross College, Oxford, and a Fellow of the Royal Meteorological Society. Address: St Cross College, St Giles, Oxford OX1 3LZ, UK.

Professor Sir Granville Beynon, formerly Head of the Department of Physics at University College of Wales, Aberystwyth, has been a leading figure in ionospheric research for many years. He has also played a major role in international cooperation in geophysics. He was President of the International Union of Radio Science (URSI), a founder member of the International Geophysical Year (IGY) and of the European Incoherent Scatter Radar System for sounding the high atmosphere (EISCAT). He is a fellow of the Royal Society. Address: Department of Physics, University College of Wales, Aberystwyth, SY23 3BZ, UK.

Professor Frank Llewellyn-Jones, CBE, took a degree in physics in 1929 at Merton College, Oxford and went on to collaborate with Professor J S E Townsend in research on properties of electrons and ions in gases. During World War II he worked at RAE Farnborough on aero-engine ignition systems and other discharge phenomena. In 1945 he was appointed Professor of Physics and Head of Department at University of Wales, Swansea, where he established research groups in ionisation physics and in electrical contact phenomena. In 1965 he was appointed Principal of the University College of Swansea and from 1969–71 was Vice-Chancellor of the University of Wales. He is the author of several books and numerous research papers. Address: Department of Physics, University College of Swansea, Singleton Park, Swansea SA2 8PP, UK.

Professor Brian David Josephson was educated at Cardiff High School and at Cambridge where he was awarded a BA in 1960 and PhD in 1964. He became a Fellow of Trinity College in 1962 and Professor of Physics in 1974. He was awarded the Nobel prize for physics in 1973 for work which he describes in the present volume. He has received many awards and honours nationally and internationally. Together with his work on superconductivity he has published on consciousness and the physical world and on the relationship between science and religion. Address: Cavendish Laboratory, Madingley Road, Cambridge CB3 0HE, UK.

Index

Aaserud F, 39
Abraham H, 350
Abrikosov A, 369
Acland H, 317
Adams G, 220
Adams J C, 289
Aharonov Y, 157
Albert, HRH Prince, 315
Allen E W, 296
Almond M, 291, 296
Amaldi E, 73
Ambegaokar V, 374
American Institute of Physics (AIP)
    Center for History of Physics, 29–43
  Archive, 40
  Catalogue of sources, 42
  Educational activities, 32
  Einstein exhibit, 32
  Newsletter, 36, 42
  Niels Bohr Library, 35
Ampère A M, 145, 150, 314
Anderson P, 369, 374
Anderson R G, 223
Andreev A, 371
Andrews T, 319
Appleton E, 295, 297, 327–30, 337–8
Apthorpe W, 238
Arago F, 273–4
Archival bias, 68
Archives for history of quantum physics, 38

Aristotle, 123, 261
Aspinall A, 296
Astbury W, 226
Asteroids, 275
Aston F, 350
Atomic collision processes, 354–6
Aurorae, 262, 266, 308
Ayscough J, 208

Babbage Institute, 42
Bachelard G, 194
Bacon F, 188, 193
Bailey V A, 357
Ball R S, 276
Baratoff A, 374
Bardeen J, 368, 375
Barnes B, 218
Barnett M, 329
Barone A, 376
Baxendell J, 309
Beck R and J, 258
Beech M, 266
Beevers C, 226, 228
Beiser A, 172
Bennett J, 207–8
Benzenberg J, 267
Berigny A, 318
Berkner L, 338
Bernal J D, 232
Bernal chart, 232
Big science, 188
Biot J B, 150, 267
Bitter F, 175

Black P J, 182
Blackett P M, 298
Blandford H, 310
Bleaney B, 85
Bloch E, 350
Boas-Hall M, 207
Bogoliubov N, 368
Boguslavski P, 275
Bohm D, 157
Bohr N, 3, 37, 69, 114, 118
Bohr Library, 35
Bonpland G, 268
Booth C, 213
Bostik, 106
Bound momentum, 160
Bournon J-L, 267
Bowles K W, 342
Boyle C, xi, 184, 201
Boyle R, 193, 207
Boys C V, 103
Bragg W H, 226, 228–9
Bragg W L, 226–7
Brandes H, 267
Braun E, 38
Breit G, 329
Bristol University, 89
Brodie B, 319
Brown J, 222
Browning J, 289
Brush S G, 113
Buerger M, 234
Buisson H, 313
Bulletin Signalétique, 12, 14
Burke J G, 262, 266
Burnett J, 223

Cailletet L, 83
Cajori F, 113
Calvert H, 222
Cambridge Engineering Laboratory, 239
Cambridge Instrument Company, *237–60*
 organisation, 241–3
Campbell R, 207, 209, 215
Carnot S, 190
Cassini G, 121
Cavendish Laboratory, 94–5

Center for History of Physics, AIP, *29–43*
Center for the History of Chemistry, 42
Center for the History of Electrical Engineering, 42
CERN, *66–77*
 and the UK, 69, *73–5*
 Provisional Council, 72
Chadburn Brothers, 212
Chadwick J, 69
Chaldecott A, 223
Chamanlal, 297
Chambers G F, 271, 285, 315
Chambers R E, 157
Chapman S, 330, 338, 357
Chappuis J, 320
Chaucer G, 99
Cherwell, Lord (F E Lindemann), 73–4, 288, 291–2
Chew V, 216
Chladni E, 267, 281
Churchill W, 73
Clarke E M, 219
Claude G, 83
Clegg J, 296
Clifton R B, 79
Cockcroft J, 73
Cocks C, 207
Cohen I B, 112, 121
Cohen M H, 372–3
Cole W, 213
Coleridge S T, 98
Collins H, 100
Collins S, 207
Comets, 264, *281–4*
Compton A H, 173
 effect, 175
Compton K T, 362
Computer simulation, 350
Condé J, 269
Conference proceedings, 19
Consensus in physics, 169
Contact potential, 173
Cook W, 362
Cooper N, 368
Cornu A, 319
Coulomb gauge, 152

Index

Coulson T, 118
Cowling T, 357
Crawforth M, 223
Crompton K, 362
Crompton R, 363
Crookes W, 351
Crystal structure, *226–36*
   photography, 228–31, 233, 235
   and reciprocal lattice, 232
Curie M, 361

Dalton J, 308
Darbishire F, 315
d'Arrest H, 283
Darwin C, 196
Darwin E, 98
Darwin H, 237
Daubeny C, 316–8
Daumas M, 212
Davenport S, 208
Davies A J, 356
Davies B, x, 126, 127
Davies D K, 363
Davies J G, 296–7
Davy H, 98
Deane P, 213
Deaver B, 369
Denning W, 284–5, 291
Descartes R, 123, 197
Deutsches Museum, 38
Devens R, 274
DeVorkin D H, x, 38, 44, 65
Dew-Smith A, 237
Dictionary of Scientific Biography, 17
Dobson G M, 288, 291, 313, 323
Doll R, 369
Druyvesteyn M, 357
DuBridge L, 174, 179
Duhem P, viii
Dumas J, 83
Duncombe P, 238
Dunkin E, 279
Dunsheath P, 238

Eastlea B, 195
Ebonite, 105
Eccles W, 329

Eckersley T, 295
Egerton A, 322
Einstein A, 114, 116, 130, 173, 194
Einstein exhibit, 32
   photo-electric equation, 172, 180
Electric telegraph, 131
Electrical apparatus, 93
Electrical inertia, 147
Electrokinetic momentum, 160
   invariance of, 161
Electromagnetic momentum, 153
Electromagnetism, mechanical
   theory of, 150
Electron microscope, *237–60*
   corrosion, 251–2
   electroplating, 251
   'Geoscan', 251
   'Microscan', 238
   slag inclusion, 253
   specimen chamber, 245
   vibration isolation, 239
Electro-tonic state, 145, 151
Elkin W, 288
Ellicott A, 268
Ellyett C, 296–7
Elster J, 180
Energy
   electrokinetic, 165
   electromagnetic, 163
   electrostatic, 163, 165
   individual potential, 164
   mutual potential, 163
   self, 163
Erman G, 275
Europe and China, 194
Evans C J, 354
Ewald P, 232
Experiment and nature, 197–8
Experimental techniques in
   ionisation physics, *359–61*
Experimental verification of
   Josephson effect, 376
Explanatory research, viii, ix, 159, 163

Fabry C, 313, 321, 341
Fairbank W, 369
Falicov L, 372

Faraday M, 17, 96, *145–7*, 156, 191, 229, 315
Fedynski V, 288
Fermi E, 36
Fermi–Dirac distribution, 171
Fireballs, 262, 264, 266
Fisher C, 373
Flamsteed J, 207
Flux,
　electric, 163
　magnetic, 163
Forman P, 195
Fourier J B, 150
Fowler A, 322
Franck J, 355
Frank-Kamenetschi D, 349
French A P, x, 111, 173
Friedrich W, 227–8
Funding of physics, 187

Galileo, 123, 306
Galle J, 283
Gauge
　auxiliary, 166
　physical, 166
　transformations, 152, 155, 158, 165–6
Gauss C F, *147–8*
Gautier A, 307
Gay-Lussac J, 314
Gee B, xi, 205, 223
Geiger H, 120
Geitel F, 180
Giaevar I, 373
Gilardini A, 363
Gilbert G K, 264, 297
Gill E, 19
Gingerich O, 45
Ginzburg V, 369
Glaisher J, 283
Glen W, 40
Goodman M, 9
Gordon W E, 342
Gor'kov L, 369
Gorler C, 85
Goudsmith S, 116–17
Gough W, 208
Gowing M, 12, 41, 75

Graaff R, Van de, 361–2
Granville Beynon W J, xi, 327
Grassman's law, 147
Greenhow J, 296
Gregory J, 208
Griffin J, 220
Gunther R, 319
Gutton A, 358

Hackmann W, 11–12, 213, 217
Hall A R, 219
Halley E, 262
Halley's comet, 298
Hallwachs W, 180
Haloes, 262
Hartley W, 319
Harvard meteor expedition, 289
　observatory, 288, 294
Harvey A P, 7
Harwell, 41, 69, 73
Hasegawa I, 269
Hassé H, 362
Hauksbee F, 207, 222
Hautefeuille P, 320
Hawkins G S, 290, 296
Haydn F, 86
Heaviside O, 154, 329
Heisenberg W, 194
Helmholtz H, von, 79, 286
Henry J, 118
Herlofson N, 294, 297–8
Hermann A, 67
Herrick E, 275
Herschel J F W, 278, 280, 307
Herschel W, 307
Hertz G, 355
Hertz H, 154, 180, 191
Hey J, 295, 298
Hind J, 277
Historians and teachers, 127
History of CERN, *66–77*
History of physics
　applied, viii
　as a means of clarification, ix, 144
　internal, vii, 185
　oral, *44–65*
　packages, *126–43*

Index

History of physics (*cont.*)
  in physics education, viii, *111-43*
History of science
  in old advertisements, 134
  in old encyclopaedias, 131
  in old newspapers, 130
  in paintings, 135-7
  in photographic records, 139
  in poetry, 135
  in popular lectures, 131
  in recordings, 139
  in science fiction, 134
History of solid state physics, 38-9
Hittorf W, 351
Hoch P, 38
Hoddeson L, 33, 38
Hodgson E, 175
Hoffmeister C, 289
Holmyard E J, 99
Holton G, 116
Hooke R, 18-19, 207, 263
Hooker W, 318
Hoskin M, 45
Houghton J, 313, 322
Houzeau A, 318
Howard E C, 267
Howard J, 322
Hubble E, 37
Huggins W, 321
Hughes A L, 174, 179
Hughes D, xi, 261, 274, 284-6
Hulbert E, 321
Humboldt F W H A, von, 268, 279, 307
Huxley L, 364
Hypotheses in physics, 90

Imoto S, 269
Industrial revolution, 188
Information, sources of historical, 44
INSPEC, *14-15*
Institute of physics, 89
Instrument makers, *205-25*
  master craftsmen, 214-15
  mathematical, 215

Instrument makers (*cont.*)
  philosophical, 206
  rewards and recognition, 220
  and scientists, 217-21
  status of, 214-15
Interaction of physics and society, 187
International Years, 330, 336-8, 345-6
Ionisation physics, *348-65*
  computer simulation, 350, 356
  critical potentials, 355
  cross sections, 356
  Debye screening, 358
  gas purification, 359
  ionic mobilities, 361
Ionosphere, *327-47*
  Appleton layer, 331
  incoherent scatter soundings, 341
  ionosonde, 334
  Kennelly-Heaviside layer, 329
  and Radar, 327
  and radio atmospherics, 328
  radio soundings, 328, 331
  term coined, 330
Irish physicists, 150
Ishawara J, 116

Jacchia J, 290
Jackson J D, 158-9, 164
Jahoda M, 77
James A N, 175
James R W, 227
Jason group, 39
Jauslin K, 274
Jodrell Bank, 296-8
Jones T, 99
Josephson B, xii, 366-8, 370-1
Josephson effect, *366-77*
Joule J P, 190
Journal der Physik, 10
Jowett B, 79
Jupiter's satellites, 123

Kahlbaum G, 315
Karlinski F, 318
Kaufmann W, 351
Kedzie R, 318

Keesing R, 181–2
Kelvin, Lord (W Thomson), 82
Kennelly A E, 329
Kennelly–Heaviside layer, 329
Kidd J, 316
Kirchhoff G, 148–9
Kirkwood D, 281
Klein M, 173
Knipping P, 227
Knowledge and 'know-how', 218–20
Koenigsberger L, 80
Kollath R, 357
Konkoly T M, Von, 287
Kowarski L, 67, 75
Kramers H, 69
Krige J, x, 66, 67, 72
Kuhn T, 193
Kundera M, 69
Kunhardt E, 357
Kurti N, x, 78, 81, 84

Laboratories, research, 188
Laboratory chemicals, 92
　culture, 91
　health risks, 99–100
　tea, 95
Ladenburg E, 321
Lambert R K, 175
Lamont J, von, 307, 309
Langevin P, 350
Langmuir I, 348
Laplace P S, de, 150
Laser history project, 39
Laue M, von, 227
Lauscher F, 325
Layton E T, 219
Leadstone S, x, xii, 169
Lecomte J, 321
Lehmann E, 321
Lenard P, 180
Le Verrier U J, 281
Levi-Cività T, 155
Lichtenburg G, 266
Liénard A, 156
Lightning, 262
Lincoln F, 102

Lindemann F E (Lord Cherwell), 288, 291–2, 313
Lindinger W, 364
Linville D, 318
Lipson H, xi, 226, 232–4
Llewellyn-Jones F, xi, 348, 349, 355–8,
Lockspeiser B, 73
Lockyer N, 283, 286
Lomax J, 47
London F, 369
London R, 105
Long J V, 253
Lorentz H A, 149, 153, 165
　gauge, 155
Lorenz L, 149
Loss of technological knowledge, 106
Lovell A C B, 284, 291, 296, 298, 299
Luessen L, 357
Lunar craters, 263

MacCallum S, 359
McCrosky R, 290
McGucken W, 287
Magnetic hyperfine coupling, 85
Magnetic storms, 308
Mann J, 207
Marconi G, 328
Marignac J C, de, 315
Mariologulos E, 318
Marshall J, 207
Martin B, 208, 210, 220
Massey H, 352
Mathematical artefacts, 159
　instrument makers, 215
　practitioners, 206
Mathematics and nature, 197
Maurer R, 178
Maxwell J C, viii, 145, 147, *150–4*, 191
Mayer J R, 286
Meadows A J, xi, 306
Measuring definition
　of scalar potential, *163–4*
　of vector potential, 160–1

Mechanical theory of
    electromagnetism, 150
Medawar P, 112
Medicus H, 175
Meldrum C, 309
Melford D, 238
Merian P, 314
Meteorites, 266–91
Meteoroids (defined), 291
Meteors, *261–305*
    ablation of, 291
    and asteroids, 275
    and bolides, 268
    and comets, *281–4*
    cosmical velocity of, 292
    hyperbolic, 289
    orbits of, 281
    and ozone, 293
    photography of, 294
    and radar, 295
    radiant of, 270–1, 280
    showers, 274
    as solar fuel, 285
    spectroscopy of, 287, 294
Meteors
    Geminids, 275, 297
    Giacobinids, 297
    Leonids, 268, 270, 273, 276, 280, 281, 295
    Lyrids, 275–6
    Perseids, 268, 275, 283, 291, 297
    Taurids, 290
Methodological pluralism, 71
Methodology, 'colonisation' by scientific, 198
Michelson–Morley experiment, 116
Miethe A, 321
Miller W, 319
Millikan R A, 97, 173–4, 180, 353
Millman P, 294
Mohr C, 352
Momentum
    bound, 160
    electrokinetic, 153, 160
    electromagnetic, 153
    invariance, 161–2
    reactionless, 162
Montanari G, 262

Moral responsibility of the scientist, 199–200
Morrison P, 34
Morrison-Low A, 223
Morrissey C, 47
Morse P, 357
Moscow Observatory, 294
Mozart, 86
Müller M, 79
Multhauf R, 218
Mulvaney J, 319
Museums, 106

Näbauer M, 369
Nagaoka H, 295
Nairn, Ramsden and Adams, 212
Naismith R, 295, 297
Nernst W, 351
Newton H A, 267, 275, 281
Newton I, 117, 193, 197, 207, 262, 264
Nicol J, 374
Niessl G, 289, 291
Noakes G, 170
North J D, 306
Nuclear physics laboratory, 71

Oatley C, 238, 243
Obsolete terms, 144
Odling W, 318
Oklahoma University libraries, 9–10
Olbers H, 275–6
Olivier C P, 284, 293
Olmsted D, 270–1
Olson B, 318
Open Systems Intercommunication, 9
Operational definition
    of energy, 159–60
    of momentum, 159–60
Operational research, 91
Öpik E J, 289, 291, 294
Oppolzer T, von, 282
Oral history, 31, 33, 34, 38, 40, *44–65*
    interviews, 45, *54–60*
    pitfalls, 47

Oral history (*cont.*)
   the product, 60
   selected readings, 64
Oscillation camera, 231
Oxford
   Bodleian Library, 5
   Radcliffe Camera, 5
   Radcliffe Science Library, 3, 5, 6
Ozone, atmospheric, *313–26*
   consolidated, 319
   'fever', 317
   and medicine, 316–18
   and meteorology, 317

Pais A, viii
Palmieri L, 318
Partington J, 313–14
Pasteur L, 111
Paterno G, 376
Pattison M, 79
Penning F, 359
Pepys S, 207
Perrin J, 350
Personal recollections, unreliability of, 115–16
Pestre D, 71, 76
Peters C F, 282
Phillips J, 316
Phillips M, 36
Philosophical Magazine, 11
Photo-electric effect, *169–83*
Photon, 114
Physical Society, 89
Physicists and experimental engineers, 253
Physics Abstracts (Science Abstracts), 12–14
Physics and the Arts, 195
Physics Briefs (Physikalische Berichte), 12–14
Physics classification, 15–18
   education, 190
   holdings catalogues, 22–8
   and literature, 195
   and masculinity, 195
   militarisation of, 189, 201
   and religion, 196
   and society, *184–201*

Physics classification (*cont.*)
   teaching, use of original data in, 118
   and technology, 190
   and war, 188
Pickering W H, 271, 292, 294
Pictet R P, 83
Piddington J, 295
Pidduck F, 357
Pierce J, 295
Pippard B, 368, 370–3
Plagiarism, subconscious, 84
Planck M, 114
Plasma physics, *348–65*
Plasticine, 102
Plavec M, 284
Poggendorff J, 17
Poisson distribution, 119–20
Popper K, 90
Posen A, 290
Potential
   scalar, 164
   vector, *144–68*
Pound R, 85
Powell C, 362
Prentice J, 297
Price D, de Solla, 220
Pritchard H, 238
Project hindsight, 192
Protestantism and capitalism, 194
Pure and applied research, 189, 192
Pyrex, 101
Pythagoreans, 197

Quasi-history, 114
Quetelet L, 275–6

Rabi I, 76
Radar, 327
Radcliffe Observatory, Oxford, 309, 317
Radiation pressure, 175
   quantisation of, 114
Radioactive decay, 119
Ramsauer C, 357
Ramsden J, 218
Ratcliffe J A, 327, 335

Rayleigh, Lord (J W Strutt), 114, 321–2
Reconstruction of theories, 144
Rediscovery, 86
Redundant explanations, 144
Reeves R, 207
Referativnyi Zhurnal, 12
Repeating experiments, 95, 103
Replication, *88–107*
Research laboratories, 188
Research libraries, *3–28*
Reynolds line, 89
Richardson O W, 173, 350
Richer J, 121
Riemann B, 152–3
Ritchmeyer F, 178
Rivé A, de la, 315
Roberts R W, 226
Roche J J, x, 127, 144, 164, 170, 174
Roemer O, 121–2
Rohrlich F, 158
Roller D H, 9
Ronayne J, 192
Röntgen W, 349
Rose M, 85
Rowell J, 375
Rowley J, 216
Royal Astronomical Society, 56
  Institution, 103, 229
  Irish Academy, 7
  Observatory, Greenwich, 277
  Radar Establishment (RRE), 258
  Society Catalogue, 11–12
Rudwick M, 47, 61
Rutherford E, 37, 97, 120, 350–1, 358

Sabine E, 307
Sadler D H, 217
Sainte-Claire Deville H, 83
Sampson R A, 322
Sarton G, 218
Saxton J, 221
Scalar potential, 149
Schaeberle J, 268
Schelling F, von, 314
Schiaparelli M, 280–1, 285

Schmidt M, 313
Schoenberg D, 368
Schonbein L, 314–15
Schottky W, 358
Schrieffer J, 368
Schrödinger, 184, 191
Schuster A, 351, 354
Science Museum, London, 7
  policy, 190
  Reference Library, London, 7
  and technology, 203, 218–19
Scientific reductionism, 198–9
  world view, 195–6
Scottish physicists, 150
Sealing wax, 101
Seidel R, 39
Seneca, 262
Shankland R 116
Shapin S, 214
Shapley H, 289
Shaw D F, ix, 3
Shellac, 104, 106
Shirkov D, 368
Short J, 217
Siemens-Helmholtz E, von, 81
Sillito R, 179, 182
Simcock A, 314
Simpson A, 207
Sinding E, 290
Skellet A, 295
Sloane, 264
Smith C G, 317
Smyth C Piazzi, 279, 309
Social history of science, vii, 137
Solid state physics, history of, 38–9
Southampton, University college, 89
Southworth R B, 290
Sparrow C M, 294
Sputnik, 120, 340
Stanjukowitsch K, 288
Stansfield D A, 98
Stansfield H, 89–90, 103
Stansfield R G, x, 88, 90, 91, 94, 96, 97, 98, 99, 100
Starr L, 362
Stewart Balfour, 309, 329
Stewart G S, 296

Stokes G, 310
Stone E, 309
Strachey R, 310
Strickland S, 271
Strutt J W (Lord Rayleigh), 114, 321
Stuewer R, 179
Sunspot cycle, 306–7
Sunspots, *306–12*
   and magnetic storms, 308, 312
Superconductors, weakly coupled, *366–77*
Support for science, 185
Survey of textbooks, 170
Sutherland L, 79
Syncretism in the concepts of physics, ix

Tait P G, 145, 319
Talyrond, 243, 245
Taylor E G R, 207–8
Taylor W H, 227
Teichmann J, 38
Telescope, Lord Rosse's, 277
Tempel G, 282
Textbooks, 118
Thackray T, 217
Thomson G, 69
Thomson J J, 112, 118, 156, 164, 180, 190, 350, 359
Thomson W (Lord Kelvin), 81, 150, 191
Thouless D, 368
Threlfall R, 105
Tilley D, 368
Tolmachev V, 368
Townsend J, 350–3, 355–6, 361–2
Trigg G, 174
Troughton E, 218
Tuttle H, 282
Tuve M, 329
Twemlow F, 319
Twining A C, 271
Tyndall A M, 362
Tyndall J, 318

UNESCO, 71

Unwin D, xi, 237, 260

Vector potential, *144–68*
   physical interpretation of, 146–7, 152–3, 156, 160–1, 165–6
Venkatareman K, 297
Verdi, 86
Vigroux M, 320
Volta-electric induction, 145

Walker S C, 274–5
Wall J, 98
Walshaw C D, xi, 313
Warnow J, 32–3, 35, 41
Warren J W, 182
Watkins F, 208
Watson F G, 289
Watson H W, 353
Watson J, 112
Watson W and Sons, 217
Watson-Watt R, 327, 330
Weart S R, x, 29, 36, 38, 65
Weber W, 148
Weiss E, 283
Weissenberg K, 232
   goniometer, 233
   photograph, 233
Weyl H, 155–6
Whewell W, 147
Whipple F L, 284, 289–90, 293, 299
Whiteside J, 264–5
Whittaker E T, viii
Wiechert E, 156, 350
Wigand A, 321
Wilson C T R, 350, 352–3
Wilson H A, 350
Wolf R, 318
Work function, 171
World of Learning, 7–9
Wright F W, 290
Wroblewski A, 121
Wyatt R J, ix, 3, 9, 22
Wyckoff R, 226, 229

X-rays, *226–36*
X-ray photography, 231

Index

Yardley K, 226
Yarwell J, 207
Younson E, 222

Youtz B, 114

Zeleny J, 350, 352